Graptolite Paleobiology

Books in the **Topics in Paleobiology** series will feature key fossil groups, key events, and analytical methods, with emphasis on paleobiology, large-scale macro evolutionary studies, and the latest phylogenetic debates.

The books will provide a summary of the current state of knowledge and a trusted route into the primary literature, and will act as pointers for future directions for research. As well as volumes on individual groups, the Series will also deal with topics that have a cross-cutting relevance, such as the evolution of significant ecosystems, particular key times and events in the history of life, climate change, and the application of new techniques such as molecular paleontology.

The books are written by leading international experts and will be pitched at a level suitable for advanced undergraduates, postgraduates, and researchers in both the paleontological and biological sciences.

The Series Editor is *Mike Benton*, Professor of Vertebrate Palaeontology in the School of Earth Sciences, University of Bristol.

The Series is a joint venture with the *Palaeontological Association*.

Previously published

Dinosaur Paleobiology
Stephen L. Brusatte
ISBN: 978-0-470-65658-7 Paperback; April 2012

Amphibian Evolution
Rainer R. Schoch
ISBN: 978-0-470-67177-1 Cloth; March 2014

Cetacean Paleobiology
Felix G. Marx, Olivier Lambert, Mark D. Uhen
ISBN: 978-1-118-56153-9 Cloth; April 2016

Avian Evolution
Gerald Mayr
ISBN: 978-1-119-02076-9 Hardback; September 2016

Graptolite Paleobiology

Jörg Maletz
Freie Universität Berlin,
Berlin, Germany

WILEY Blackwell

This edition first published 2017
© 2017 by John Wiley & Sons Ltd

Registered office
John Wiley & Sons Ltd, The Atrium, Southern Gate, Chichester, West Sussex, PO19 8SQ, UK

Editorial offices
9600 Garsington Road, Oxford, OX4 2DQ, UK
The Atrium, Southern Gate, Chichester, West Sussex, PO19 8SQ, UK
111 River Street, Hoboken, NJ 07030-5774, USA

For details of our global editorial offices, for customer services and for information about how to apply for permission to reuse the copyright material in this book please see our website at www.wiley.com/wiley-blackwell.

The right of Jörg Maletz to be identified as the author of this work has been asserted in accordance with the UK Copyright, Designs and Patents Act 1988.

Library of Congress Cataloging-in-Publication data applied for

ISBN: 9781118515617 (hardback)
ISBN: 9781118515723 (paperback)

A catalogue record for this book is available from the British Library.

Wiley also publishes its books in a variety of electronic formats. Some content that appears in print may not be available in electronic books.

Cover image: courtesy of Jörg Maletz
Cover design: Wiley

Set in 9/11.5pt Trump Mediaeval by SPi Global, Pondicherry, India

Printed in Singapore by C.O.S. Printers Pte Ltd

10 9 8 7 6 5 4 3 2 1

Contents

List of Contributors

Denis E. B. Bates
Aberystwyth, United Kingdom

Anna Kozłowska
Polish Academy of Science, Institute of
Paleobiology, Warszaw, Poland

Alfred C. Lenz
London, Ontario, Canada

Jörg Maletz
Freie Universität Berlin, Berlin, Germany

Jan Zalasiewicz
University of Leicester, Leicester, United Kingdom

Yuandong Zhang
Nanjing Institute of Geology & Palaeontology,
Nanjing, China

Preface

Graptolite specialists would certainly regard their fossils as the most "sexy" fossils of the world, even though they may not be the most popular ones in the view of the general public. However, graptolites are among the most useful fossils found in the geological record, as every geologist and paleontologist working in the Paleozoic time interval would tell you. Our research cannot compete with the public attention that the dinosaurs generate, and not even with the trilobites or ammonites – favourites among the fossil collectors. We do not look for the big fossils or produce the reconstructions that Hollywood uses for its movies to scare, but also fascinate, its audiences. The general public looks at fossils and paleontologists in a more Indiana Jones fashion, or, if you will, compare us with Dr Alan Grant, the paleontologist of the *Jurassic Park* movies. Nothing could be further from reality, as most paleontologists are not working with the prehistoric beasts that stimulate the fantasies of the moviegoers. Paleontology is more commonly a detailed investigation of tiny objects, fossils that most people will never look at or even recognize as such. It is the microfossils, often less than 1 mm in size and barely visible to the naked eye, that earn us our life and reputation in the scientific world, and also in companies using geological information in their exploration of natural resources.

Fossil graptolites early on earned themselves the reputation of being extremely useful for dating purposes and thus important for geological exploration. Even James Hall in the 1840s recognized this potential when he prepared one of the earliest monographic works on graptolites. Graptolite research is not a hobby for specialists, for people sitting in their offices, identifying fossils and putting them into small boxes. Our research has been motivated by the need for a biostratigraphical framework for rock successions, as exemplified by the graptolite biostratigraphy established for the Australasian Bendigo and Castlemaine goldfields by Thomas Sergeant Hall in the 1890s, and extended and revised by William John Harris and David Evan Thomas in the 1930s. At the time the Australian state of Victoria was one of the major regions of the world for gold production, and the precious gold was hosted in the Paleozoic rocks in which our favourite fossils, the graptolites, were also found. Ballerat, Bendigo and Castlemaine, among others, became famous names as the most productive goldfields of Australia and of the whole world, even though we as paleontologists know these names mainly from the modern regional Australasian chronostratigraphy (the Bendigonian and Castlemainian Stages of the Ordovician System) and the fossil faunas we investigate.

Hydrocarbon source rocks, particularly "hot shales", are a more recent area of interest, since our modern world relies upon hydrocarbons as an energy source and much more. Without hydrocarbons our world would be quite different, with no gas for our cars or heating systems, or plastic for so many purposes. In particular the Silurian hot shales in Iraq, Saudi Arabia and North Africa, but also in China, North America and Europe, are the source of the hydrocarbons that modern petroleum companies exploit at the moment. Here, as graptolite specialists, we are asked to help with exploration and provide expertise to search for the most productive layers.

Our work as specialists is not restricted to the scientific "ivory tower", but has important implications for the modern world. We are not working isolated in our research offices and labs, but are integrated into a larger world. Personally, we may see ourselves as the scientists and we may not be interested in the commercial application of our research, but this we cannot ignore entirely.

Our input in biological aspects and the evolution of life on our planet should also not be ignored. Graptolites are now known to be one of the longest-living groups of organisms, and the extant genus *Rhabdopleura* is often regarded as a living fossil. This term – introduced by Charles Darwin – should not be taken too literally, as it is wrong in every case. Fossils are remains of dead animals, even though we may be able to refer some fossils (e.g. Pleistocene organisms) to extant species. However, we are not able to identify a fossil graptolite specimen and refer it without doubt to an extant species. Still, with our fossils, we connect the modern world to a time lost in the mist of the geological past. More than 500 million years of evolution and our favourite organisms are still around. They survived extinction events and ecological catastrophes of many kinds. What does this tell us about their ways of life? What was their origin? Where will it end? The questions of complexity of life, of the evolution of coloniality as a means of communication among individuals and of help in the survival of a group instead of an individual – all this can be and needs to be explored. More than 150 years of research on graptolites lies behind us as graptolite specialists, and many important questions have been answered, but much is still to be learned from them. Graptolites have not yet provided answers to all our questions, but hopefully they will give us a few more hints in the future.

Jörg Maletz
Berlin, Germany
August 2016

Acknowledgments

When I was first asked by Mike Benton and Jan Zalasiewicz to write a book on graptolites, I was very uncertain and reluctant, as this would be a major and risky undertaking. Would anybody be interested in reading a book on graptolites? There are not too many books on this topic, and the last one I remembered was the Palmer and Rickards volume with its beautiful photo plates, for which I also provided a few photos, but this was a long time ago and much has happened in our scientific field since that time. I know of very few books for a more general audience, except for the ones in German by Rudolf Hundt, arguably one of the most strange and unusual people who worked on graptolites. As a self-made man with a geological background, Hundt was the most published person in German graptolite research, but the scientific community did not accept him or his work for a long time. However, his aim was to educate the general public and to show scientific work in an understandable and relatable way. Thus, Hundt – and many others working at the time, when graptolite research was not popular in Germany – should be thanked as they kept the torch alight, and now much scientific material can be found in German museum collections that otherwise would never have been collected.

From a practical point of view, graptolite collections in natural history museums and geological institutions guided my way, and for many years the curators provided the material upon which my research is based. Thus, many people need to be mentioned here: Per Ahlberg, Mats Eriksson, Kent Larsson, Anita Löfgren (Lund University, Sweden), Tom Bolton, Jean Dougherty, Michelle Coyne, Ann Thériault (GSC, Ottawa, Ontario, Canada), David. L. Bruton, Franz-Joseph Lindemann, Elisabeth Sunding (Natural History Museum [PMO], Oslo, Norway), Douglas H. Erwin, Mark Florence (Smithsonian Institution, Washington, USA), Una C. Farrell (University of Kansas Museum of Invertebrate Paleontology, Lawrence, Kansas, USA), Christina Franzén, Jonas Hagström (Naturhistoriska Risksmuseet, Stockholm, Sweden), Birgit Gaitzsch (TU Bergakademie Freiberg, Germany), Michael Howe, Paul Shepherd (British Geological Survey, Nottingham, UK), Frank Hrouda (Museum für Naturkunde, Gera, Germany), Ed Landing (New York State Museum, Albany, New York, USA), Paul Meyer (Field Museum, Chicago, USA), Hermann Jaeger, Christian Neumann (Museum für Naturkunde, Berlin, Germany), Bernard Mottequin (Royal Belgian Institute of Natural Sciences, Bruxelles, Belgium), Michael Ricker, Eberhard Schindler (Senckenberg, Frankfurt/Main, Germany), Matthew Riley (Sedgwick Museum of Earth Sciences, Cambridge, UK), Andrew Sandford, Rolf Schmidt (National Museum of Victoria, Australia), and Linda Wickström (SGU Uppsala, Sweden). A special mention goes also to the German Science Foundation (DFG) for the support of my research over the years.

I should not forget the many graptolite specialists, without whom a book like this would not have been possible: Denis E.B. Bates (Aberystwyth, UK), Christopher B. Cameron (Montreal, Canada), Chen Xu (Nanjing, China), Roger A. Cooper (Lower Hutt, New Zealand), Robert Ganis (Southern Pines, North Carolina, USA), Dan Goldman (Dayton, Ohio, USA), Juan Carlos Gutiérrez-Marco (Madrid, Spain), Anna Kozłowska (Warszaw, Poland), Alfred C. Lenz (London,

Ontario, Canada), Kristina Lindholm (Kävlinge, Sweden), David K. Loydell (Portsmouth, UK), Michael J. Melchin (Antigonish, Nova Scotia, Canada), Charles E. Mitchell (Buffalo, New York, USA), John F. Riva (Québec, Canada), Michael Steiner (Berlin, Germany), Anna Suyarkova (St. Petersburg, Russia), Blanca A. Toro (Cordoba, Argentina), Petr Štorch (Praha, Czech Republic), Alfons H.M. VandenBerg (Melbourne, Australia), Wang Xiaofeng (Wuhan, China), Jan Zalasiewicz (Leicester, UK), and Zhang Yuandong (Nanjing, China). You all provided me with photos, information, discussions on many subjects, and in the preparation of the various chapters of this book, not necessarily supporting my personal views, but always willing to help me. I know many of you since I gave a talk at the 3rd International Conference of the Graptolite Working Group of the IPA in Copenhagen, Denmark, organized by Merete Bjerreskov (Kopenhagen, Denmark) in 1985. I was just a diploma student from Germany at the time, trying to make the best of it and finding friends and colleagues for life. I will never forget the situation when Prof. En-Zhi Mu, one of the greatest graptolite specialists of all time, shook my hand and said: "See you at the next graptolite conference in Nanjing". Graptolites have been part of my life ever since and I am deeply grateful that this research gave me so much pleasure and enjoyment.

Obviously, my family have supported all my efforts for so many years, and helped me to survive in the sometimes difficult "science environment" that I choose to live in. Especially, I must not forget to thank my friend Ralf Kubitz, who has dealt with me for decades now, and without whom I might not have been able to finish this project.

1

Graptolites: An Introduction

Jan Zalasiewicz and Jörg Maletz

What are graptolites? To many geologists, they are somewhat scratch mark-like markings on rocks that represent one of the more strange fossil groups, lacking the ferocity of the dinosaurs, the smooth elegance of the ammonites or the charisma of the trilobites. And yet, observed closely, they represent one of the most beautiful, mysterious and useful of all of the fossil groups.

Their beauty is often concealed by the unkindness of geological preservation, all too many specimens being crushed by the weight of overlying strata, or distorted by the tectonic forces that raise mountains. They are also, simply, too small for casual human observation. Many are smaller than a matchstick, and their tiny shapes can appear as mere scratch-like markings on the rocks. Others are quite large, with some umbrella-shaped colonies in the Ordovician measuring about 1 m in diameter, and some stick-like straight Silurian monograptids measuring more than 1 m in length.

But there are – more commonly than one might think – those specimens that have managed to resist the twin pressures of burial and tectonics, perhaps because a rigid mass of pyrite (fool's gold) crystallized within their remains, or because they were encased in chemically precipitated calcium carbonate or silica before they were deeply buried. These, when looked at through a hand lens, or, better, a stereo microscope, reveal a rich diversity of extraordinary, other-worldly geometric patterns, finely engineered for purposes that we still, for the most part, can only guess at. The precision of their construction, and the distinct architectures shown by different species are, of course, key to their identification (Figure 1.1) and hence to their use by geologists.

The exquisite morphological detail can, in some specimens, extend to the finest scale of observation, where minute parts of these fossils, magnified

Graptolite Paleobiology, First Edition. Jörg Maletz.

Figure 1.1 Images of well-preserved graptolites, showing the complexity and beauty of their construction. (A) *Archiclimacograptus* sp., obverse view, SEM photo, Table Head Group, western Newfoundland, Canada. (B) *Dicranograptus irregularis*, obverse view, relief specimen, Scania, Sweden. (C) *Spirograptus turriculatus* (Barrande, 1850), proximal end, SEM photo, Kallholn Shale, Llandovery, Dalarna, Sweden. Scale indicated by 1 mm long bar in each photo.

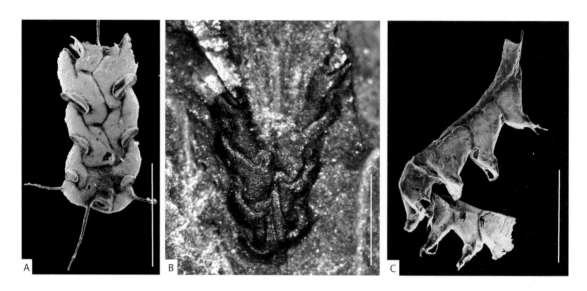

hundreds of thousands of times by an electron microscope, show traces of their original molecular architecture, relics of the biological processes that built the entire fossils but also remain largely mysterious.

Biology

Graptolites are biological enigmas of the first order. They were all colonial, and seemingly obliged to be so. A few colonies went down to just a handful of individuals, while some had thousands. They are represented today by the colonial pterobranch hemichordate *Rhabdopleura*, which, through modern taxonomic analyses, is now regarded as lying within the graptolite clade (Chapter 2). *Rhabdopleura* comprises bottom-living colonies (Figure 1.2E) that share a pattern of behaviour with corals and bryozoans. They are animal architects constructing the "homes", the collagenous tubes, in which they live. One of the major differences, however, is that their housing constructions are formed from an organic compound, not from minerals like the calcium carbonate used by the corals. *Rhabdopleura* is most closely related to the cephalodiscids (order Cephalodiscida), a second, less well organized and not truly colonial group of pterobranchs forming their tubaria from organic material in a very similar fashion (Figure 1.2 F, G).

Thus, graptolites built the robust, easily fossilizable constructions, or more precisely their tubaria, while the architects themselves, the delicate and perishable zooids of the colony, were almost never preserved in the fossil record, and we know of them only through their living representatives. The discovery of that evidence, in the 1980s (Chapter 2), in the form of the "fuselli" and "cortical bandages" with which the graptolites, quite literally, wrapped their homes, is one of the classic paradigm shifts in the whole of paleontology. Moreover, in the intricacy, complexity and integration of these homes, which were not skeletons, the planktic graptolites far surpassed the

Figure 1.2 Pterobranchs and their housing constructions (tubaria). Extant *Cephalodiscus* (A, B, F, G) and *Rhabdopleura* (C–E) to show the zooids (A–D) and their tubaria. Illustrations after Sars 1874 (C, D), Lester 1985 (B), Dilly et al. 1986 (A), Emig 1977 (F), and M'Intosh 1887 (G). Illustrations not to scale. [(A) adapted from Dilly et al. (1986) with permission from John Wiley & Sons. (B) adapted from Lester (1985) with permission from Springer Science + Business Media.]

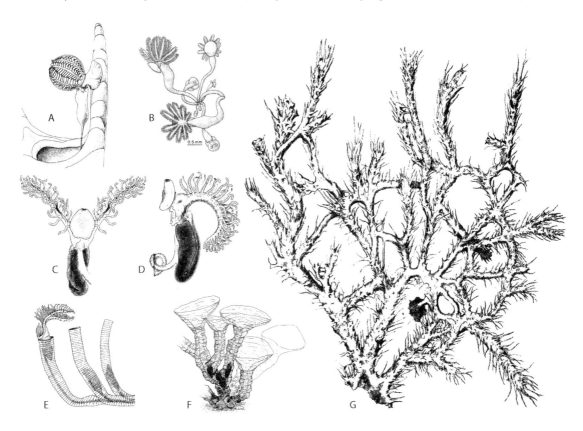

often crude and untidy constructions of the living, benthic taxa (Chapter 8), especially those of the encrusting forms.

Analysis of the command-and-control systems by which the graptolite zooids, acting cooperatively, carried out these scarcely believable constructional feats is in its infancy, while the implications for graptolite evolution, and, more widely, for understanding the evolution of animal behaviour, have scarcely been examined at all. There must be implications here, too, for the extremely rapid evolution shown by the graptolites, or, to be specific, by the planktic graptoloids (Chapter 9). Again, these implications have yet to be seriously examined. We are, in a very real sense, at the beginning – we trust – of a new phase of graptolite research.

Evolution

The planktic graptolites in particular provide splendid examples of evolution (Chapter 7). Their evolutionary changes can be followed, often stratum by stratum, through the geological column. In Darwin's concept of "descent with modification", they show clear changes in graptolite species assemblages and morphology through successions of strata and also, importantly, provide the basis for biostratigraphy.

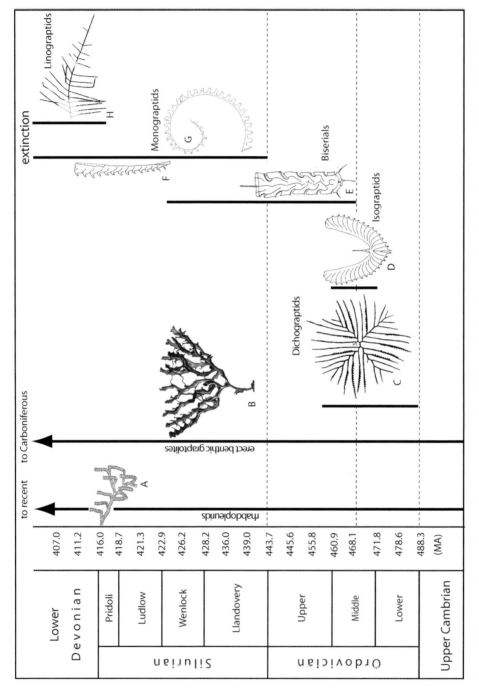

Figure 1.3 Large-scale evolutionary changes in graptoloids. (A) Encrusting benthic graptolite, *Rhabdopleura normani* Allman, 1869. (B) Benthic dendroid, *Dictyonema cavernosum* Wiman, 1896. (C) Multiramous *Goniograptus thureaui* (M'Coy, 1876). (D) Two-stiped, reclined *Isograptus mobergi* Maletz, 2011d. (E) Biserial graptolite, *Archiclimacograptus* sp. (F) Straight monograptid *Monograptus priodon* (Bronn, 1834). (G) Coiled monograptid *Demirastrites* sp. (H) Secondarily multiramous *Abiesgraptus* sp. Graptolite illustrations not to scale.

The overall pattern of change (Figure 1.3) has been clear since Lapworth's day: the change from the many-branched early forms that, already by the Lower Ordovician, settled into myriad forms of two- to four-branched dichograptids, including the classic "tuning-fork" species or pendent didymograptids (Chapter 10). Early in the Ordovician there was the development of graptolites with two "back-to-back" branches, the biserial graptolites that dominated faunas from then on, and into the early Silurian, with some then reverting wholly or partly to a uniserial state, such as the V-shaped dicellograptids or the Y-shaped dicranograptids (Chapter 11). Following the end-Ordovician crisis when graptolites nearly became extinct, the monograptid graptolites arose. It is somewhat counterintuitive that this morphology, seemingly so simple, took so long to appear. Single-stiped graptoloids, though, had been around since early Ordovician times and evolved several times independently, as can be seen in the Lower Ordovician genus *Azygograptus* (see Beckly & Maletz 1991) and the Upper Ordovician *Pseudazygograptus* (see Mu et al. 1960). The monograptids, liberated of the need to involve another stipe in their construction, rapidly evolved a dazzling – and often highly complex – range of overall forms and thecal shapes, including many variations on the spiral theme, and developed secondary branches in some cases.

There were many other innovations. At least twice in their history, graptolites found means to largely replace their solid-walled living chambers with elaborate, delicate meshworks: the archiretiolitids of the Ordovician (Chapter 11) and the retiolitids of the Silurian (Chapter 12). The latter represent the peak of graptolite complexity, at least as far as the architecture of their living chambers is concerned, and their study is a highly specialized endeavour, even within the specialist world of graptolite paleontology.

The evidence that is preserved is that of the graptolite tubaria, collected from various levels in strata in various parts of the world. Sampling by paleontologists reflects only tiny fragments of the ancient world of the Early Paleozoic. These fragments may be more or less representative of that world, but much evolution must have taken place in regions where strata were not preserved, or have not yet been recovered. Given this, what can be said about the patterns of evolution, when looking more closely?

One can look, most simply, for micro-evolutionary species lineages. Those that we recognize, of course, are all inferred, by linking morphological resemblance across successive stratigraphic levels. There are a number of seemingly clear examples, particularly well seen in those lineages where morphological change seems more or less unidirectional, and where ancestor–descendant relationships seem clear. One example is the evolution of the triangulate monograptids (genus *Demirastrites*) by elongation of the thecae, a tendency that found yet greater expression in the bizarre rastritids (genus *Rastrites*) that evolved from the triangulates (Chapter 13). There are a number of such examples, and some of these show remarkably rapid rates of morphological change when placed against a numerical timescale. The selective pressures that led to such morphological changes, and the biochemical mechanisms that controlled them, remain largely unknown.

At a larger scale, the origins of the major groups of graptolites and the architecture of the evolutionary tree have been the focus of much recent attention. In particular, there have been serious attempts at cladistic analysis (Chapter 7) that seeks to compare morphological characteristics between different groups, without reference to stratigraphic level, in order to extract information on evolutionary relationships. Advances have been made, and the origins of a number of the major graptolite groups have been traced by these means. There remain outstanding questions, but the outlines have become clearer. This is despite the patchiness of the sampling in time and space, and despite the fact that many of the key evolutionary steps involved subtle changes to the earliest-formed parts of the colony – parts that are only rarely preserved in sufficient detail to extract useful information. There is still much work to do to solve the remaining mysteries.

Stratigraphy

In a practical sense, the mechanisms that drove and shaped graptoloid evolution might be thought immaterial. The graptoloids, through the ~100 Ma of their existence, from the beginning of the Ordovician to midway through the Devonian, provide to geologists a biostratigraphical zonation that is among the best in the stratigraphic column (Chapter 6). This zonation continues to be refined, by the ever-more-precise characterization of individual graptolite taxa, by better constraints on their stratigraphic ranges, and by improved correlation between the graptolite successions in different parts of the world.

The graptolites continue to underpin much of the geological timescale of the Early Paleozoic. The fine time resolution that they provide complements and, arguably, still overshadows such well-established biozonations as those provided by benthic macrofossils such as brachiopods, and by the conodonts, the acritarchs and, more recently, by the chitinozoans.

Graptolite biostratigraphy remains highly effective (Figure 1.4) despite the fact that graptolites were, for the most part, restricted to offshore/deepwater settings, being rare, poorly diverse or absent in shallow shelf environments (Chapter 4). Furthermore, even within these deeper water settings, they occur almost exclusively in the "graptolite shale" facies that accumulated under anoxic conditions, being generally absent from the intervening "barren beds" that accumulated when oxygen (and a burrowing biota) reached the deep sea floor. This may reflect a preservational bias, as graptolites probably flourished in general under normal marine conditions, when the sea floor as well as the sea surface was oxygenated. In these conditions the organic tubaria have much less chance of preservation because of scavenging by bottom-dwelling organisms (see Chapter 5).

Major advantages of the graptolites as biostratigraphical index fossils include their size relative to microfossils, such that preliminary identifications may be made in the field, and the distribution of the living (and dead) colonies through transport by marine currents into regions where they may not have lived, but which enhance their value to the biostratigrapher, particularly in rock successions where no other fossils can be found.

Furthermore, biostratigraphical assignments in practice are often made on the basis of a small amount of material, perhaps only a handful of incomplete specimens. Indeed, in some cases a single fragment may be enough to establish the presence of a biozone. This reflects the extraordinary morphological complexity and diversity of these fossils, which can make even fragments commonly distinctive and identifiable to species level. It also helps that the graptolites, unlike palynomorphs, were only very rarely reworked into younger strata, because they rapidly became brittle and friable after burial.

Hence graptolites have been key to the unravelling of the geological structure of many regions where strata of Early Paleozoic age dominate (Chapter 6). For instance, the Southern Uplands of Scotland were famously interpreted by Charles Lapworth in the mid-19th century as comprising multiple repetitions of strata, and with more refinement from the 1970s on as one of the best examples in the world of a fossilized accretionary prism. The structures of the Welsh Basin, too, and of parts of the Appalachians and other mountain ranges around the world, have been deciphered with the help of graptolites.

Going beyond "abstract" regional studies, graptolites have been key to resolving major economic deposits hosted within strata of Early Paleozoic age, such as the Bendigo goldfields of Australia. Today, they are key to working out the structure of some of the world's most important oil source rocks (in the Middle East and north Africa, for example) and more recently in the identification of shale gas horizons.

The material of which graptolite tubaria were made, formerly termed periderm, a term that is no longer used by a number of graptolite workers because it is not a "skin", also has its uses. Originally transparent, it progressively changed its colour on progressive deep burial and heating, from straw-yellow to orange to brown and finally to black, which becomes "shinier" (i.e. has progressively greater reflectance) on further burial and heating (see Chapter 5). In this way, graptolites can be used as a kind of geothermometer, to determine the highest temperatures that buried rock strata once reached, and therefore to determine the history of the hydrocarbons that they contained.

Ecology

In exploring the ecology of the graptolites (Chapter 4), there is much still to study. The benthic graptolites have clear analogies with such filter-feeding organisms as sponges, bryozoans and others, and indeed the ecology of the living pterobranchs themselves may be studied.

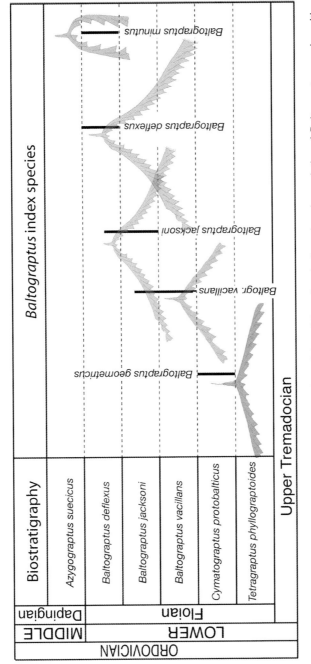

Figure 1.4 Graptolite biostratigraphy of the Floian, Lower Ordovician of Scandinavia, based on the evolution of *Baltograptus* species with subhorizontal (*B. geometricus*) to pendent (*B. minutus*) habit, as an example of biostratigraphical subdivision of an evolutionary lineage. Graptolite illustrations not to scale.

However, the planktic graptolites have no close analogues in the modern plankton, and it helps little here that virtually all of the evidence that we have of them is that of their robust homes or housing constructions, the tubaria, with no evidence left of the animals themselves. Because of this, there have arisen great differences in opinion on how the planktic graptolites may have swam, fed, reproduced, and so on. There have been ingenious attempts at reconstructing the life mode of graptolites, both physically, for instance by the use of model graptolites placed in wind tunnels to simulate ocean currents, or by computer models placing "electronic graptolites" in "electronic water".

Exploration of the world of the planktic graptolites is hindered because so little of the other zooplankton, with which they would have shared the water column and competed, and which they would likely have preyed upon and been the prey of, have not been preserved. Graptolites are the only consistently preserved macro-zooplankton from the early Paleozoic, although accompanied by various microfossils (Figure 1.5): chitinozoans (possibly the eggs of some as yet unknown creature), acritarchs, radiolarians, conodonts, scolecodonts and, in places, some of the more robust planktic arthropods and scattered problematica. Almost nothing has remained of a likely host of jellyfish, arrow-worms, comb-jellies, and delicate euphasiid-style arthropods that would clearly have influenced the life of graptolites. Even the benthic graptolites are poorly known in this respect, as many of their remains are preserved after they were transported from the shallow sea floors on which they lived into deeper water strata.

These other planktic organisms almost certainly played some part in the rollercoaster, "boom-and-bust" evolutionary history of the graptolites, where rapid changes in morphology, disparity and diversity were punctuated by periodic crashes in species numbers, followed by re-radiation from surviving stocks (Chapter 7). These changes were clearly linked to oceanographic changes, and perhaps with the supply of oxygen to deep waters. Counterintuitively, graptolite diversity typically decreased with increases in oxygenation, suggesting that these organisms might have adapted to, or preferred, low-oxygen conditions. When the low-oxygen waters shrank, then graptolites commonly became more prone to extinction. But this general link of graptolite diversity patterns to ocean oxygenation may not have been mediated through such direct cause and effect. Levels of ocean oxygenation broadly reflected global climate, then as subsequently, and the graptolites may have been sensitive to other parameters, as yet unknown, that were more directly linked to climate. A major factor affecting the evidence is that, as noted above, graptolites are much more easily preserved in black shales formed on oxygen-starved and hence scavenger-free sea floors (Chapter 5); nevertheless, some of the most notable reductions in graptolite species diversity seem to be clearly associated with major changes in climate and ocean chemistry.

The last of the bursts in graptolite diversity in the Lower Devonian was the final one, and the planktic graptolites disappeared forever, while the benthic dendroids persisted until the Carboniferous, and a few encrusting pterobranchs remain with us today (Figure 1.3) although these last have left very few traces of their long existence in the stratigraphic record. The disappearance of these extraordinary organisms from the sea has bequeathed to us a magnificent fossil record, and some of the greatest puzzles in paleontology.

Paleogeography

While graptolites are often cited as ideal zone fossils because of their wide geographical distribution, a more nuanced analysis shows that they, like modern zooplankton assemblages, can be resolved into assemblages controlled by oceanographic setting, water temperature and other factors (Chapter 4). Thus there has been, for instance, the recognition of continental (Laurentian) and oceanic assemblages around the North American continent (Figure 1.6A), and of low-latitude and high-latitude faunas (Figure 1.6B), while development of the latter concept has helped refine paleoclimate reconstructions of the late Ordovician, in indicating the shifting location of such features as the Arctic Front.

A continuing debate among graptolite paleontologists has been the relative importance of depth control versus control by lateral separation of water masses of different properties. Because planktic graptolites as fossils are, by definition,

Figure 1.5 Organisms associated with graptolites. (A) Silicified ostracod carapace. (B–C) Phosphatic brachiopods. (D–E) Conodont elements. (F) *Obruchevella*, silicified microbial organism. (G) Radiolarian. (H) Chitinozoan. (I) Phyllocarid fragment in shale. Specimens from the Middle Ordovician of eastern Canada.

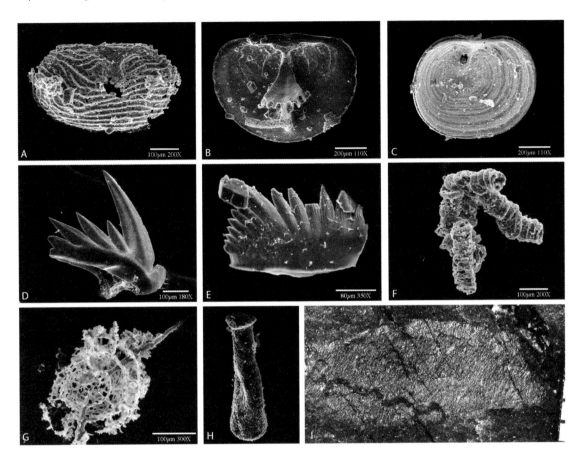

never found in their life habitat, but have fallen through the water column to rest, in death, on the sea floor, this has been a difficult question to resolve, with interpretations favouring both models, or variations of them, having been suggested in recent years.

Colony Shapes

The planktic graptoloids, in particular, developed an enormous diversity in colony shapes (Chapters 9–13), far outcompeting the benthic taxa, which had stricter constructional limitations on their tubaria imposed by life on the sea floor. Benthic graptolites either encrusted the sediment surface in various ways or reached upwards into the water column, their stipes and branches forming erect, bushy or fan-shaped colonies (Chapter 8). Throughout their development they were more affected by the local environmental conditions of currents and sediment input than were their planktic cousins. Thus, their shapes tended to be irregular as is seen in recent bryozoan colonies and even in corals, growing towards the light, but being unable to overcome their limitations. These two groups in some ways provide the closest living analogues to the numerous fossil benthic graptolite taxa of the Paleozoic.

With the emergence of the planktic graptoloids, a new chapter of graptolite colony design and construction was opened. Their intricate thecae and

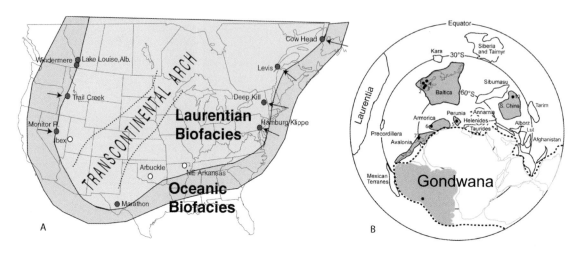

Figure 1.6 Paleogeography of graptolites. (A) Biogeography of Laurentia showing oceanic and shallow water, epicratonic (Laurentian) biofacies (based on Maletz et al. 2005a, Fig. 4). (B) Biogeographical distribution of the Floian dichograptid *Baltograptus* in high latitudes (from Goldman et al. 2013). The areas in grey show the distribution of *Baltograptus* species. (Map based on Egenhoff & Maletz 2007, Fig. 2.)

elaborate branched colonies are only understandable in terms of their free-swimming lifestyle. This was a true revolution in many aspects and brought about a fundamental change in the organization of graptolite colonies. The graptolites threw off the limitations imposed by their old benthic lifestyle and evolved a freestyle architecture unrivalled by any other organisms (Chapter 9). The colony shapes became more regular and typically highly symmetrical, since maintaining themselves in the water column was much more easily achieved with a more symmetrical and balanced design. This was acquired through close integration and coordination of the individual zooids and the thecae forming the individual stipes, so that no single stipe could grow faster than another. Initially, there arose umbrella-shaped swimming colonies like that of *Rhabdinopora*, the closest relative of the benthic *Dictyonema*.

Slightly later in the evolution of the group, still in early Ordovician times, a great reduction in the number of stipes took place and led to many additional changes in colony design. Four-, two- and even single-branched designs quickly evolved in the Early to Mid Ordovician, with a variety of branching patterns developing in the colonies. While the thecal development became simplified

in one respect, losing some of the more complex developmental aspects of the early dendroid graptolites (see Chapter 10), at the same time the thecae became morphologically more complex, with the development of a wide range of modifications of the thecal apertures.

Seemingly small constructional changes may have led to enormous modifications of the colonies, as shown by the example of the development of cladial (secondary) branching. Such many-branched colonies evolved many new designs: for example, xiphograptids of the genus *Pterograptus* (Chapter 10), the beautifully curved *Nemagraptus* (Chapter 11), and the spectacular cyrtograptids (Chapter 13) with their regularly developing single cladia (Figure 1.7A).

One of the greatest evolutionary breakthroughs in the transformation of their colonies was the construction of the biserial colony of the Axonophora (Chapter 11). Complex growth patterns within the proximal thecae enabled development into a biserial colony with two stipes growing parallel to each other in a back-to-back fashion (Figures 1.7 F, G). This design dominated the Ordovician and early Silurian graptolite communities, with the emergence of a multitude of new genera and species. This seemingly simple scheme became elaborated by spines, flanges,

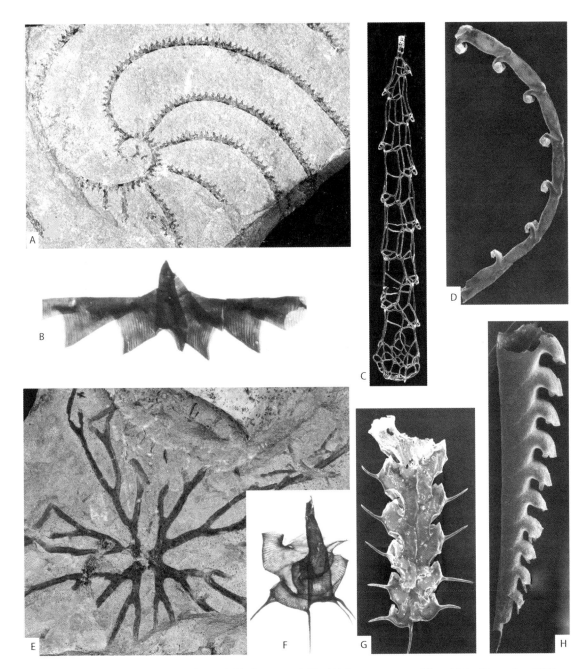

Figure 1.7 Colony variation in Ordovician and Silurian graptoloids. (A) *Cyrtograptus multibrachiatus* Bjerreskov, 1992, Arctic Canada. (B) *Expansograptus abditus* Williams and Stevens, 1988, Cow Head Group, western Newfoundland, Canada. (C) *Neogothograptus balticus* (Eisenack, 1951), glacial boulder, Northern Germany. (D) *Streptograptus galeus* Lenz & Kozłowska, 2006. (E) *Paradelograptus norvegicus* (Monsen, 1937), Fezouata Biota, Morocco. (F) *Amplexograptus maxwelli* (Decker, 1935). (G) *Dicranograptus nicholsoni* Hopkinson, 1870, Viola Limestone. (H) *Monograptus priodon* (Bronn, 1835), Arctic Canada. Illustrations not to scale. Photos by A.C. Lenz, J.C. Gutiérrez-Marco and J. Maletz.

genicular rims, and modifications of the thecal shape – there is an astonishing range of such variations. But still, this was not the end, as external complex systems of bars forming "fences" around the colonies were explored by the archiretiolitids. Later, and independently, the Silurian retiolitid graptolites evolved a way of simultaneously lightening and strengthening their tubaria with the help of a system of fine meshes laid down on very thin fusellar layers (Figure 1.7C).

During the Silurian the monograptids (Figures 1.7D, H), with a single stipe growing straight up from the sicula, started to dominate the faunas (Chapter 13). The biserial Axonophora diminished and eventually vanished in the mid-Silurian. Another complete reorganization of the construction of the colonies of planktic graptolites happened, but this time in a less dramatic fashion. The new design had appeared already in the basal Silurian, but tracing its origins has proved difficult. Initially, the monograptids were inconspicuous elements in the basal Silurian graptolite faunas and their evolution only slowly gained speed, before they rapidly became the overwhelmingly dominant component of graptolite faunas, with new colony and thecal designs.

The end of the reign of planktic graptolites in the world's oceans came not unexpectedly. After many smaller and larger extinction events in the Silurian, the diversity of the planktic graptolites reduced and their colony designs became severely limited and generally conservative. Only a few straight monograptids survived into the Lower Devonian, even though a short-lived bloom of multiramous taxa (*Abiesgraptus*, *Linograptus*) briefly recreated a diversity that had long been lost. Thus, the end came slowly and was barely noticeable: the world's oceans lost a group of organisms that long represented a major player in the ecosystem, while other groups positioned themselves and became ready to take over.

While unsurprising, the end of the planktic graptolites, their final extinction, is still not explained in detail. The main problem in understanding the interconnections between fossil groups remains the lack of sufficient data. Many groups of organisms do not leave fossil remains and are thus untraceable in the fossil record. The main players in the marine ecosystems of the Paleozoic likely have not been discovered and, without knowledge of these, our explanations are incomplete and questionable. However, we may reasonably postulate that the graptolites, after major blooms in the Ordovician and Silurian, had competitors in their marine ecosystem that eventually outcompeted them and led to their demise. With such little understanding even of the graptolites' lifestyle, it is hard to gain a better insight into their interactions with other groups of organisms, to reconstruct the environment in which they flourished and from which they were finally expelled. However, one intriguing point to note is that, since the disappearance of the planktic graptolites, no other marine group of organisms has evolved into a similar type of macro-zooplankton that was able to leave evidence of its presence. Modern planktic organisms are usually small, grow extremely fast and are present in large numbers but do not sacrifice either time or energy to construct a preservable housing system.

What remains of the graptolites, and their cousins within the pterobranchs, the cephalodiscids, are a number of small colonial and pseudocolonial, benthic organisms. Their origin can be traced back to the Mid Cambrian, when colonial pterobranchs were already a common component of the ecosystem. Thus, the benthic pterobranchs were either more successful or just lucky to survive into our modern days. They had a very successful time during the Late Cambrian to Carboniferous, but then their diversity diminished and most taxa extinguished, leaving the modern cephalodiscids and rhabdopleurids as the sole survivors. During their early days the benthic graptolites were one of only a few known groups of colonial organisms inhabiting the shallow water marine environments. They flourished in these shallow waters, but competition, probably with the bryozoans, which developed a similar lifestyle, may have been too strong for them in the end. Pterobranchs can be found in many environments today, from coastal sandy beaches to deep-water Arctic and Antarctic regions, but they are usually only a small, inconspicuous component of the faunas. It is said that rhabdopleurids can only be found close to marine research stations, but this may be regarded as an inside joke, as this is the only place where scientists search for them. Looking back, we can say that graptolites have been around for the last 500 million years, much longer than mankind will ever last on this planet, and have accompanied the evolution of life until today. They had their heyday long ago, but, if left undisturbed, may quietly last another 500 million years – a symbol of hope for the future of our planet.

History of Research

As is the case with many fossil groups, misunderstanding and incomplete knowledge of graptolite anatomy and comparison with unrelated taxa led scientists to many unusual interpretations about graptolites (Figure 1.8A, B, E). The example of the synrhabdosome interpretation of Ruedemann, with the – as it turned out – nonexistent "floats" and other hypothesized structures of these "supercolonies", is only one example of such misunderstanding, but certainly will not be the last. It all started with the genus *Graptolithus* erected by Linnæus (1735), a term used for supposedly

inorganic markings on rocks. We now recognize the *Writing in the Rocks* – the title, later, of a popular book on graptolites (Palmer & Rickards 1991) – as paleontological reality, and we can now understand this kind of "writing" and interpret it in a paleobiological sense. However, it took more than 250 years for scientists to gather the information we have at hand today, and to relate the fossil graptolites to the living pterobranchs.

Many scientists have devoted their time to the enigmatic and elusive graptolites and provided us with their insights and interpretations (Chapter 15). What counts at the end is the result, while the many blind alleys that were pursued along the way ultimately helped set us on the path to our modern understanding. Even what we now perceive as errors along the way were useful and even important, as the example of Nancy Kirk and her extrathecal tissue model of graptolites shows. Here a suggestion that was shown to be wrong in hindsight produced enormous amounts of discussion, new research and fruitful ideas. The recent understanding of graptolites as a largely fossil group of pterobranch hemichordates would not

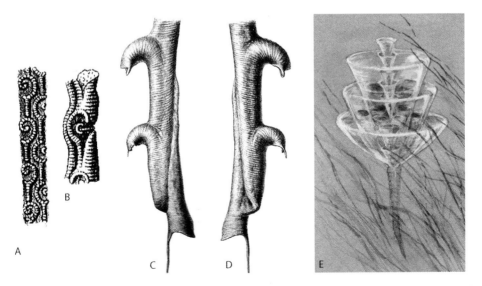

Figure 1.8 (A, B) Reconstruction and interpretation of indeterminable biserial graptolite fragment from Richter (1853, pl. 12, Figs 12, 13). (C, D) Excellent illustration by Georg Liljevall (ca. 1895) for Gerhard Holm, *Monograptus priodon* (Bronn, 1835), proximal end in obverse (C) and reverse (D) views (published in Bulman 1932a, pl. 1, Figs 10, 11). (E) Erroneous reconstruction of biserial graptolite with complex float structure by Franziska Zörner-Bertina (ca. 1950) for Rudolf Hundt, unpublished (original at Naturkundemuseum Gera, Germany; provided by Frank Hrouda, Gera).

have been possible without her provocative and thoughtful suggestions.

Outlook

As a short overview of graptolites and their development, this chapter is just the starting point for the reader to explore an unusual but fascinating group of organisms. Generally thought to be extinct, graptolites are recently recognized to include a few extant living members and provide deep insight into more than 500 million years of biological evolution on our planet. Surprisingly little change happened during this long time interval to the benthic taxa that represent the graptolites, and their closest ancestors from the Ordovician times have apparently not changed much at all. However, this is an inference from their tubaria, the housing construction, since information on the soft-body anatomy of the fossil taxa is not available. Will it be possible in the future to find fossilized graptolite zooids? Perhaps this is destined to be forever a vain hope, but graptolite workers will go on searching for direct clues to the architect and builder of some of the most remarkable constructions in the history of life on this planet.

2 Biological Affinities

Jörg Maletz

All organisms on this planet are in some way interconnected and phylogenetically related. This is why a single tree of life can be used to describe biological evolution on our planet, as Charles Darwin, Ernst Haeckel and others realized in the 19th century. Some organisms can be more easily related to other groups, while it is difficult to relate others. When we look at the birds and its early member *Archaeopteryx lithographica* from the Solnhofen Limestone of Germany, we can still imagine the problems associated with the comparison of living organisms and fossils. As paleontologists we remember the discussion on the feathered dinosaurs and whether they are related or not to the modern birds, the only living organisms with feathers. The discussion ended with a valuable solution and we can now recognize birds as feathered dinosaurs, talking about "non-avian" and "avian" dinosaurs.

The further we go back in time – in geological history – the more difficult it is to compare the fossil organisms with our modern world. The Ediacaran world would have been extremely strange to us, and only the Cambrian explosion brought to light most of the modern phyla. During the Cambrian explosion the graptolites emerged, first as small and inconspicuous animals, probably from some worm-like organisms similar to modern enteropneusts, but soon the pterobranchs with their typical housing construction, the tubarium, must have originated. We know very little of this time of early evolution, but fairly recently we realized that the graptolites are not that strange and easily incorporated into our zoological system. *Rhabdopleura* and *Cephalodiscus*, two extant members of a very old lineage of organisms have provided us with a glimpse into the distant past. They provide the key to our modern understanding of the supposedly long extinct phylum of the Graptolithina.

Graptolite Paleobiology, First Edition. Jörg Maletz.
© 2017 John Wiley & Sons Ltd. Published 2017 by John Wiley & Sons Ltd.

Graptolites as Organisms

Graptolites are an unusual group of organisms, and in the past there were considerable difficulties in trying to compare them with other (modern) groups of animals. There has long been speculation on their phylogenetic relationships, but initially this did not have any scientific basis. For a very long time graptolites were considered extinct and a comparison with extant animals was rarely attempted. Originally, graptolites were even interpreted as inorganic markings on rocks by Linnæus (1735, 1768), later as cephalopods (e.g. Walch 1771; Bronn 1835; Geinitz 1842), and were identified as an odd group of bryozoans (e.g. Ulrich & Ruedemann 1931) or as polyps and hydroids (Hisinger 1837; Murchison 1839; Portlock 1843; Geinitz 1852). Thus, the graptolites long rested among various groups of benthic, colonial organisms before their real phylogenetic relationships were finally realized and understanding of these organisms became settled. They are now placed among the Hemichordata (Beklemishev 1951a,b), where they are included as the subclass Graptolithina (Mitchell et al. 2013; Maletz 2014a). The extant pterobranch *Rhabdopleura* is now regarded as the only extant member of the graptolites. It survived from the early Paleozoic until today, apparently with little change in the nature of its tubarium, and may be regarded as a living fossil. Tubaria comparable in all details with those of the extant *Rhabdopleura* have been chemically isolated from limestones of Ordovician age, showing a long existence of the group during which little or no changes were produced in their housing construction. Nothing, however, can be said about the anatomical changes of the soft-bodied organisms producing this construction, as there is no fossil record available.

The Hemichordata include three groups: the Enteropneusta, the Pterobranchia and the rare Planctosphaeroidea. This relationship, based initially on the tripartite body of the living members (Figure 2.1A, C), is supported through recent DNA analyses (e.g. Halanych 1995; Swalla & Smith 2008; Cannon et al. 2009, 2013, 214). However, the exact phylogenetic relationships between the two main groups of the Hemichordata, the Pterobranchia and the Enteropneusta, remain uncertain. The Pterobranchia may either represent a sister group of the Harrimaniidae and originate from within the Enteropneusta, or they represent a sister group of the Enteropneusta (e.g. Cameron et al. 2000; Winchell et al. 2002; Cannon et al. 2009; Osborn et al. 2012). Peterson et al. (2013) found microRNA support for a monophyly of the Enteropneusta and interpreted the Pterobranchia as a sister group of the Enteropneusta. The data also supported a monophyly of the Hemichordata.

Hemichordata

When Bateson (1885) introduced the term Hemichordata, he meant to show a close relationship to the Chordata. He suggested a number of characteristics indicating this relationship, including the presence of a notochord, the assumed precursor of the backbone of the vertebrates. A notochord is possibly already present in some early Cambrian fossils such as *Pikaia* and *Haikouella*, but these interpretations are tentative and impossible to verify from the fossil record. Hence, the relationships of these fossils are controversial. The stomochord of the Enteropneusta has been compared to the notochord of the chordates, but according to Newell (1952) and Ruppert (2005) may be no guide to our understanding of hemichordate relationships. Harmer (in M'Intosh 1887) initially described a notochord in *Cephalodiscus*, but this notion received considerable criticism (e.g. Masterman 1897) and this feature is now termed the stomochord to indicate its independent origin.

The Hemichordata include largely extant, marine organisms with an elongated, worm-like body. They are benthic animals with a body subdivided into three distinct portions: the proboscis or head, the collar, and the trunk regions (Figure 2.1A). The proportions and shape of these parts can be modified considerably according to the lifestyle of the animals. In the Pterobranchia (Figure 2.1C), the body is small, with a rounded, short trunk, modified internal organs including a U-shaped gut, and one or several pairs of ciliated arms for food gathering on the collar. Pterobranchs either live as individuals or as colonial organisms in a housing construction, the tubarium, actively secreted from glands on the proboscis (Figure 2.1B).

Figure 2.1 Adults of Enteropneusta and Pterobranchia. (A) The enteropneust *Saccoglossus* sp. (based on Bulman 1970a). (B, C) *Cephalodiscus* sp. (adapted from Lester (1985) with permission from Springer Science + Business Media). (B) Part of colony with zooids sitting at the openings (apertures) of the tubarium. (C) Single zooid at the apertural spine of a tubarium, showing the main body parts.

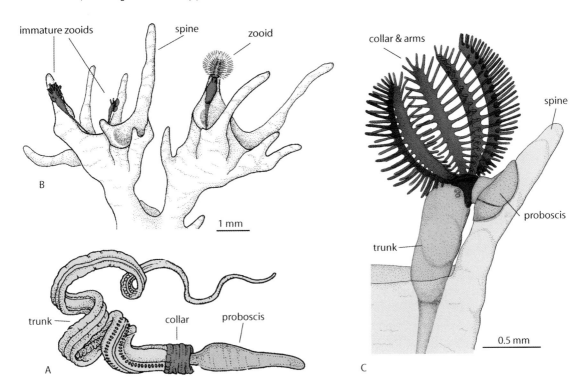

Hemichordates have rarely attracted scientific interest in the past, as they are inconspicuous in most faunas, and scientists have rarely looked at their taxonomy and ecology. However, they have recently been considered as close to the origin of the deuterostomes, and thus of the early evolution of the vertebrates (e.g. Gerhart et al. 2005; Röttinger & Lowe 2012) and in line to the ancestry of man. Halanych (1995) interpreted the Hemichordata as a sister-group to the Echinodermata, with which they form the Ambulacraria (Figure 2.2), more recently supported by Philippe et al. (2011).

The Hemichordata include three main groups: the Enteropneusta (acorn worms), Pterobranchia and Planctosphaeroidea. Of these, the Planctosphaeroidea (Figure 2.3H) are the most curious and unexplored of all hemichordates. Originally described by Spengel (1932), the genus *Planctosphaera* and its only species *Planctosphaera pelagica* are known from a small number of specimens and have been observed alive only once (Hart et al. 1994). The taxon is now understood as a gigantic larval form, possibly of a deep-sea enteropneust (Damas & Stiasny 1961; Scheltema 1970), but positive evidence for this suggestion is still lacking. The anatomy is similar to the *Tornaria* larvae of the enteropneusts, but the specimens are much larger, up to 22 mm in diameter, while the typical *Tornaria* larvae of many enteropneusts (Figure 2.3B) are generally only ca. 1 mm in diameter.

The acorn worms or Enteropneusta are bottom-dwellers and live as infaunal or epifaunal elements in all marine regions. The ontogeny of the Enteropneusta is variable and ranges from a direct development in the Harrimaniidae to taxa with an indirect development through a planktic tornaria larva (see Figure 2.3A–G) in the Ptychoderiidae and Spengeliidae. The exact development of the

Figure 2.2 The phylogenetic relationships of the Hemichordata (based on Maletz 2014a).

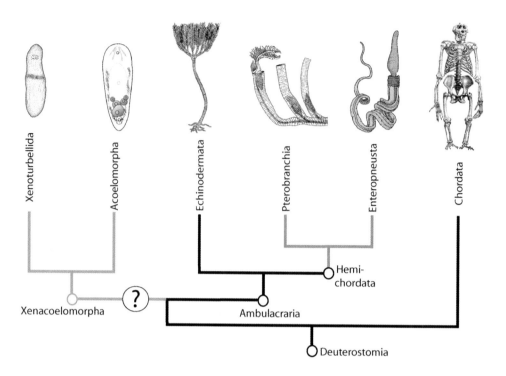

recently discovered deep-sea enteropneusts of the Torquaratoriidae is unknown. Asexual reproduction is present in *Balanoglossus capensis* (Gilchrist, 1908, 1923) and may be found in other taxa. Juvenile specimens of this species are able to shed pieces of their tails which develop into new individuals.

In recent years, with the increasing interest in marine biodiversity, the Enteropneusta are investigated in some detail and new species have been recognized. Many of the new taxa that have been described in the last decade show surprising ecological adaptations, providing evidence of a wide ecological and geographical distribution. Enteropneusts seem to be present in every imaginable marine environment and have clearly entered numerous ecological niches during the long geological timespan of their existence.

Unfortunately, little is known of the evolution of this group, as enteropneusts are rarely preserved in the fossil record (Maletz 2014b). They possess a slender and highly fragile body without a preservable covering which is only preserved under very special circumstances. The oldest taxon referred to the Enteropneusta is *Spartobranchus tenuis* (Walcott, 1911b) from the middle Cambrian Burgess Shale of British Columbia, Canada. Caron et al. (2013) recently described this species in some detail and recognized that it is associated with a burrow formed from fibrous material. They suggested that this material may represent a precursor to the tubarial material of the pterobranchs.

Mazoglossus ramsdelli Bardack, 1997 (Figure 2.4) is another one of the few fossil species of acorn worms. A number of specimens have been found in the marine "Essex Fauna" of the Mazon Creek Biota of Carboniferous age of North America. The specimens are poorly preserved in siderite concretions as flattened outlines. Organic material is not found and the material can be recognized only as a slightly paler shadow on the sediment surfaces. As the specimens clearly show a tripartite body outline, they can be identified as enteropneusts, but further anatomical details are not available. The few Mesozoic records of fossil enteropneusts

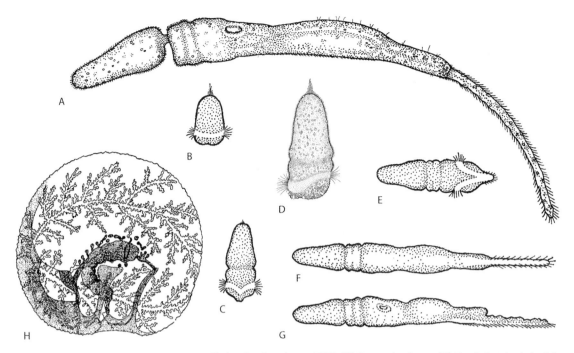

Figure 2.3 Ontogeny of *Saccoglossus horsti* (after Burdon-Jones 1952). (A) Burrowing larva. (B) Newly hatched planktic larva. (C–D) Settling larva in different views. (E) Post-settlement creeping larva. (F–G) Burrowing larva in ventral (F) and lateral (G) views. (H) *Planctosphaera pelagica* (after van der Horst 1936). Illustrations not to scale.

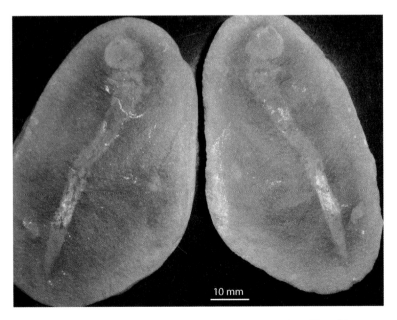

Figure 2.4 *Mazoglossus ramsdelli* Bardack, 1997. Counterparts, Mason Creek Biota. The white spots on the specimen may be an effect of recent weathering.

include a number of poorly preserved specimens up to 60 cm long. They have been referred to the genera *Megaderaion* and *Mesobalanoglossus*, but also do not show much detail and cannot be compared with modern taxa (Maletz 2014b).

Modern acorn worms can be found as miniature interstitial organisms in sandy regions, as shallow water infaunal organisms in beach areas, or as epifaunal elements in deep-water regions, where they search the sediment surface for organic matter. The size of their worm-like bodies ranges from less than 1 cm to more than 2.0 m (*Balanoglossus gigas* from Brazil). One species, *Saxipendium coronatum* Woodwick and Sensenbaugh, 1985, has been discovered at the Galápagos Rift hydrothermal vents. Here, these "spaghetti worms" live at nearly 2500 m water depth attached by the posterior part of the trunk. The anterior part of the worm drifts in the water, sampling the water column for food. The animals are often more than 20 cm long and about 0.5–1.0 cm wide, of yellow-white colour and very fragile. They occur in large masses of specimens, covering rocky surfaces around the geothermal vents.

Enteropneusts have been discovered in yet more varied environments. One miniature species, *Meioglossus psammophilus* Woorsae et al., 2012, less than 0.6 mm long, exists as a meiofaunal element in sandy, marine environments, where it went undiscovered until recently (Worsaae et al. 2012). The specimens were recovered from reefal sands at a water depth of up to 15 m off Belize and Bermuda. More such miniature enteropneusts probably remain to be discovered, since these small organisms are easily overlooked.

Shallow-water enteropneusts usually live as infaunal elements and are recognized mainly through their fecal remains found in tidal flat regions (Figure 2.5). They live in simple U-shaped to complex burrows (see Stiasny 1910), the walls of which may be coated by a slimy material to stabilize the tubes in the soft sediment. According to van der Horst (1934), the coiled burrow of *Saccoglossus inhacensis* is lined by a thin layer of clear sand that is easily seen against the dark surrounding muddy sediment.

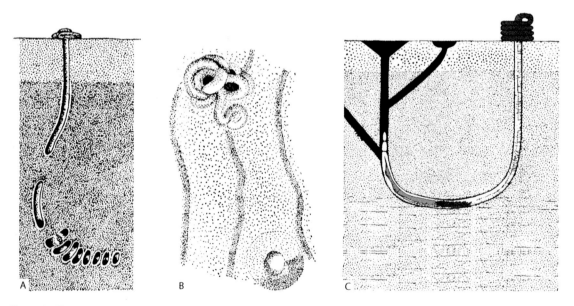

Figure 2.5 Enteropneust burrows and traces (after Hyman 1959, Fig. 50). (A) *Saccoglossus inhacensis*, spiral burrow. (B–C) *Balanoglossus clavigerus*, funnel opening and fecal coil as surface indication of burrow (B) and scheme of U-shaped burrow (C). Illustrations not to scale.

Modern deep-sea enteropneusts commonly form complexly meandering traces (Smith et al. 2005). They may also have formed some of the typical deep-water trace fossils of the Phanerozoic for which the producer has never been found. Support for this was provided by Halanych et al. (2013), who showed that modern acorn worms form transparent tubes, possibly via glands or specialized organs at the base of the collar, to which they seem to be attached. These tubes are stable for up to several weeks and thus may be preservable in the fossil record.

Pterobranchia

The Pterobranchia as the second important group of the Hemichordata are mostly known only to a small group of paleontologists and biologists. The paleontologists call their pterobranch fossils "graptolites", but the relationship to the extant pterobranchs – even though often suggested – has only recently been verified through a combined cladistic analysis of extant pterobranchs and fossil graptolites. Mitchell et al. (2013) differentiated the Pterobranchia into two groups, the individual to pseudo-colonial Cephalodiscida and the truly colonial Graptolithina. Both share many similarities of the zooidal anatomy (Plate 2), but also the habit of secreting a housing – the tubarium (Figure 2.6) – from an organic material similar to the chitin of arthropods. This material is secreted from glands on the proboscis of the animal and is highly durable in the fossil record. The oldest remains of graptolite tubaria are found in the Middle Cambrian (Harvey et al. 2012). These small fragments preserved indications of the fusellar construction and do not differ at all from much younger material. The chemical composition of the tubarium material has never been

Figure 2.6 Tubaria of extant and fossil Pterobranchia. (A) *Rhabdopleura compacta* Hincks, 1880, extant, part of small colony showing fuselli, SEM photo. (B) *Streptograptus sartorius* (Törnquist, 1881), fossil, infrared photo showing fuselli. (C) *Pseudoglyptograptus vas* (Bulman & Rickards 1968), fossil specimen in reverse view in shale, showing the median septum and the interthecal connections (common canal) of the two stipes of an axonophoran graptolite, vague striation indicates the fusellar construction. Scale bar indicates 0.1 mm in (A–B) and 1 mm in (C).

successfully analysed and the composition remains unknown (see Sewera 2011). The fossil material consists of an aliphatic polymer (Gupta et al. 2006; Gupta 2014), but originally it may have been an organic compound, possibly a scleroproteic material, most likely collagen (see Towe & Urbanek 1972; Bustin et al. 1989).

Chemically isolated graptolite specimens may be hundreds of millions of years old, but under exceptional conditions their tubaria may remain preserved (Figure 2.6B) in ultrastructural detail and retain their original flexibility. Thus, the existence of pterobranch hemichordates can be documented from the presence of their typical tubaria or pieces of these in many sedimentary rocks. Their general construction and composition has not changed since Cambrian times, as can be seen in pterobranch fragments found in the Middle Cambrian Kaili Formation of China (Harvey et al. 2012). Even though consisting of small fragments, the development of fusellar structure clearly indicates a pterobranch origin for the material.

Allman (1869) first described a representative of the Pterobranchia as *Rhabdopleura normani* and referred the taxon to the Polyzoa (now Bryozoa or moss animals). It was clearly recognizable as a colonial organism through the dark rod (the stolon) visible in the tubarium, connecting the individual zooids. At the same time Michael Sars (1868) had found a similar organism, initially named *Halilophus mirabilis*, but later described under the name *Rhabdopleura mirabilis* by his son (George Ossian Sars, 1872, 1874). The second genus of the Pterobranchia, *Cephalodiscus*, was introduced by M'Intosh (1882, 1887) and was instantly recognized as related to *Rhabdopleura*. M'Intosh (1882) identified the free-roaming zooids of the tubarium as one of the main differences to *Rhabdopleura*, and also noted a number of other differences in the anatomy of the zooids.

The biological descriptions of the extant pterobranchs concentrated on the anatomy and interconnection of the zooids and less on the secretion of the tubaria, by contrast with the study of the fossil members, the graptolites. These are only known from the fossilized remains of their tubaria, and nothing is known of the anatomy of the zooids of these common fossils. Experiments show that the zooids of modern pterobranchs decay to an unidentifiable organic mass within a few days of death (Briggs et al. 1995), while the tubaria persist for much longer. Thus, the chance of preservation of the delicate and small pterobranch zooids is negligible. The few records of fossil zooids (e.g. Bjerreskov 1978; Rickards & Stait 1984; Durman & Sennikov 1993; Loydell et al. 2004) refer to mineralogical differentiations and replacement at places around and in the tubaria, where zooids may be expected, but these did not show zooidal shapes or anatomical details. A tentacle-bearing, tube-dwelling, soft-bodied organism, *Galeaplumosus*, from the early Cambrian Chengjiang Formation of China, was assumed to be an early pterobranch (Hou et al. 2011), although this interpretation has been challenged recently. Maletz (2014b) interpreted the single available specimen as a fragmentary fossil of unknown origin. If indeed a pterobranch, *Galeaplumosus* would be the oldest available specimen, but better material is needed to confirm this. Maletz (2014b) also questioned the interpretation of *Herpetogaster* (Caron et al. 2010) as a possible basal pterobranchs, and considered this claim as unproven.

The record of the extant pterobranchs *Rhabdopleura* and *Cephalodiscus* was preceded by the description of many fossil members of the clade under the general term "graptolites", but a link between them was not made. Linnæus (1735, 1768) first introduced the term *Graptolithus* for a number of features then understood as inorganic markings on rock surfaces. The term was quickly translated to "graptolite" and became the word used for these fossils. It was probably Wahlenberg (1821) who first realized that some of the materials of Linnæus were true fossils. Walch (1771) had earlier described and illustrated graptolite remains as fossils, but referred them to the "orthoceratites" (straight cephalopods in modern terminology), a common practice at the time. That graptolites were related to the pterobranchs was recognized by Kozłowski (1947) following detailed work on chemically isolated fossil specimens.

Communality and Coloniality

All Pterobranchia share a number of characteristics showing their phylogenetic relationships. This can be seen in the anatomy of the zooids, but also – and more clearly at least for the paleontologists – in the development of the tubaria. The zooids as the individual animals of the pterobranchs live in close connection to each other in a common structure, the tubarium, but the interconnection of the zooids differs between the two groups. In the Cephalodiscida (Plate 2A–B) the tubarium is quite variably formed, ranging from isolated tubes with a single opening from where the zooids can crawl out to filter-feed, sitting on the apertural spines at the openings (Figure 2.1B), to a communal tubarium with one or more large internal spaces in which the zooids roam freely. Communal tubaria possess numerous openings from which the zooids emerge for feeding, but the zooids may not be constantly connected to a single opening and may wander independently from opening to opening.

The individual tubes of the communal tubaria of the cephalodiscid subgenera *Orthoecus* and *Idiothecia* form large masses with a distinct organization, either as low meadows of individual tubes or as branched, bushy colonies with occasional lateral connections between the complexly structured branches. In *Acoelothecia* the whole tubarium is reduced to a mass of individual interconnected bars, forming a meshwork in which the zooids are protected. The genus *Cephalodiscus* includes species with a complex bush-like tubarium with an interconnected inner space and numerous apertures. These apertures are provided with spine-like protrusions from which the zooids feed.

The zooids of all *Cephalodiscus* species appear to be separate at the mature stage, but asexually produced juveniles and immature zooids live connected to the sucker of a mature adult in the tubarium (Plate 2D–E). Thus, a kind of pseudocolonial lifestyle has to be considered for *Cephalodiscus*.

The Graptolithina, as the second and more important group of the Pterobranchia, include truly colonial communities in which all zooids (Plate 2F–H) are interconnected for life through their stolon system. The genus *Rhabdopleura* is the only extant representative of the group and may serve as a key to understanding the graptolites as fossil organisms. *Rhabdopleura* bears numerous interconnected zooids in a strictly organized colony (Figure 2.7A). The zooids are not

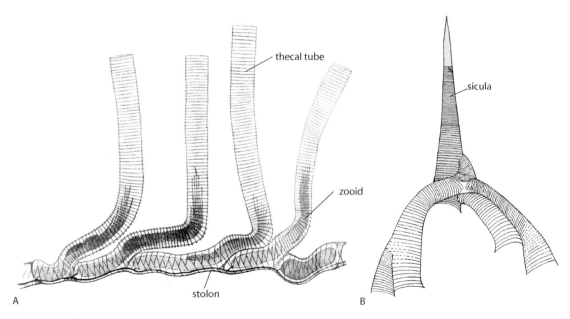

A

B

Figure 2.7 (A) *Rhabdopleura normani*, part of colony, showing the stolon (black line) at the base of the thecal tubes and the zooids retracted into the basal parts of their tubes (based on Schepotieff 1907a, pl. 22). (B) *Jenkinsograptus spinulosus*, proximal end with sicula and a number of successively formed tubes; the stolon is not preserved (adapted from Bulman 1970a, Fig. 52, with permission from the Paleontological Institute). Illustrations not to scale.

free to roam in the tubarium as are *Cephalodiscus* zooids. Each zooid has its own little living chamber, in a way similar to the situation in the Bryozoa. In the Bryozoa, however, the housing construction is a dermal development and not built via a secretion from a special gland as in the graptolites.

All zooids of a single graptolite colony are descendants from the founding zooid, the sicular zooid, of the colony. The development of the extinct taxa of the Graptolithina (Figure 2.7B) followed the style seen in *Rhabdopleura*. The sicula as the first tube of the colony is sexually produced. The remaining thecal tubes are formed as a succession of interconnected tubes. As the zooids have not been preserved in the fossil record, some details, however, have to be interpreted from the tubaria alone. Every graptolite tubarium possesses a connecting feature, named the common canal, through which the stolon is thought to have connected the individual zooids. Thus, the common canal can be taken as evidence for the colonial development of the graptolites and the indication

that the zooids were not entirely separate individuals, but formed a true colonial organism.

Ontogeny and Astogeny

The ontogeny of an organism describes its development from the fertilization of the egg to formation of the final, mature organism, but in a colonial organism the growth pattern is more complex and includes the ontogenies of the individual organisms (zooids in the Pterobranchia) and the growth of the whole colony, often termed astogeny (basically the ontogeny of the colony). The ontogeny of the individual zooids therefore has to be distinguished from the astogeny of the colony.

All Pterobranchia possess a complex double cycle of sexual and asexual reproduction, in many aspects similar to the reproduction of corals and bryozoans (Figure 2.8). The sexual reproduction produces planktic larvae developing into the founder zooids for new graptolite colonies.

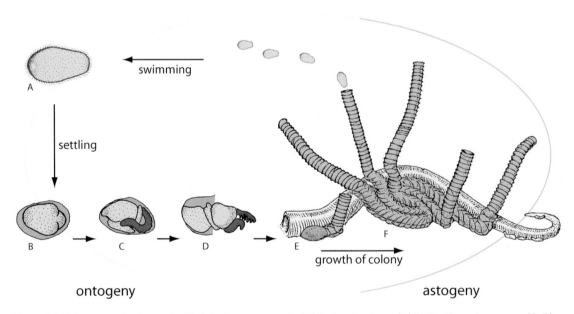

Figure 2.8 Ontogeny and astogeny in *Rhabdopleura compacta*. (A) Swimming larva. (B) Settled larva in cocoon. (C–D) Developing zooid. (E) Juvenile colony with dome and thecal tube of sicular zooid. (F) Larger colony with five thecal tubes. Illustrations not to scale.

Subsequent growth of the colony is by asexual budding. The graptolite astogeny includes the asexual budding of zooids from the stolon of the sicular zooid, and the subsequent and sequential budding of later zooids. This development is known only in the extant *Rhabdopleura*. The stolon is preserved, however, in many dendroid graptolites, and may serve as an indication of the coloniality of the organisms (e.g. Saunders et al. 2009). In *Cephalodiscus*, all asexually budded zooids originate from the base of the stalk (the sucker) of the mother zooid and separate when mature to live their independent lives (Lester 1985), and a stolon system does not exist.

The initial sexual reproduction occurs through the fertilization of eggs by male sperm. Eggs are produced by female zooids of the Pterobranchia and the fertilization apparently is organized within the tubaria. The details of this part of the reproduction are unclear. The colonies of *Rhabdopleura* bear male and female zooids in various numbers and fertilization may happen within the colony, as the zooids are restricted in their movement.

The fertilized eggs of *Rhabdopleura normani* are brooded in the coiled tubaria of the female zooids (Lester 1988a). Up to seven eggs have been counted in a single female tube. The eggs remain in the brood chamber until the developing embryos reach the stage at which they can swim, after about 4–7 days. Then, they push past the younger embryos and the female zooid and swim away. After a maximum of a few hours they settle on a suitable spot to start building a new colony. At 400–450 µm long, the embryo swims via the ciliae of the densely ciliated epidermis of the body. The larva settles within 24 hours after being released and attaches itself with the posterior part of the body, where the stalk will develop subsequently. It surrounds itself completely with a protective cocoon, the dome, in which the metamorphosis to the zooid takes place (Figure 2.8B–E). The dome is formed from the same material that will later be used to secrete the tubaria, but is structureless. The transformation of the larva within its protective cocoon takes 7–10 days, during which it relies on the stored nutrients. After finishing the metamorphosis, the zooid produces

a hole in one side of the dome and starts secreting the first thecal tube by adding regular fusellar rings (Lester 1988b). The second zooid of the colony originates as an asexually produced bud from the base of the stolon, the contractile stalk of Stebbing (1970a), as is seen in *Rhabdopleura compacta*. The developing zooid breaks through the wall of the dome and starts secreting its own tube. Subsequent origination of zooids is from the growing stalk of the colony.

The astogenetic growth of the *Rhabdopleura* colony is through a monopodial pattern, in which a permanent terminal zooid exists at the tip of the branch and new zooids are budded subsequently from the stalk (stolon) behind this zooid (Figure 2.9A). Schepotieff (1905, 1907a) described this growth of a colony of *Rhabdopleura* bearing a permanent terminal bud with its tube termination in a closed pointed tip. Lankester (1884) did not support this interpretation, but indicated a distally open tube for the terminal zooid, which is more likely in light of the growth of the colony. The growing tube behind the terminal zooid is closed at certain intervals by a diaphragm to produce separate chambers for the developing zooids. The zooids later break through their chamber wall and start to produce their own housing tubes, forming a distinct growth unconformity (Figure 2.9A) at the base of the erect tubes.

The asexual reproduction of the colonies in the extinct taxa of the Graptolithina can only be inferred from the tubarium development, where a common canal connects all zooidal tubes. However, anatomical details are not available for any fossil pterobranchs and the interpretation is entirely based on the comparison with extant rhabdopleurids. The fossil tubaria indicate a sympodial budding in which each zooid at one point represents the terminal bud (Figure 2.9B). Each zooid in turn leaves a small opening through which the next budding zooid emerges to secrete its thecal tube. Thus, a permanent terminal bud as in *Rhabdopleura* (Figure 2.9A) does not exist. The nature of the transition from the monopodial growth in rhabdopleurids to the sympodial growth in most derived graptolites is unknown.

Figure 2.9 Monopodial and sympodial budding. (A) Monopodial growth, fragment of *Rhabdopleura normani* colony with growing end and permanent terminal zooid (based on Lankester 1884, pl. 39). (B) Sympodial growth, reconstruction of dichograptid branch with zooids; each zooid is the terminal one at one time (zooids reconstructed from Schepotieff 1906, pl. 25).

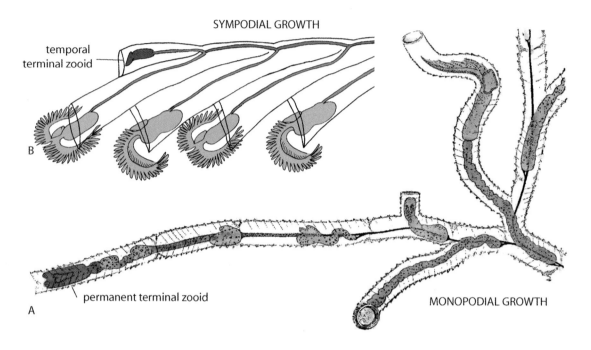

Cephalodiscida

The Cephalodiscida, as the least advanced group of the Pterobranchia, live in large communities as pseudocolonial organisms (Harmer 1905; Lester 1985). These communities are probably generated from a single sexually produced mother zooid. In many tubaria, a number of interconnected zooids of various sizes and growth stages (Figure 2.10A) can be found attached to the base of the stalk of the mother individual (M'Intosh 1887; Harmer 1905). Mature zooids appear to be able to reproduce asexually, whether they are sexually produced (mother zooids or founding zooid) or produced by asexual budding and subsequent separation. Dilly (2014) discussed the reproductive biology of *Cephalodiscus nigrescens* and *Cephalodiscus gracilis* and, contrary to earlier descriptions, indicated that the tubaria are inhabited by joint zooids and not by individuals. However, his illustrations indicate that he regarded the budding individuals associated with

their mother zooid as a colonial organism and did not consider the separation of the zooids after maturation. The zooids of the individual subgenera and species of *Cephalodiscus* generally resemble each other in their anatomy (Figure 2.10B) and the differences in their tubaria (Figure 2.11) strongly exceed the soft-body anatomical differences.

Even though the Cephalodiscida in general secrete a tubarium, the genus *Atubaria* appears to be free-living as naked individuals on corals. The *Cephalodiscus* tubaria can be encrusting, compact or even branched, dendroid in shape, and may recapitulate many shapes also known from the colonial Graptolithina. In most taxa, the individual dwelling tubes of the zooids are entirely separate. However, a communal dwelling is formed of interconnected tubes or other three-dimensional constructions for the protection of the zooids in other taxa. Openings may be smooth and straight or bear a variety of elaborations, from robust ventral apertural lobes to long apertural spines.

Figure 2.10 The zooids of *Atubaria* and *Cephalodiscus*. (A) *Cephalodiscus* sp., pseudocolony of mature and associated juveniles (adapted from Lester 1985 with permission from Springer Science + Business Media). (B) *Cephalodiscus dodecalophus* M'Intosh, showing internal anatomy (from Schepotieff 1907b, pl. 48). (C) *Atubaria heteroplopha* Sato, 1936 (adapted from Komai 1949, Fig. 1, with permission from The Japan Academy). Illustrations not to scale.

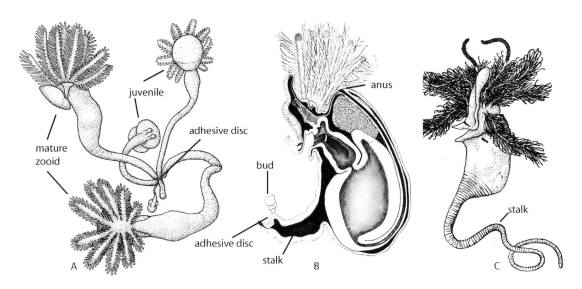

The tubaria of the Cephalodiscida are quite variable in their construction, and the differentiation of the extant subgenera of the genus *Cephalodiscus* depends on these differences. They vary from encrusting (*Cephalodiscus graptolitoides*) to erect (*Cephalodiscus dodecalophus*) forms with multiple branchings, including lateral connections resembling the dissepiments and anastomosis of the Graptolithina. The openings in the communal tubaria (called ostia in many publications) are either simple or adorned with single or multiple spines. The tubarium is dissolved into a complex mesh of spines and bars in *Acoelothecia*, in which the zooids roam freely and without separate housing.

The evolutionary relationships of the extant subgenera are unknown and the relationships of the constructional features of the tubaria have not been investigated. It may be assumed that the complete individual tubes of *Cephalodiscus* (Figure 2.11A) are the basic concept of the tubarium in the Cephalodiscida. Complete, initially closed tubes exist in the subgenera *Orthoecus* and *Idiothecia*. In *Orthoecus*, the nearly straight and parallel-sided tubes are initially connected by a mesh of rods and free distally (see Dawydoff 1948). The individual tubes originate from a basal surface, a hardground or rock surface and build a meadow. In *Idiothecia* the individual tubes are enclosed partly in a thick development of cortical tissue and form erect structures, often with branching stipes in which the apertures are oriented in every direction.

The species *Cephalodiscus calciformis* (Emig, 1977) has the most unusual shape of the tubaria (Figure 2.11B). The individual tubes of the tubarium are interconnected and inhabited by the zooids and their attached asexually developing buds. The openings are wide funnel-shaped structures, not found in any other pterobranch. The zooids are comparable to the zooids in other species of *Cephalodiscus*, but little detail is known of their development or interconnection. The tubarium construction of the Cephalodiscidae in many aspects shows features later used for the construction of the tubaria of the Graptolithina. Erect *Cephalodiscus* tubaria show branching and lateral connection of "stipes", even though the zooids do not work in a colony, but are individuals.

Few possible fossil cephalodiscids have been described, preserved only in the form of their tubaria. However, although in extant taxa the

Figure 2.11 Tubarium shapes in extant Cephalodiscidae. (A) *Cephalodiscus nigrescens* (adapted from Hyman 1959, Fig. 54). (B) *Cephalodiscus calciformis* (adapted from Emig 1977, Fig. 1). (C) *Cephalodiscus* (*Orthoecus*) *rarus* (from Andersson 1907, pl. 2, Fig. 6). (D) *Cephalodiscus* (*Idiothecia*) *levinseni* Harmer, 1905 (after Harmer 1905, pl. 2, Fig. 11). (E) *Cephalodiscus sibogae* (from Ridewood 1907, Fig. 2). Illustrations not to scale.

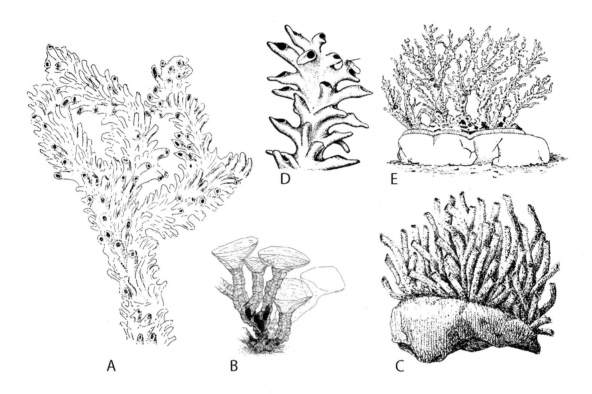

differentiation of the Graptolithina and the Cephalodiscida can be made easily, it is nearly impossible to separate them in fossil material, especially when the specimens are flattened and poorly preserved, as is usually the case.

Rickards and Durman (2006) referred the Upper Cambrian genus *Aellograptus* and its type species *Aellograptus savitskyi* to *Cephalodiscus*, and thus synonymized *Aellograptus* with *Cephalodiscus*. Other potential cephalodiscid genera can be found among the Ordovician taxa *Eocephalodiscus*, *Melanostrophus* and *Pterobranchites*. Of these, only *Eocephalodiscus* has unanimously been referred to the cephalodiscids, although to its own family (Eocephalodiscidae: Kozłowski 1949). The genus *Melanostrophus* was long known from very fragmentary and poor material. It has been the focus of long debate, but Eisenack (1937) first recognized fusellar structures to confirm its pterobranch rela-

tionships. Zessin and Puttkamer (1994) discussed new material of *Melanostrophus* and erected the family Melanostrophidae for this taxon. Mierzejewski and Urbanek (2004) identified *Melanostrophus* as a "*Cephalodiscus*-like taxon" based on the investigation of isolated fragments. *Pterobranchites* is known from small fragments and its tubarium construction is unknown. Some similarity may be seen to the encrusting *Cephalodiscus graptolitoides*.

The most interesting species included in the Cephalodiscidae is the strange *Atubaria heterolopha* (Sato 1936; Komai 1949). The species was found as a couple of isolated zooids, found feeding on a colony of the hydrozoan *Dycoryne conferta* in Sagami Bay, Japan. The specimens were very actively moving their arms up and down, as they were very agitated when collected. The specimens are anatomically very similar to other cephalodiscids, but do not possess an adhesive disc at the end of the stalk (Figure 2.10C). All 43 specimens collected were female individuals without indication of growing buds. However, juveniles without completely developed arms were among the material. A tubarium or comparable housing construction was not found associated with the specimens, and the interpretation is that they belong to a taxon that does not produce a tubarium at all.

Graptolithina

The recognition of *Rhabdopleura* as an extant, benthic graptolite (Mitchell et al. 2013) did not come as a surprise to most graptolite specialists, although it enhanced direct comparison of fossil graptolites with their extant relatives. It was the result of a meticulous investigation of graptolite tubaria. Andres (1977) had compared benthic Ordovician graptolites with modern Pterobranchia and concluded that graptolites most probably evolved from a *Rhabdopleura*-like ancestor. Crowther and Rickards (1977) and Crowther (1981) came to the same conclusion. The results were based on the recognition of the basic fusellar construction of the tubaria and the secondary addition of cortical bandages in both groups. This led to the long-accepted reconstruction of graptolites in which rhabdopleurid zooids were included (Figure 2.12A).

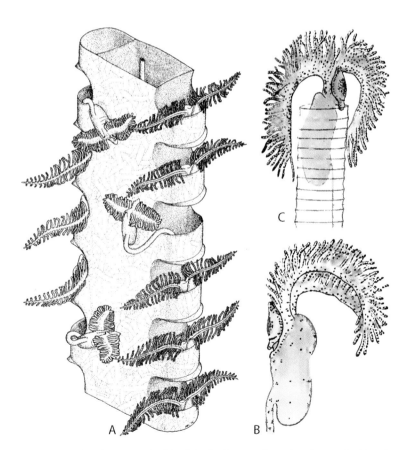

Figure 2.12 (A) Graptolite zooids in a fragment of a biserial tubarium, surface showing cortical bandages (reconstruction based on Crowther & Rickards 1977). (B–C) *Rhabdopleura normani* zooids (based on Schepotieff 1906, pl. 25). Body of the zooids is ca. 1 mm long without arms.

This recognition of a combination of fusellar construction and late stage cortical additions in the tubaria of graptolites and pterobranchs was key to understanding the supposedly extinct class of the Graptolithina. Thus, after a long period of speculation, the fossil graptolites are now safely nested in the Pterobranchia and we can improve our understanding of their life-style and ecological importance. Even after more then 200 years of research, however, undisputed fossil graptolite zooids have never been found, and the anatomical comparison with the extant *Rhabdopleura* remains only inference.

Outlook

The graptolites are now recognized as a clade of the Hemichordata, but this is certainly not the end of biological and paleontological research about the phylogenetic relationships of the group, and a number of questions are still open. Why and how did the Pterobranchia develop their housing construction, the tubarium? Why are graptolites colonial organisms and what factors were involved in generating coloniality in this group, but not in others? Certainly, coloniality had an advantage for them, but apparently it did not last, at least for the planktic taxa that are extinct. Are we in for more surprises?

3 Construction of Graptolite Tubaria

Jörg Maletz, Alfred C. Lenz and Denis E. B. Bates

Terminology is a necessary but often annoying part of any scientific work. It is the language that we have to learn before we can communicate with others on a certain topic. Graptolites are no different in this respect from other organisms, living or fossil. In the past, we understood graptolite fossils – the tubaria – as the skeletal remains of these ancient organisms. In modern understanding, the graptolite tubarium is more similar to the housing of a bee's hive or a wasp's nest. However, it is constructed from a secretion of the organism itself and not from foreign material. The zooid, the animal that occupies this housing construction, but is not fixed to it, even though the stolon restricts its movements, secreted it from special glands. The connection between a graptolite zooid and its housing construction thus differs considerably from that of the colonial bryozoans, or from the well-known non-colonial barnacles. In both, the construction is an integral part of the animal like an external skeleton and the animal is unable to separate from it and move around freely.

A hermit crab looks for an empty shell and occupies it. The crab is able to move into another home if it is too small – similar to a human, but it does not construct the home by itself. Graptolites are one step further advanced: they are the architects and the inhabitants of their homes and they are very good at architecture. To the advantage of the paleontologists and geologists, graptolites used their abilities to create fashions and designs that are easily recognized and used for geological purposes, especially for dating Lower Paleozoic rock sequences (biostratigraphy), where they are unrivalled. It is unfortunate that the builders of the graptolite tubaria are virtually unknown, except by comparison with the few modern benthic taxa (see Chapter 2). So

Graptolite Paleobiology, First Edition. Jörg Maletz.
© 2017 John Wiley & Sons Ltd. Published 2017 by John Wiley & Sons Ltd.

we have to infer their morphology from the extant *Rhabdopleura*, and can only extrapolate as to how the zooids of the extinct planktic graptolites may have looked and how they constructed their tubarium.

Naming the Tubarium Features

The terminology of the graptolite tubarium has been developed over centuries of research, although some of these terms have changed considerably since the first mention of these animals. Certain terms were introduced early on, but later were abandoned when we learned that they were inappropriate to describe those particular features. For example, the tubarium of the graptolites – the rhabdosome in older literature – has been termed an exoskeleton. However, it does not represent an exoskeleton, a dermal cover with the attachment of muscles on the inside as is known from insects and other invertebrates. Mitchell et al. (2013) and Maletz et al. (2014) suggested use of the term tubarium instead of the coenecium (for the pterobranchs) and rhabdosome (for the fossil graptolites), as it was done in the past (see Bulman 1955, 1970a) to show that both terms describe the same morphological feature. Lankester (1884) originally introduced this term for the housing construction of extant pterobranchs. He considered the terms coenecium or zoenecium inappropriate as they describe the differently formed housing constructions of bryozoans.

Graptolites have been described by many specialists from various countries and in numerous languages, and a reflection of this is seen in our modern graptolite terminology. Terms vary from country to country, and words in many languages may have to be combined or translated to form a coherent language in which we as scientists can communicate our knowledge. A number of papers provide the early basis for our graptolite terminology (e.g. Wiman 1893a,b, 1896a; Törnquist 1894), but over time, thanks to the recent study of isolated, three-dimensionally preserved graptolites by means of scanning and transmission electron microscopes, many new terms have been introduced and are found scattered in numerous papers describing graptolite faunas. For a long time, the compilation of graptolite terminology was based on the two editions of the *Treatise on Invertebrate Paleontology* (Bulman 1955, 1970a). A new edition is on the way now, and an up-to-date, much expanded glossary has been published recently (Maletz et al. 2014). In this chapter only the most basic and important characteristics are described and illustrated, while relevant features concerning individual graptolite groups may be found in the later taxonomic chapters.

Construction Material

All graptolites, by comparison with the living *Rhabdopleura*, produced an organic housing construction from glandular regions on their cephalic shield. It is this housing construction, the tubarium (Lankester 1884; Maletz et al. 2014), that is preserved in the fossil record, and is the key to the taxonomy of all graptolites. The tubarium construction (Figure 3.1A) of the planktic graptolites differs considerably from that of the benthic taxa, but a number of important homologous characteristics can be found to connect all extant and extinct taxa. All graptolite tubaria are formed from two main building materials, the fusellum and the cortex (Kozłowski 1938, 1949) (Figure 3.1B). The fuselli comprise the fundamental material forming the tubarium walls, collectively called the fusellum. They are secreted mainly as halfrings or full rings to form a simple tube. The fuselli are stacked in series one upon each other, secreted by the zooids like lines of bricks on a chimney. Each fusellus bears at least one suture connecting it to the next-formed fusellus, but in fusellar half rings, two sutures are present (Kozłowski 1938). The zooid secretes a single fusellus by starting at one point and, while excreting the organic material, moves in a circle to create a tube. Usually, two halfrings form a complete ring and, thus, two oblique sutures are formed on what is called the dorsal and ventral sides of the tube. The sutures of the

Figure 3.1 (A) Uniserial tubarium of *Neocolonograptus lochkovensis* (Přibyl, 1940) showing fusellar construction (based on Urbanek 1997a, Fig. 49). (B) Fuselli and cortex (based on Kozłowski 1938, Fig. 2). (C) Biserial tubarium of *Diplacanthograptus spiniferus* (Ruedemann, 1912), showing increasing width of fuselli towards thecal apertures (based on Mitchell 1987). Illustrations not to scale.

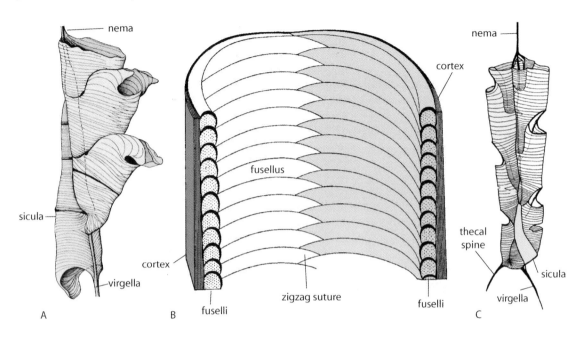

individual fuselli are often perfectly aligned in a zigzag pattern in the tubes, showing masterful construction by the graptolite zooids. Only in some earlier graptolite taxa is this regularity missing and the sutures are randomly distributed (Andres 1977, 1980). Perfectly regular zigzag sutures can also be seen in the creeping tubes of the modern *Rhabdopleura*, whereas the erect zooidal tubes possess irregular sutures (Plate 2C).

The cortex, often called cortical tissue or cortical bandages, is secondary material smeared on the surface of the fuselli, but it represents an integral part of most colonies and probably helps to support and stabilize them. Cortical tissue is generally laid down on the fusellar tissue in the form of thin bandages, like paintbrush strokes in a painting (Crowther & Rickards 1977). It is usually found as ectocortex on the outside of the tubarium, but sometimes, and in lesser amounts, as endocortex on the inside of the tubes. In the cephalodiscids (order Cephalodiscida), the cortex can include loose masses of spongy material surrounding the

individual and separate tubes of the zooids, and thus include more material than the individual thecal tubes. Cortical bandages can be formed as patternless strokes of material, but in many taxa the cortex bears a distinct surface ornamentation, best seen in the Retiolitidae (see Chapter 12).

The ultrastructure of the fusellar and cortical material provides us with some information on the secretion of the tubarium (Figure 3.2), but it can only be investigated with the transmission electron microscope (TEM) and scanning electron microscope (SEM). Each fusellus is composed of three layers, a basal layer formed from a sheet of granular fabric, followed by a three-dimensional meshwork of fibrils, and covered by a sheet of granular fabric. Thus, the internal part is much less dense than the bounding layers (Figure 3.2A). The fibrils have a diameter of 20–70 nm, but this is variable depending on the taxon and the position within the individual secreted layer. The development is identical in fossil and extant pterobranchs, but some differences exist between the various groups.

Figure 3.2 Ultrastructure of graptolite tubarium. (A) *Desmograptus micronematodes* (Spencer, 1884), SEM, fuselli, with thicker ectocortex below, and thinner endocortex above. (B) *Gothograptus nassa* (Wiman, 1895), SEM photo, bandages on genicular hood, with pustules. (C) *Dendrograptus*? sp., TEM section through two fuselli and bandages of the ectocortex. (D) Stacked fuselli with endo- and ectocortex (based on Bates & Kirk 1986a, Fig. 1). Photos provided by D.E.B. Bates.

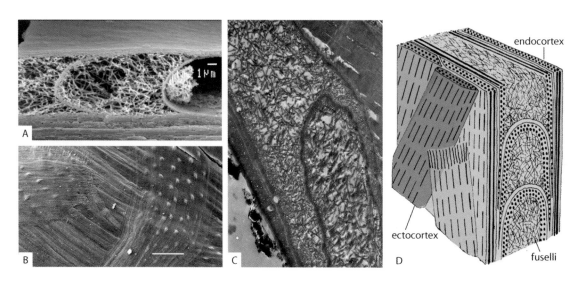

Tubarium Design

We have to presume that the earliest pterobranch secreting a tubarium was very similar to the extant *Cephalodiscus*, a pseudocolonial organism with a number of modern taxa and a very poor fossil record. The tubes of *Cephalodiscus* in taxa with tube-building capacity are closed at the base and possess a single opening at which the zooid can be seen sitting to gather food. However, a number of species construct communal tubaria in which the mature zooids can move around independently. These look very similar to certain tubaria formed from colonial Graptolithina and the differentiation is only possible based on the anatomy of the zooids. Thus, in the fossil record a separation of pseudocolonial Cephalodiscida and primitive colonial Graptolithina is often impossible. In the colonial Graptolithina, the zooids are interconnected for life and have a much more restricted mobility. The general pattern of tube-building, however, is identical.

Rhabdopleura constructs a tubarium with sometimes hundreds of zooids, covering a considerable area (Figure 3.3A). The tubarium shows an encrusting colony with many irregularly distributed branching points forming multiramous colonies. The early Graptolithina were similar to the extant *Rhabdopleura* (Mitchell et al. 2013) and produced encrusting colonies with variable shape, but by the Mid Cambrian, erect colonies with a bushy or tree-like shape appeared (Figure 3.3C). From these, the planktic graptoloids originated in the basal Ordovician and initially still possessed a multitude of thecate branches (Figure 3.3B).

Tubarium size

Theoretically, the size of a graptolite colony is unlimited, and colony diameters of nearly a metre have been found in planktic forms in the Lower Ordovician, including thousands of zooids. Others have colonies with just a few zooids and can barely be seen with the naked eye. The size is unfortunately often estimated from fragmented colonies, since large ones usually break during extraction from the rocks. The holotype of *Holograptus deani* is one of the largest known nearly complete specimens described so far. Elles and Wood (1902) stated that the length of the main stipes of this specimen is about 40 cm, indicating a possible

Figure 3.3 (A) *Rhabdopleura normani* Allman, 1869, large encrusting colony, showing highly irregular placing of thecal tubes and branching points (from Ridewood 1907, Fig. 6). (B) *Clonograptus flexilis* (Hall, 1865), planktic graptoloid with dichotomous branching (after Lindholm & Maletz 1989, Fig. 2F). (C) *Dendrograptus fruticosus* Hall, 1865; note the massive stem and slender stipes showing dichotomous branching (from Hall 1865, pl. 19, Fig. 8). Illustrations not to scale.

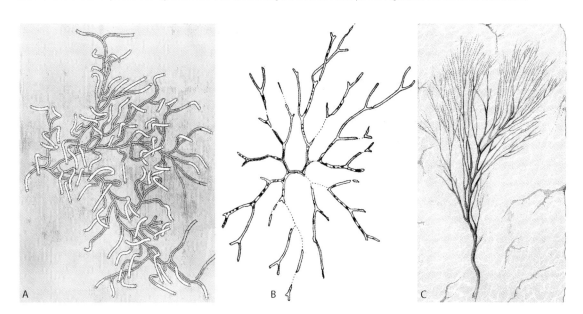

diameter of the reconstructed specimen of about 80 cm. Pritchard (1892) found a fragmentary specimen of *Temnograptus magnificus* (now *Paratemnograptus magnificus*) with an estimated diameter of about 100 cm in which the preserved part already showed a diameter of 75.75 cm. Maletz et al. (1999) illustrated a similar specimen as *Clonograptus* sp. cf. *C. multiplex* with a diameter of about 50 cm. The longest stick-like colony ever recorded is about 1.45 m long and belongs to the straight monograptid *Stimulograptus halli* (Loydell & Loveridge, 2001). As the proximal end is not preserved, a length of more than 1.5 m must be estimated. Even at the most distal preserved part of the colony, it is only 2.5–3.0 mm wide. Compared with the known growth rate of the extant *Rhabdopleura*, the specimen might have been at least 25 years old when it died.

It is more difficult to estimate the colony size of benthic graptolites, as these are usually preserved as transported fragments. Bouček (1957) illustrated fragments of *Pseudodictyonema giganteum* Bouček, 1957 with a length of more than 15 cm. He described the specimen as a small fragment of a much larger colony. Thus, a diameter of at least 30 cm is possible. Usually, however, complete colonies with a diameter of a few cm dominate the benthic dendroid communities.

Branching style

Branching is one of the main ways of modifying the shape of the graptolite colonies. Branching invariably takes place at the growing points of the colony in the form of dichotomous branching, where thecae are added to increase the length of the stipes, and therefore the size of the colony. The genus *Clonograptus* shows typical dichotomous branching (Figure 3.3B). At the tip of the stipes, branching occurs. The resulting two new stipes form a more or less constant angle to the direction of the previous branching division, and none of the new stipes follows the direction of the previous stipe. A distinct lateral appearance of the new stipes can be found in a number of taxa (*Trochograptus*, *Schizograptus*), but branching invariably occurs at the tip of the stipes.

A branching mode with considerable constructional differences can be found in cladial branching

Figure 3.4 Cladial branching in *Cyrtograptus*. (A) *Cyrtograptus perneri* Bouček, 1933, SEM photo. (B–F) Several stages in the development of a cladium, showing the secondary nema as the leading rod of the cladial stipe (based on Bulman 1970a, Fig. 65). (G) *Cyrtograptus radians* Törnquist, 1887, Thuringia, specimen with numerous cladia and fragment of *Monoclimacis* sp. on top. Photos provided by A.C. Lenz and M. Schauer. Illustrations not to scale.

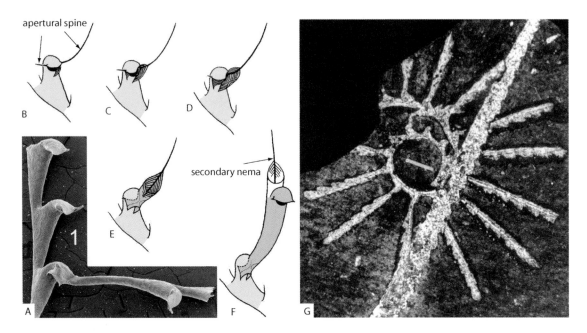

(Elles & Wood 1911). This mode of branching was first found in Silurian monograptids (*Cyrtograptus, Diversograptus, Linograptus*), but is actually widely distributed in certain Ordovician taxa. Cladial branching is a secondary branching of the stipes. It is not formed at the tip or the growing end of a stipe, but is possible at any position along the stipe. A stipe formed through cladial branching starts at the aperture of a theca and is produced by a mature zooid. This branching is thus formed at some distance from the growing end of the stipe and leads to a considerable secondary modification of the colony shape. Small flanges of fusellar material are produced by the mother zooid at two sides of the thecal apertures and subsequently are enlarged to form a thecal tube (Figure 3.4A–E). In all monograptids a secondary nema or pseudovirgula is first secreted and the newly developed stipe follows the lead of this secondary nema. The cladial branching may theoretically be present at any theca, but in most taxa a distinct position along the main stipe is maintained. Therefore, the position of the cladia can be used for taxonomic purposes

and to determine the various species of the genus *Cyrtograptus* (see Chapter 13).

Cladial branching first appears in some Ordovician taxa, and is known from chemically isolated material of *Pterograptus* (Skwarko 1974; Maletz 1994b). Other Ordovician taxa developing cladia are the axonophoran genera *Nemagraptus*, *Amphigraptus* and *Tangyagraptus* of the Dicranograptidae (Finney 1985). The cladia in all Ordovician genera differ from the cladia in Silurian monograptids in the lack of a secondary nema on the dorsal side of the cladial stipes.

Colony shapes

The colony shapes of graptolites vary considerably and are controlled by numerous factors, from ecological and ecophenotypical to genetic control and teratological modifications. Benthic graptolites invariably show ecologically determined shapes of their tubaria. The colony forms are influenced considerably by the development of the surfaces over which they grow in encrusting taxa, but also by water currents and interactions with other

Figure 3.5 (A) *Palaeodictyota* sp. showing the typical anastomosing stipes of a conical, benthic graptolite colony (adapted from Bulman 1970a, Fig. 21-3 with permission from the Paleontological Institute). (B) *Dictyonema* (*Dictyonema*) *elongatum* Bouček, 1957, fragment showing fine dissepiments in a benthic dendroid graptolite (based on Bouček 1957, Fig. 20C). Reconstructions not to scale.

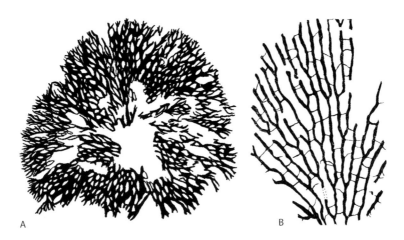

A   B

organisms living in the surrounding area. Thus, the colonies are often highly irregular and do not show the pronounced regularity of the colonies of planktic taxa. Colony shape in planktic graptolites is considerably less influenced by ecological factors, and highly symmetrical colonies are usually developed. Irregularities may be more commonly produced by growth anomalies, parasitism, or accidental death of individual zooids.

Multiramous sheet-like encrusting, fan-shaped or bushy colonies are common in benthic graptolites, with the main variation of the stipes being based on the density or closeness of the stipes. Often stipes are kept at a certain distance by secondary development of dissepiments or of anastomosis (Figure 3.5) of the individual stipes, a temporal touching and connection of the stipes, and sometimes with an exchange of thecae.

Planktic graptolite colonies show less intraspecific variation, but a higher degree of interspecific variation in comparison with the benthic taxa. Early planktic forms are still similar to their benthic ancestors and have a cone-shaped colony (*Rhabdinopora*), but more open umbrella-shaped colonies rapidly arose. The graptolite colony shape is often considered in relation to the position of the sicula, the first theca of the colony. The orientation of the sicula with the aperture downwards and the nema upwards is used as a standard orientation and the direction of the stipe growth relates to this position, but should not be regarded as representing the orientation of the living colony in the water column. The shape of the colonies is then described from pendent to scandent (Figure 3.6A), following the suggestion of Elles (1922).

Scandent colonies are especially characteristic, but can be formed in two different ways. They mostly involve only two stipes, but scandent triserial (with three thecal series, *Pseudotrigonograptus minor*: Fortey 1971) and quadriserial (with four thecal series: *Phyllograptus*, *Pseudophyllograptus*: Chapter 10) tubaria can be found in the Ordovician. The most important group is the Axonophora (Chapter 11), in which the two thecal series are positioned back-to-back, a condition called dipleural (Figure 3.6C). The biserial, dipleural development originates in the Middle Ordovician from a reclined ancestor that can be identified as an *Isograptus* species (see Figure 3.6B).

Biserial, monopleural colonies (Figure 3.6D) are found in the Glossograptidae (see Chapter 10). Here the two thecal series are positioned side-by-side, enclosing the sicula and nema between the thecal series and obscuring the proximal development. A partial monopleural development is found in the genera *Bergstroemograptus* and *Kalpinograptus* (see Chapter 10).

Figure 3.6 (A) Pendent to scandent colony shapes (adapted from Bulman 1970a with permission from the Paleontological Institute). (B) *Isograptus victoriae*, reclined two-stiped colony. (C) *Pseudamplexograptus distichus*, scandent, dipleural colony. (D) *Paraglossograptus proteus*, scandent, monopleural colony. Shading is used in (B–D) to show the disposition of the two stipes in these taxa. Illustrations not to scale.

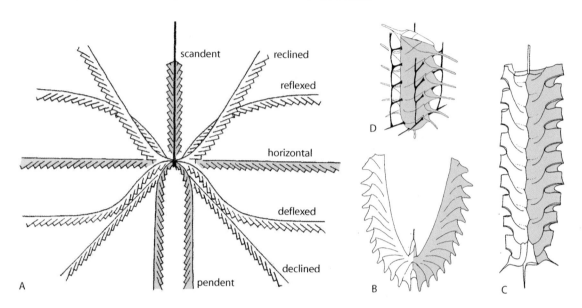

Thecal Tubes

The pterobranchs secrete simple tubes, the thecae in which they spend most of their lives. As the graptolites (suborder Graptolithina) are clonal, colonial organisms, they form these thecae initially as identical housing constructions in a repetitive mode. The thecae along the stipes of a single colony are identical. However, later in the evolution of the group, a change to an often gradual, but sometimes also instant change in thecal style along the stipes can be noted (see Chapter 13: Monograptidae). The thecae are the main building blocks of the graptolite colony and define the shape. Along the stipes, the individual thecae are interconnected through internal openings, representing the space from which the next zooid emerges to build its own tube, showing the clonal nature of the colonies. These interconnected initial parts of the thecae are called the "common canal" (Figure 3.7A).

In most cases the thecae show a distinct serial arrangement and thecal overlap, expressed through an interthecal septum (Figure 3.7A), the

wall common to the mother and daughter thecae. This interthecal septum is often constructed through the joint efforts of both thecal zooids. The thecal overlap is characteristic of most graptolite taxa, but many Silurian monograptids show no thecal overlap at all. Otherwise, the thecal overlap may change along the stipe in many dichograptids and sinograptids, and a distally increasing thecal overlap is common (e.g. *Nicholsonograptus*) (see Chapter 10).

The thecae are differentiated into the protheca and the metatheca, but the differentiation is not always straightforward and is a subjective feature, since a break in fusellar structure is not present. In general a protheca may be described as the part of a theca before the insertion of the interthecal wall (Figure 3.7A).

Autothecae and bithecae

Wiman (1895) differentiated autothecae, bithecae and stolothecae in benthic graptolites, an observation dubbed the triad budding system or the Wiman rule by Kozłowski (1949). Cooper and Fortey (1983) identified the stolotheca of Wiman

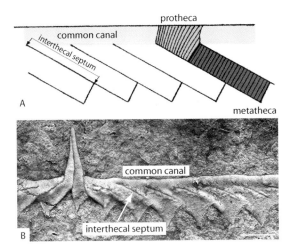

Figure 3.7 (A) Terminology of dichograptid thecae. (B) *Expansograptus holmi* (Törnquist, 1901), obverse view, showing thecal style of relief specimen, common canal and interthecal septae visible on the stipe. Scale indicated by 1 mm long bar in (B).

(1895) with the prothecal part of an autotheca and simplified the thecal notation to include only two thecal types. The differentiation is based on the presence of larger and smaller thecae in many benthic and some early planktic (Anisograptidae) taxa. The reason for this differentiation of thecae is unknown, and a number of hypotheses have been proposed. Hutt (in Palmer & Rickards 1991) amongst others suggested a sexual differentiation, with the bithecae housing male and the autothecae the female zooids. Kirk (1973) postulated that the smaller bithecae were the home of cleaning individuals, and compared the situation with that in the bryozoans. Modern *Rhabdopleura* colonies can include male and female zooids, but all thecal tubes are constructed in an identical fashion and a separation is impossible. Thus, there is no help from modern graptolites to provide an explanation for the presence of autothecae and bithecae.

The authothecae and bithecae of benthic graptolites are usually organized in highly regular patterns with each autotheca provided with an associated bitheca, except for the branching points at which the bitheca is replaced by an autotheca representing the first theca of a new stipe. The bithecae are positioned on the side of the autothecae, but they are found alternating on the stipes.

One bitheca is present on the left side of the stipe and the next bitheca is on the right side of the stipe. A breakdown of this regularity can be found in the early planktic Anisograptidae, where a number of taxa are known to have successive bithecae on one side or successively lose the bithecae until only the first bitheca, the sicular bitheca, is found as a leftover (Lindholm 1991). The development of bithecae can be fairly complex, in which the bithecal apertures open into the autothecae or possess elaborate apertural modifications (Bulman 1970a).

The Wiman rule (Figure 3.8A–C) originally describes the interconnection of the thecal types in dendroid (benthic) graptolites, which is considered to be the external expression of the internal stolon system connecting the zooids (see Chapter 2). Saunders et al. (2009) described and illustrated in detail the stolon development and the surrounding thecal tubes for *Desmograptus micronematodes* (Spencer, 1884). At each introduction of a new theca, the previous theca produces a new autotheca on one side and a bitheca on the other side (Figure 3.8). Thus, the point of insertion of the next autotheca appears as a triad with the metatheca of the previous autotheca in the centre. All auto- and bithecae are produced on the sides of the stipes in bithecate taxa, but in younger graptolites with the development of only a new autotheca, the origin of this theca is on the dorsal side of its mother theca (Figure 3.8E). The lateral origin and subsequent growth onto the dorsal side of the stipe by the autothecae is called the plaited overlap by Lindholm (1991).

Dicalycal thecae

At the points where a stipe branches, a slight modification of the normal thecal development occurs. In dendroid graptolites the bitheca is replaced by an autotheca and the theca producing two autothecae is termed the dicalycal theca (Figure 3.8A). Maletz (1992a) describes in detail the theoretical concept behind this type of development and the complex quadri- to biradiate proximal development of the Anisograptidae.

Elles (1897) introduced the first system of thecal notation for the graptolites. Cooper and Fortey (1982) revised the system to its current form. The thecal notation is fairly simple and straightforward in two-stiped taxa (Figure 3.9), but complex in

Figure 3.8 Thecal development. (A) Thecal diagram showing triad budding and the presence of a dicalycal theca (based on Cooper & Fortey 1983). (B) Stipe fragment with bithecae, dorsal view, showing lateral origin of thecae. (C) Stipe fragment with bithecae in lateral view. (D) Stipe fragment with plaited overlap, but lacking bithecae. (E) Dichograptid stipe fragment without bithecae and dorsal thecal origins. Illustrations not to scale.

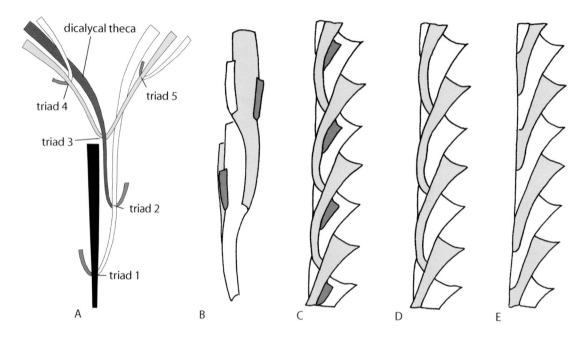

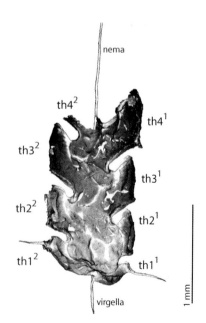

Figure 3.9 *Archiclimacograptus* sp., showing the thecal notation for biserial axonophorans.

multiramous forms due to the number of orders of stipes, and is rarely used for complex multiramous species. The thecal notation system used by Legrand (1964, 1974) for the Anisograptidae has been abandoned.

Seriality

The thecal tubes in the rhabdopleurids are quite irregularly developed, and no consistent orientation or development can be noted. However, a distinct regularity was achieved in these colonial, clonal organisms, and regularity was a highly valued concept for derived graptolites. This also includes the seriality of the thecae. All thecae of a single branch show the same orientation and open in the same direction (Figure 3.8C–E). This is quite a change from the random direction and orientation of the thecae of earlier, benthic taxa.

Thecal morphology

The variability of the autothecae of the graptolites is quite astonishing (Figure 3.10), if we consider that the underlying concept is a simple

Figure 3.10 Thecal styles in planktic graptolites. (A) *Monograptus priodon* Bronn, 1835, hooked thecae.
(B) *Pseudostreptograptus williamsi* Loydell, 1991b, thecae with cupulae (arrows) and branched lateral apertural spines.
(C) *Proteograptus opimus* (Lenz & Melchin, 1991), hooked thecae showing biform development, with distal thecae bearing paired lateral lobes. (D) *Paraorthograptus pacificus* (Ruedemann, 1947), geniculate thecae with spines. (E) *Neodiplograptus tcherskyi tcherskyi* (Obut & Sobolevskaya, 1967), biform thecae in biserial graptoloid, proximally thecae are geniculate, distally straight, outwards inclined. Various magnifications. Photos provided by A.C. Lenz and D.K. Loydell.

parallel-sided to aperturally widening tube formed from fusellar halfrings. Differentiations and modifications are introduced at various places on the thecal tubes and lead to significant or even dramatic changes. In extreme forms it is impossible to recognize the simple tubes in the complexly modified thecae (see Chapter 13: Monograptidae). The thecal profiles have been named from some typical biserial genera such as climacograptid, glyptograptid and orthograptid shapes, but these terms should be used sparingly. The "climacograptid thecae" named after the genus *Climacograptus*, are characterized by a pronounced geniculum (Figure 3.10D) and/or a genicular flange in many species, but other modifications can be found on the geniculum.

The old "thecal types" are in general based on a number of independently changing characteristics and do not represent a precise concept. In the Silurian, hooked and hooded thecae on monograptids were common (Figure 3.10A, C) and a differentiation is often difficult as the outline of both types is identical, so may only be possible in chemically isolated material.

Thecal isolation

In many graptolites the apertural parts of the thecae are completely isolated, and do not closely adhere to the stipe. These thecae are called isolate as they form free tubes like the erect tubes of a *Rhabdopleura*. In extreme forms the isolated thecal tubes can be more than 10mm long, as is

Figure 3.11 Thecal style. (A) Isolated thecal apertures in *Psigraptus arcticus* Jackson, 1967, isolated specimen, Erdaopu, Jilin, China (photo provided by Zhang Yuandong). (B) *Sinograptus typicalis* Mu, 1957, showing prothecal and metathecal folding. Scale indicated by 1 mm long bar in each photo.

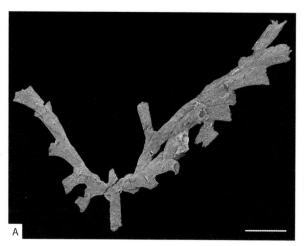
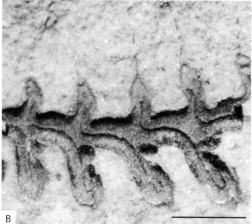

seen in specimens of the Lower Silurian *Rastrites maximus* (see Chapter 13). In this species the initial parts of the thecae, the prothecal parts, are more bulbous and the metathecae are completely separated. In other taxa only an apertural isolation of the thecae can be noted. A good example is the Upper Tremadocian species *Psigraptus arcticus* (Figure 3.11A), an unusual reclined anisograptid.

Thecal folding

Thecal folds have been discovered in a number of taxa, but the details of their growth are still uncertain. These features are generally termed prothecal folds (Figure 3.11B), based on the position of the folds on the thecae. The Sinograptidae (see Chapter 10) is a family of Middle Ordovician graptolites in which prothecal folds are the determining feature, even though a number of species without clearly developed prothecal folds can be observed. Prothecal folds have been discovered in planktic graptolites but are unknown in benthic taxa. The reason for such complexities is unknown. The presence of prothecal folds in the genus *Pseudisograptus* and in many axonophorans may be unrelated to the prothecal folds in the Sinograptidae, as their development and growth direction is quite different.

Colony Growth

The graptolites – known at least from the extant *Rhabdopleura* – are suggested to have a complex cycle of sexual and asexual reproduction and growth (see Chapter 2). The main visible part of the reproduction in the fossil record is the asexual reproduction of the clonally developed zooids reflected in the construction of their tubaria. Sexual reproduction yielding the sicular zooid is represented by a small part of the colony only, the sicula (Figure 3.12). Due to the clonal, colonial development of graptolite colonies, we have to differentiate the ontogenetic growth of the individual zooids and the secretion of their tubes from the astogeny, the growth of the colony. Actually, the growth of the zooids is known only from the few extant members (see Chapter 2). Therefore, the ontogeny and astogeny of fossil graptolites has largely been assembled from the secretion of their tubaria.

Each one of the graptolite colonies starts with the secretion of the prosicula by the sicular zooid. The sicula is either a cocoon-shaped feature, the dome (Figure 3.12A), without a primary opening as in the rhabdopleurids, or a conical prosicula with a primary opening at which the secretion of the metasicula with its fusellar rings starts, as

Figure 3.12 Ontogeny and early astogeny. (A) Dome of *Rhabdopleura compacta* Hincks, 1880 (based on Stebbing 1970b, Fig. 3). (B–D) *Rectograptus gracilis* (Roemer, 1861), sicular development, showing the individual parts (based on Kraft 1926). (E) *Monograptus* sp., sicula with first theca (based on Kraft 1926). (F–G) *Rectograptus gracilis* (Roemer, 1861), proximal ends in reverse (F) and obverse (G) view (based on Kraft 1926). Illustrations not to scale.

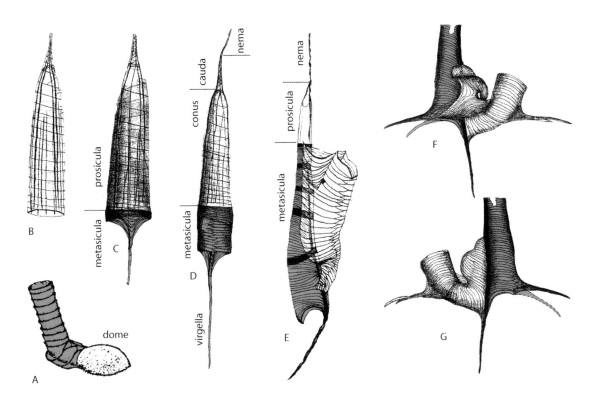

in all derived graptolites (Figure 3.12B–G). The homologization of the two types of prosicula is still uncertain as no true transitional forms exist, and both show considerable constructional differences (Mitchell et al. 2013).

The dome of *Rhabdopleura* (Figure 3.12A) is secreted as a featureless membrane from glands on the ventral epidermis of the larva, encapsulating itself in this cocoon in which it metamorphoses undisturbed into the first zooid of the colony. The zooid then resorbs a hole into one end of the dome and starts secreting the first thecal tube (Lester 1988b). The dome in place of a prosicula is known from a few benthic graptolite taxa, notably from the genus *Epigraptus* (Kozłowski 1971; Andres 1977).

The typical sicula of the derived benthic graptolites is a cylindrical to bottle-shaped construction with a basal attachment disc, with a primary opening

(see Chapter 8) through which the sicular zooid emerges to secrete the metasicula to complete its housing. The prosicula is not structureless, but bears a distinct helical line indicating its secretion as a strip of organic material, similar to the later secretion of the tubarium. Thus its construction differs considerably from the dome of early graptolites and of the surviving extant taxon *Rhabdopleura*.

The sicula in planktic taxa is essentially similar to the sicula of derived benthic taxa, but bears no indication of an attachment to the substrate. The prosicula is conical, with a helical line (Figure 3.12B). The metasicula (Figure 3.12C–G) is constructed from fusellar halfrings (Kraft 1926). The prosicula can be differentiated into the conus and cauda following Hutt (1974a), but both parts are only discernible in excellent preservation (Figure 3.12D).

Longitudinal rods (Figure 3.12B) are generally present on the outside of the prosicula and are formed before the secretion of the metasicula. The number of longitudinal rods varies considerably, but may be consistent for a given species. Williams and Clarke (1999) noted the paired development of the longitudinal rods in Lower Ordovician species. Nothing is known of the value of the longitudinal rods to the prosicula, but they may have provided strengthening.

Metasicula

The metasicula is invariably formed from fusellar halfrings, as are all later thecal tubes. A differentiation of the dorsal and ventral side of the sicula is possible only when the origin of the first theca is visible. This side is – by definition – the ventral side (see Maletz 1992a, p. 298). In many cases the sicular aperture is provided with a rounded lip, the rutellum. The same term is also used for the ventral lips of later thecae. The rutellum can be modified considerably and evolves into the virgellar spine in the axonophorans and a number of other smaller groups (Maletz 2010a). Interestingly, the virgellar spine evolved independently a number of times, and the Pterograptidae and the genus *Phyllograptus* developed a dorsal virgellar spine.

Nema

The cauda of the prosicula merges into the nema (Figure 3.12), a rod-like feature at the tip of the sicula in all planktic taxa. The presence of a free nema at the tip of the sicula is universally regarded as the indication of a planktic lifestyle, as the tip of the sicula is also the attachment point in dendroid graptolites. The nema was fairly short in early planktic taxa, but appeared to increase in importance and was adorned with a variety of modifications in later graptolites (see nematularia). In the Dichograptina and Sinograpta the nema was free and often bore a nematularium at the end, which is rarely preserved.

The Axonophora included the nema inside the biserial colony. Usually, the nema was incorporated and integrated into the colony design and formed the central guide for the growth of the tubarium. It was attached to the ventral thecal walls or connected to the thecae through thickened lists. In rare cases it meandered freely within the tubarium. The nema often projected far beyond the distal end of the colony; sometimes it was longer than the whole colony. The Monograptidae incorporated the nema in a different way into their tubaria. They used the nema as the dorsal rod to which the thecae were attached. Thus, monograptid stipe fragments can often be differentiated from dichograptid ones through the presence of a nema.

Origin of the first theca

The colonial growth of the graptolite colony starts with the origination of the first post-sicular zooid, budding from the sicular zooid (Figure 3.12E–G). Its place of origin varies through time and can be regarded as an important characteristic for the taxonomy and evolutionary understanding of the group. The origin of the first post-sicular theca (th1[1]) is through a foramen (opening) in the pro- or metasicula. The position of the foramen changes from a position in the median part of the prosicula in early taxa, to a position in the lower part of the metasicula in the youngest taxa. Another change can be seen in the development type of foramen: as a resorption foramen in Ordovician to Lower Silurian taxa, while during the early evolution of the Silurian monograptids a change to a primary foramen can be seen. The formation of the sinus and lacuna stages (see Chapter 13) is well known from isolated Upper Silurian monograptids, but the evolutionary origins have not been traced sufficiently.

Only parts of the proximal development of biserial axonophorans can be seen from flattened shale material or even relief specimens, as certain parts of the development are covered by the later growth of the thecal tubes. Thus, growth series of chemically isolated material (Figure 3.13) are necessary and important in the understanding of the proximal development or proximal astogeny of these graptolites. In higher magnifications these specimens may show the individual growth lines (Figure 3.12F–G), providing the evidence for the interpretation of the astogeny used for taxonomic and evolutionary studies (see Mitchell 1987; Maletz et al. 2009; Melchin et al. 2011). For many species, however, the proximal development is unclear due to unfavourable preservation of the available material.

Figure 3.13 The astogeny of *Dicranograptus nicholsoni* Hopkinson, 1870, Viola Limestone, JM26. (A) Sicula with indication of th1¹ origin. (B) Downward growth of th1¹ visible. (C) Th1¹ nearly complete. (D) Sicula with th1¹ before secretion of apertural spine. (E) sicula with incomplete first two thecae. (F) Proximal end with th1¹ and th1² complete. (G) Specimen with six thecal pairs and indication of distally diverging stipes. All specimens in reverse view, except for (D), which is in obverse view. Scale indicated by 0.5 mm long bar in each photo.

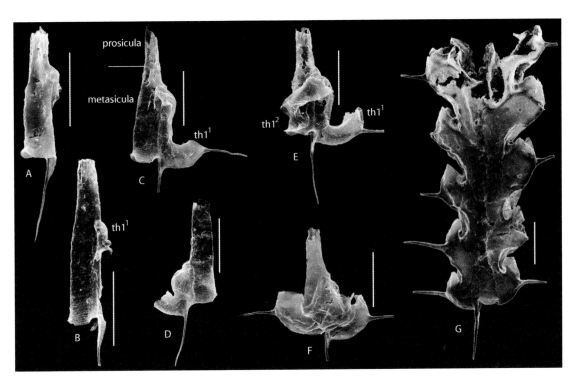

Extrathecal Developments

Numerous extrathecal developments have been recognized in graptolites, and are here discussed in relation to their position on the colony and the precise development. Many of these features are known from few specimens or from poor and fragmentary material.

Nematularia

Nematularia are all features related to the secretion of the nema and are regarded as distal modifications or additions to the nema (Figure 3.14). Nematularia occur at the distal ends of many planktic graptolite colonies, and may come in the form of much thickened width, heart-shaped two- or three-vaned, bifurcating or multifurcating branches, or spiralled. Potentially all planktic graptolites may have produced nemal vanes or other types of nematularia, but these are rarely preserved in the fossil record. In the past the nematularia have been interpreted as gas-filled chambers or flotation devices, but this is now discredited. Nematularia are common in some taxa, but their fragility prevents a detailed analysis in most cases. The oldest nematularia are found in Tremadocian anisograptids (e.g. Jackson 1974) and nematularia are common in the Glossograptina (Chapter 10). A consistency in the presence of nematularia has not been observed, but this may be due to the fragility of these features. They may easily break off before the graptolites are embedded in the sediment. All nematularia found in chemically isolated material are planar constructions, formed from fusellar material.

Bulman (1947, p. 64) described and illustrated the three-vaned nematularia of *Pseudoclimacograptus scharenbergi*, but erroneously referred them to

Figure 3.14 Nematularia. (A) *Pseudoclimacograptus scharenbergi* (from Bulman 1947, pl. 9). (B–C) *Archiclimacograptus decoratus* (Harris & Thomas, 1935), distal end of flattened colony with fragmented nematularium, Table Head Group, western Newfoundland. (D) Isolated complete nematularium of *Archiclimacograptus decoratus* (Harris & Thomas, 1935) showing growth lines. (E) *Cystograptus vesiculosus* (Nicholson), section through part of nematularium (based on Urbanek, Koren & Mierzejewski 1982, Fig. 5). Scale indicated by 1 mm long bar in each photo.

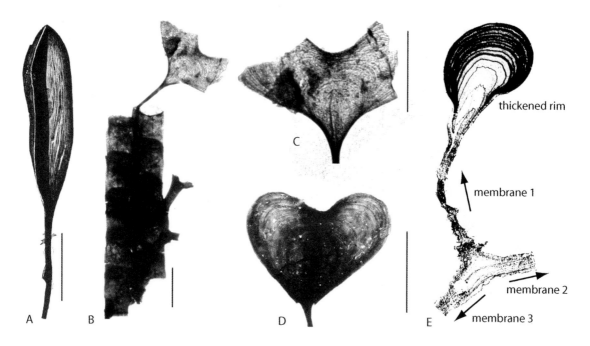

Glyptograptus brevis at the time. According to Bulman (1947), the material shows fine "growth lines" approximately parallel to the outline that can be identified as fuselli or microfuselli. Mitchell & Carle (1986) discussed the material and the possible secretion, preferring a pterobranch model for their interpretation of its secretion. Their model, however, predicts a secretion and expansion of the nematularium from the inside (see Mitchell & Carle 1986, Fig. 4), not comparable to the secretion of the tubarium by a pterobranch model. Three-vaned nematularia exist also in the Silurian *Cystograptus* (Jones & Rickards 1967) and may be the model for the interpretation of most nemal vanes. Urbanek et al. (1982) investigated chemically isolated material of *Cystograptus vesiculosus* from the South Urals. The authors confirmed the presence of a single-walled construction, built exclusively of fusellar tissue and lacking cortical bandages. They did not find an extension of the nema in the nematularium, which is present in

the nematularia of *Pseudoclimacograptus scharenbergi* (see Rickards 1975, Fig. 46). Extensive nematularia can be found in some Retiolitidae, but have rarely been seen in isolated material (Lenz & Kozłowska-Dawidziuk 2001). Maletz (2010b) illustrated the slender three-vaned nematularium of *Plectograptus robustus*. Müller and Schauer (1969) and Müller (1975) discussed the nematularia in some detail. They may be interpreted as helping to maintain a stable position in the water column or as a flotation device, but no proven use can be established.

A terminal vane with fusellar construction and a strongly thickened rim at the distal end of the nema in an abnormal specimen of *Orthoretiolites hami* Whittington, 1954 was described by Bates and Kirk (1991). A heart-shaped vesicle at the end of the nema is the characteristic feature of *Archiclimacograptus decoratus* (Harris & Thomas, 1935). It may be overgrown by the advancing colony, but is invariably present in larger specimens. The few chemically

Figure 3.15 Proximal webs and membranes. (A) *Loganograptus kjerulfi* Herrmann, 1882, Christiania, Norway (from Herrmann 1882, pl. 2, Fig. 12), showing a large proximal membrane. (B) ?*Loganograptus kjerulfi* Herrmann, 1882, juvenile with initial membrane growth (from Herrmann 1882, pl. 1, Fig. 1). (C–D) *Didymograptus murchisoni* Beck in Murchison, 1839, inside views of specimen with extensive proximal membranes (identified as *Didymograptus pakrianus* in Jaanusson 1960, pl. 1, Figs 5–6). (E) *Cyrtograptus* sp., Cape Phillips Formation, Arctic Canada, dorsal view of proximal membrane. Illustrations not to scale.

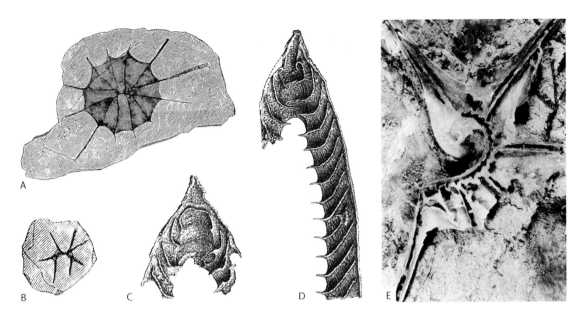

isolated specimens (Figure 3.14B–D) indicate a secretion from fusellar material, since growth lines are easily recognizable. The nematularium starts with the addition of fusellar material at the tip of the nema, forming a flat, oval-shaped feature. Two separate lobes then grow outwards from the oval structure, and eventually a heart-shaped nematularium is constructed. In some species these nemal vanes or nematularia can be overgrown by the advancing thecal series, even though originally they were constructed in advance of the growing colony.

Proximal membranes

Proximal web structures are important features of many early planktic taxa. They have been generally interpreted as supporting the colonies, helping to stabilize the specimens in the water column and retard sinking. Proximal webs are most common in the Dichograptidae (Hall 1865; Herrmann 1882, 1885), especially in the genera *Dichograptus*, *Loganograptus* (Figure 3.15A–B) and in certain

Tetragraptus species, but various other taxa may bear these features. The membranes in *Dichograptus* may be formed from fusellar material as the growth increments are visible under special conditions, but for most taxa the origin and development is unclear, since isolated material does not exist. In many species, the juvenile specimens do not possess web structures or membranes, indicating that these are an adult or mature feature of the colonies. A special form of proximal membrane can be found in *Didymograptus murchisoni*. Jaanusson (1960) described specimens under the name *Didymograptus pakrianus* in which considerable proximal overgrowth forms around the proximal end and covers the initial thecae. The membrane forms a larger communal cavity for the proximal zooids, but does not close off the thecal apertures (Figure 3.15C–D).

A considerable thickening of the proximal stipes can be seen in many mature specimens of the multiramous taxa *Clonograptus* and *Adelograptus*. The thickening is known only in flattened specimens and may either be a real thickening of the stipes or the development of lateral membranes to widen the stipes. Bates et al. (2011) described secondary thickening of the tubarium for many other taxa. The authors interpret the development of extrathecal cortex as a gerontic feature, aimed at reinforcement of the colony, possibly to stabilize the colony orientation in the water column in planktic taxa.

Proximal membranes have been discovered also in the Silurian cyrtograptids (Figure 3.15E), but are either rare and unusual, or difficult to preserve in the fossil record. Lenz (1974) described a single *Cyrtograptus* specimen with an extensive membrane development that extended along the spiral proximal part of the colony and formed a web around the cladial stipes. Underwood (1995) illustrated a similar, but even more complete membrane in a flattened specimen in shale.

Additional nemal constructions

A number of Ordovician and Silurian axonophoran graptolites developed paddle-like structures from the nema along the lateral sides of the colony. These are termed scopulae and have rarely been described from isolated material. Bates (1987) illustrated a nemal vane in a specimen of *Phormograptus sooneri* Whittington 1955 from the Upper Ordovician Viola Limestone of North America. It consists of a bladed fusellar membrane with a thickened rim formed from cortical bandages. Bates and Kirk (1991) described lateral scopulae in *Orthoretiolites hami*. They are similar to the distal nemal vanes found in retiolitids (see Maletz 2010b: *Plectograptus robustus*).

Lacinia

The term lacinia was used first for the meshwork of lists in the Archiretiolitidae of Bulman (1955, 1970) (now part of the Lasiograptidae: Chapter 11). This meshwork originates from thecal spines and is developed as an increasingly complex construction surrounding the colony. The spines have a fusellar core surrounded by a concentric development of cortical bandages. A similar development can be seen in a number of Glossograptidae (see Chapter 10), where the genus *Paraglossograptus*

possesses four ladder-like constructions at the edges of the colony, supported by lateral apertural thecal spines.

The lacinia can easily be misinterpreted as a retiolitid clathrium. However, the clathrium is secreted on the surface of a membrane only. The thin secondary membrane of the retiolitids, the ancora sleeve (see Chapter 12), is usually not preservable in the fossil record and only the complex meshwork of lists secreted on its surface is found, providing beautiful images of the Silurian Retiolitidae.

Ancora development

A special development can be found in the Silurian Retiolitidae (Chapter 12), the ancora umbrella and ancora sleeve, consisting of a secondary membrane surrounding the tubarium and additional list structures. Its growth starts from the virgella and is outlined as a regularly developed suite of lists laid down on a fusellar membrane. In earlier taxa only the ancora umbrella is present as branched lists surrounding the proximal end of the colony. In most Retiolitidae, however, the whole colony is covered by the secondary membrane of the ancora sleeve. The ancora sleeve may be ornamented by numerous lists of the reticulum, often irregularly placed, on either the inside or the outside of the membrane. The main lists outlining the colony are called the clathria, and also outline the internal tubarium on the surface of which no reticulum is laid down.

Outlook

The terminology of the graptolite tubarium has changed considerably through the last 150 years and – among others – one of the central terms, the use of rhabdosome for the graptolite housing construction, has been abandoned recently. As in terminologies for all groups of organisms, the terminology of graptolites is influenced by the authors working with them and their interpretation of the phylogeny and evolutionary relationships of the group. The supposed relationship of the graptolites to the bryozoans was the reason why we used so many terms borrowed from bryozoan terminology to describe graptolites. We identified the housing construction as a skeletal development and, thus, were misled for a long time. Now a first step has been made to unite the extant pterobranchs with the graptolites and understand them in a biological way. Therefore, a more suitable terminology had to be developed and it will certainly be adjusted in the future to the needs of the taxonomists among the graptolite researchers.

4 Paleoecology of the Pterobranchia

Jörg Maletz and Denis E. B. Bates

The Paleozoic Era was a time interval that differed considerably from our modern world, and in most aspects its ecosystems are quite difficult to imagine or to reconstruct. Thus many questions remain open and little light can be shed on the ecological needs and interactions of the graptolites. Initially, there was no life on land, even though the graptolites would not have cared, but the marine ecosystem and the food chain was quite different since many of the important modern players only appeared much later in geological history. So what were the conditions under which graptolites lived? What were their needs? How did they survive in the Paleozoic oceans?

The planktic graptoloids were probably the most unusual plankton that our planet has ever seen – huge colonial organisms swimming in the water column above their benthic cousins, the dendroids. They were able to move actively through the water for feeding and only sank down to the sea floor when they died. Graptoloids developed into numerous species with quite a variety of colony shapes, probably reflecting special ecological responses to their environment and their acquired lifestyle.

The graptolites might have been the slowest growing planktic organisms that we know of, if we consider the growth estimates based on the few extant members. If they were able to live for many years in the oceans without being regarded as food for other organisms, then the implication is that there were few predators and they were not too successful in collecting graptolites as food items and thereby diminishing their numbers. Otherwise, this long lifecycle might have been disadvantageous, especially if it included slow reproduction rates also. However, the evidence indicates that graptolites might have been among the first larger organisms venturing into the deep oceans and thus at the time probably had very little competition. Were the huge expanses of the open Paleozoic oceans still relatively lifeless? We might forget the possibility of the existence of large siphonophores and pelagic tunicates in the Paleozoic oceans, but these did not leave any trace in the fossil record. Thus, life might have been more complex than we can estimate.

Mode of Life

All graptolites are marine organisms and appear to be restricted to normal marine conditions, as they have never been found in extreme environments. The same can be said for the closely related Cephalodiscida, who also share the wide geographical and ecological distribution of the graptolites in the world's oceans. Even their sister group, the Enteropneusta, are invariably marine organisms and explored a wide variety of environmental conditions. None of these expanded their ecological niche to cover brackish or freshwater environments.

Modern *Rhabdopleura* may thrive under poor light conditions and even hide under empty shells, but members of the genus appear in all marine regions from very shallow water in coastal and even intertidal environments to a water depth of many hundreds of metres in the modern world's oceans. *Rhabdopleura* is found in tropical regions, but also close to the poles in Arctic and Antarctic waters; thus it is not restricted by temperature barriers.

While the only extant graptolite *Rhabdopleura* is a benthic organism, fossil graptolites can quite easily be differentiated into two main groups based on their particular lifestyle. There are the benthic taxa, fixed to the sea floor and the planktic ones, floating or moving freely in the water column as an ancient macro-plankton (Figure 4.1). A further differentiation may be seen in the encrusting taxa like *Rhabdopleura* and the erect-growing, bushy to tree-like forms of many derived benthic graptolites, looking like bryozoans moving slowly back and forth in the water currents.

Benthic Graptolites

The lifestyle of benthic graptolites may be understood by investigating the environment in which modern *Rhabdopleura* lives. However, fossil benthic graptolites were much more diverse in the constructional style of their colonies, and consequently in their ecological needs. They ranged from small encrusting colonies to erect bushy or tree-like colonies of considerable size. Others formed conical or fan-shaped tubaria, similar to those of some modern bryozoans. They probably lived in shallow-water regions all over the world, where they flourished, but are rarely preserved in the fossil record. Erdtmann (1976) described a Silurian fauna including dendroid graptolites from the dolomitic Mississinewa Shale of Huntington, Indiana. The graptolites were apparently living under low oxygen conditions within the photic zone at a depth of less than 60m in an inter-reef area, as Erdtmann (1976, p. 246) estimated from the frequency of benthic algae. Erdtmann described this graptolite fauna as a pioneering community with a short lifespan that was affected by cyclic burial events. Thus, the preservation is due to rapid entombment by the carbonate mud from the reef flanks, and explains the *in situ* preservation of the specimens.

All benthic graptolites need a firm substrate on which to anchor. Modern pterobranchs are often found on corals or other marine organisms, or are attached to rocks (Figure 4.1D), but attachment to other organisms has been noted. A species of the extant genus *Rhabdopleura* was discovered in intertidal areas on Fiji, where it lives in the cavities of corals and seals these cavities with tubarial material to stay hydrated even under intertidal conditions (Dilly & Ryland 1985). *Sphenoecium johanssoni* from the Middle Cambrian of Närke, southern Sweden, was found to be growing on the shells of the brachiopods *Acrothele* and *Dictyonina*, and in rare cases attached to the remains of trilobite shells (Bengtson & Urbanek 1986). The material was then transported into deeper water regions with an anoxic environment and is now preserved in black shales, but the animals certainly did not live there. The species is probably identical to material described as

Figure 4.1 Lifestyle of graptolites. (A) *Dictyonema cavernosum* Wiman, 1896b, benthic, erect colony (based on Wiman 1896b, pl. 1) (from Erdtmann 1986a). (B) *Rhabdinopora flabelliformis* (Eichwald, 1840), planktic graptoloid. (C) *Tetragraptus*(?) *norvegicus* (Monsen, 1937), planktic graptoloid with free nema, chemically isolated specimen (from Maletz 2011d, Fig. 1A). (D) *Cephalodiscus* (*Orthoecus*) *rarus* Andersson, 1907, specimen attached to a piece of rock (from Andersson 1907, pl. 2, Fig. 7). Illustrations not to scale.

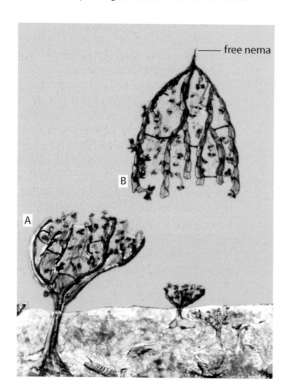
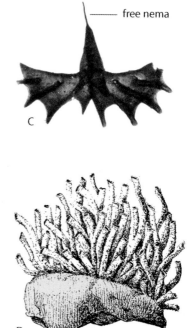

Dendrograptus (now *Sphenoecium*) *mesocambricus* from the Middle Cambrian of Krekling, Norway (Öpik, 1933). The type specimen of *Sphenoecium mesocambricus* is attached to an organophosphatic brachiopod shell, which – due to preservational aspects – is difficult to see in the centre of the colony and was overlooked in the original description (Figure 4.2A). The association with numerous head and tail shields of agnostid trilobites indicates the transportation of the colony into the black shale environment and provides information on the exact age of the material (Figure 4.2B–D). Attachment of benthic graptolites can even be seen in the ornamentation of the attachment disc, as is shown by Bates and Urbanek (2002) from chemically isolated material of *Mastigograptus* aff. *tenuiramosus* (Walcott, 1883). The illustrated specimen shows ridges in the attachment disc that originated from the ribs of the calcitic shell to which the specimen was originally attached. All known attachment structures of benthic erect graptolites are flat-bottomed and indicate attachment onto a firm surface. There is no evidence of a root system like that in modern plants in any benthic graptolite, and it would be unlikely to exist if the construction of the tubarium from the cephalic shield according to the pterobranch model is considered.

Planktic Graptolites

Obviously, the life mode of a planktic graptolite is much more complex and difficult to understand than the lifestyle of the benthic species. The main problem is the lack of any extant

Figure 4.2 Early graptolites and their associates. (A) *Sphenoecium mesocambricus* (Öpik, 1933), holotype, showing poorly preserved phosphatic brachiopod in centre. (B) *Goniagnostus nathorsti* (Brögger, 1878), cranidium. (C) *Goniagnostus nathorsti* (Brögger, 1878), pygidium. (D) *Leiopyge calva*, cranidium. All specimens from Krekling, Norway, Middle Cambrian *Goniagnostus nathorsti* Biozone.

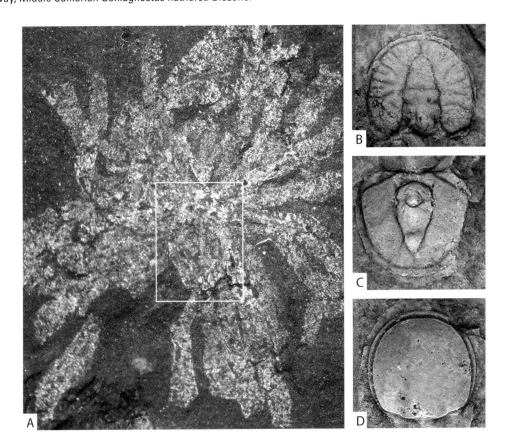

planktic graptolites for comparison and the preservation of most graptolites as flattened films of organic material in dark shales. Usually, this material shows indication of transport and fragmentation of the graptolites, and thus does not provide precise information on the lifestyle and life history of these organisms, which has to be interpreted from faint fossil indications. Even in 1868 Nicholson suggested a free-floating lifestyle for the graptolites, but this was not universally accepted and discussions for a long time considered that attachment to seaweed with long and flexible nemata was more probable, following the ideas of Lapworth. Lapworth (1897, p. 254) described his concept with the words "Das Polyparium musste von dem tragenden Objekt herabhängen wie eine Glocke am Ende eines Strickes" [*The polyparium (tubarium) had to be suspended from the supporting object like a bell on a rope*]. This concept of an epiplanktic lifestyle attached to seaweed (Figure 4.3B) has long been rejected (e.g. Bulman 1964; Kirk 1969), but is still found in many paleontological textbooks (Maletz 2014a). It is quite unrealistic to suggest that the large (often reaching a metre in diameter) colonies of the multiramous Ordovician dichograptids are attached by a long and delicate nema to seaweed for support. These colonies would have been too massive for a single point attachment and the connection would easily have been broken through current action and water turbulence.

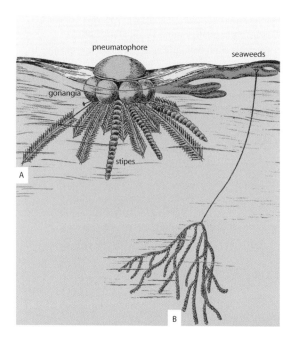

Figure 4.3 Misleading concepts of epiplanktic graptolites. (A) Colony of *Orthograptus quadrimucronatus* (Hall, 1865), showing synrhabdosome with floating vesicle ("pneumatophore"), gonangia and tubaria ("stipes" or "rhabdosomes"). (B) *Pterograptus elegans* Holm, 1881, attached to seaweed with long nema. Illustration modified from Maletz (2014a).

Synrhabdosomes

Ruedemann (1904) created the term synrhabdosome for the compound colonies or supercolonies of mostly biserial (axonophoran) graptolites, based on his interpretation of the colonies of *Orthograptus quadrimucronatus* (in Ruedemann 1895). He described this "supercolony" as based on a pneumatophore, a central disc, gonangia and the stipes (Figure 4.3A). The individual stipes (the tubaria in newer literature) were connected by the funicle, a branched structure without thecae. This interpretation was based on the descriptions of Hall (1865) and is now known to be based on an incomplete understanding of graptolite colonial development, but has long persisted in the paleontological description of graptolites. There is no fossil evidence for the pneumatophore and the gonangia as described by Ruedemann in graptolites. The funicle

connecting the individual tubaria is a misleading interpretation of the proximal ends of the multiramous *Clonograptus flexilis* in which the proximal stipes often do not show the thecae due to their orientation (see Maletz 2014a).

Modern interpretations suggest a free-living, holoplanktic lifestyle for the planktic graptolites (Figure 4.1B) in the water column (e.g. Kozłowski 1971). This is supported by the investigation of the sicula of the planktic graptolites, with their prosicula differentiated into conus and cauda. The slender cauda is the base of the nema and is not present in the benthic graptolite, in which a tubular prosicula is provided with a flat attachment site.

Locomotion

As the idea of an unattached planktic lifestyle for graptolites gained acceptance, the question of the "how" came to light. How could a graptolite colony have moved through the water column? What mechanism did it use to propel itself around in the ocean? Was there really active movement possible, or were graptolites just moved passively by water currents and eddies? These questions are difficult to answer, as we do not have any living examples of planktic graptolites today. Graptolites would be included within macro-plankton due to their size. Their colonies range from less than 1 mm in juveniles to a metre or more in length or diameter for some large graptolites. Modern plankton, however, are usually much smaller and consists of zooplankton and phytoplankton. Zooplankton includes the jellyfish, which may reach diameters of several metres and are referred to as macro-plankton. These have a large body, but do not produce a skeleton or housing construction like the graptolites. They usually have a maximum life expectancy of several months, and thus may grow and reproduce quickly as an adaptation for a successful life in the dangerous modern seas, where predators are common and organisms need to be either fast or common and reproducing quickly to survive.

From all considerations, our existent marine plankton provides a poor analogue for the Paleozoic graptolites. Planktic graptolites were an unusual type of macro-plankton, not comparable to any existing forms. They were large colonial organisms with tiny zooids filling the individual chambers or thecal tubes of the tubaria. Thus, the living tissue of the graptolites (if the pterobranch model is correct) probably made up only a small portion of the whole colony and the zooids would have had a high amount of "dead weight" to transport through the water column. The similar-sized macro-plankton of today, the jellyfish with their high proportion of jelly material and without a skeleton or massive housing construction, can move by changing their body shape without the restriction of a rigid skeleton or other inflexible structure. They thus have a considerable advantage that may explain their success. They do not need to spend enormous amounts of time and energy secreting a housing structure that then limits their activities.

The pteropods (Thecosomata), small marine snails, bear shells of generally less than 10 mm in diameter. Thus, they are more similar in size to the zooids of the Pterobranchia, but differ in being individuals and not bound to each other as the colonial zooids of the pterobranchs. They possess a thin shell and a relatively large body. The pteropods form special paired wing-like extensions called parapodia, typically expressed by *Limacina antarctica* (Figure 4.4C), one of the peculiar "sea butterflies". Closely related taxa of the Gymnosomata, the "sea angels", even lost their shells completely. Larger gastropods have relatively massive bodies in comparison with their shells, but they are benthic organisms and do not swim in the water column.

Melchin and DeMont (1995) considered the modes of movement that were possible for a graptolite colony from a constructional point of view. The authors assumed that graptolites did not use a method of movement that is unknown in modern organisms, and that they did not rely on passive movements or drifting. Comparing the various methods that modern planktic organisms use, they considered that rowing with muscular appendages would be the only possible mode due to constructional restraints. They postulated wing-like lateral extensions of the cephalic shield as the most likely method, and compared their model with the pteropod swimming wings (Figure 4.4). There are two different ways in which rowing extensions could have been formed. Either they may be constructed from enlarged arms (Figure 4.4A) or from separate wings developed independently as new features from the cephalic shield of the zooids (Figure 4.4B).

Rigby and Rickards (1989) investigated the mobility of Ordovician graptoloids and used water tank experiments with life-sized models. They were able to show that many colony shapes are designed to move passively or actively in a spiral fashion through the water column. The spiralling increases considerably the water column sampled for food by the individual zooids compared with a straight vertical fall (Figure 4.5). It could also be the reason for the many spiral-shaped monograptids in the Silurian (e.g. *Spirograptus*, *Torquigraptus*, *Cyrtograptus*, and many others). The model works best for multiramous tubaria, but can be adapted for single-stiped ones also. The colony shapes have considerable influence on the feeding efficiency of graptoloid colonies, and biserial colonies apparently have a much higher feeding efficiency than planar multiramous taxa (Rigby 1991, 1992). Kirk (1969) postulated that vertical movement of the tubaria

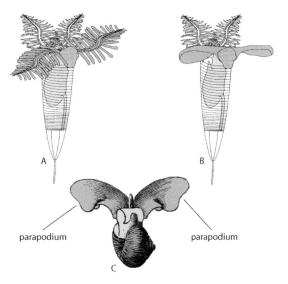

Figure 4.4 (A–B) Zooid model of Melchin and DeMont (1995). (C) *Limacina antarctica*, an Antarctic pteropod (after Woodward 1854, pl. 14, Fig. 41).

Figure 4.5 Vertical movement of graptolite colonies (after Rigby & Rickards 1989, Fig. 4).

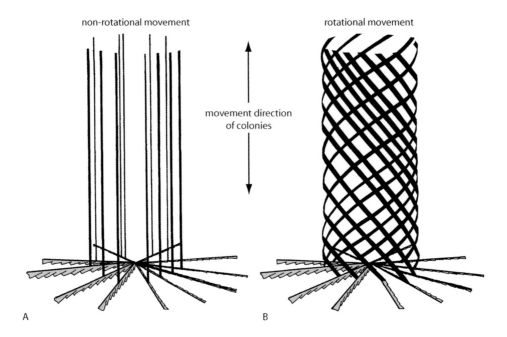

through the water column was the result of ciliary action on paired arms, similar to those of *Rhabdopleura*, the movement being in a diurnal cycle similar to that of other plankton.

The interpretations of Rigby and Rickards (1989) and Rigby (1991, 1992) unfortunately do not answer the question of how the locomotion of the graptolite colonies was accomplished. They merely provide the theoretical aspects of a graptolite colony moving through the water column and show the recognizable adaptations of the tubarium shapes for active movements.

Graptolites and Sediments

Graptolites can be found in nearly every marine sediment type and are not restricted to a narrow limit of sedimentary facies (Figure 4.6). They are most commonly found, however, in fine-grained sediments, in black shales, more rarely in greenish to grey shales, and sometimes also in coarse sandstones. Thus, the restriction to the characteristic "graptolitic black shales" (Ruedemann 1911) is not true, even though these are the types of sediments

from which most graptolites are collected. Graptolites are often common in shales as flattened films of organic material, or as pyritic internal casts of three-dimensionally preserved specimens. In the black anoxic shales they are often associated with small phosphatic brachiopods and rare conodonts. Organic-walled microfossils like chitinozoans and acritachs may also be common in the sediments. The presence in these sediments does not, however, indicate that graptolites were living under anoxic conditions. They may have been able to survive low oxygen conditions for a while, but were living in oxygenated waters, either above the bottom anoxic zone, or were transported into the anoxic environment, as were the remainder of the faunas.

Despite the idea of fragility due to the delicate appearance of their tubaria, graptolite colonies were highly durable. They are often still flexible when freed from the sedimentary matrix after hundreds of millions of years. Therefore they could be transported for long distances by turbidity currents without being completely broken. Current-oriented graptolite fossils have been found in the E-horizon of the Bouma sequence in turbidites and in intercalated pelagic black shales (Hills & Thomas 1954; Moors 1969, 1970; Cooper 1979a). They may

Figure 4.6 The graptolite biofacies: schematic distribution of lithofacies and graptolite biofacies in a sedimentary basin bounded by a carbonate platform (after Podhalańska 2013).

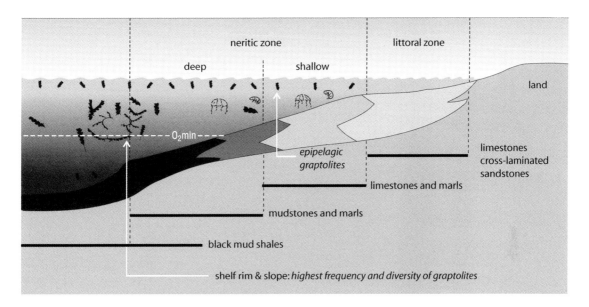

also be present in the A and C horizons of turbidites (Finney & Berry 1997; Cooper et al. 2012). Parallel orientations of elongated graptolite specimens can be regarded as an indication of current transportation (Ruedemann 1897; Hundt 1936; Moors 1968).

Graptolites preserved in carbonates are the most precious fossils for a paleontologist as they are usually preserved in three dimensions and often chemically isolatable from the rock. This is the material that provides the best information on the structural organization of graptolites. Holm (1895) was among the first to chemically isolate graptolites from limestones and describe them in detail. The Retiolitidae (see Chapter 12) are now recognized as a spectacular group of graptolites with a highly complex development of the colony from numerous thin rods (Bates et al. 2005). These graptolites are best studied from isolated material as the flattening of specimens on shale surfaces covers many important details and makes the material unrecognizable.

Patterns of Occurrence

In many successions, graptolites are either restricted to individual layers or the frequency of specimens varies quite considerably from layer to layer, even in fairly monotonous sedimentary successions. There is no even distribution of graptolites in any succession. Graptolites may be found as monospecific associations of large specimens, probably indicating a mass mortality event, or in populations of all growth stages. Individual siculae and juveniles may be associated with mature specimens, or only juveniles of a kind are found (e.g. Moors 1968; Pannell et al. 2006). Blooms of a single species may look different from the higher diversities of time-averaged faunas on shale surfaces. In many layers, diverse graptolite faunas indicate a considerable time interval during which the specimens most probably collected on the sea floor, indicating the result of time-averaging, but this effect may be difficult to separate from mass mortalities of diverse communities.

The uneven distribution and relative abundance of graptolite species in most sedimentary layers and the presence of specific faunal elements may be used to infer a faunal differentiation based on water depth (Egenhoff & Maletz 2007; Cooper et al. 2012). Thus it is quite plausible that a differentiation of graptolite faunas by depth gradient was present in the Paleozoic oceans. Rigby (1993a) investigated the size distribution and interpreted survivorship curves of

Figure 4.7 (A) Length-frequency relationship of *Orthograptus quadrimucronatus micracanthus* of sample SM X23260 (after Rigby & Dilly 1993, Fig. 14). (B) Survivorship curve for *Orthograptus quadrimucronatus micracanthus* (after Rigby 1993a, Fig. 9).

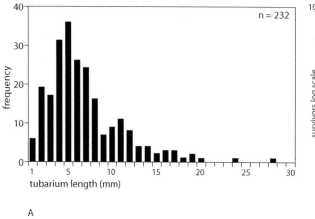

A

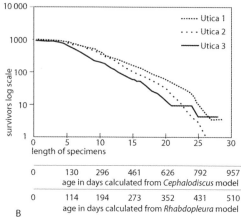

B

faunas from the Utica Shale of eastern North America (Figure 4.7). She concluded that the investigated species *Orthograptus quadrimucronatus micracanthus* and *Amplexograptus praetypicalis* lived in monospecific shoals and arrived on the bottom in showers at different times.

On many sediment surfaces, graptolites are found in diverse associations of a variable number of species. Cooper et al. (2012) counted 13 species on a shale surface of only 90 cm² from the Yapeenian of New Zealand, and noted that this is nearly half of the species known from the time interval. In this example all species belong to the planktic graptolites. However, examples of association of planktic and benthic graptolites on the same sediment surface are not uncommon. Hall (1865) described diverse graptolite faunas from the Lower to Middle Ordovician of Quebec, eastern Canada. The faunas are often highly diverse and include accumulations of benthic and planktic graptolites at many levels, of which Maletz (1997a) only discussed the biostratigraphy of the planktic taxa. The planktic graptolites are exclusively present in many other successions, however, or benthic faunal elements are very rare. The Elnes Formation of southern Scandinavia includes rich graptolite faunas (e.g. Maletz 1997b; Maletz et al. 2011), but in general no benthic graptolites. Maletz and Egenhoff (2005), however, found a few benthic graptolites of the genera *Dendrograptus*

and *Acanthograptus* in a single siltstone layer in the Engervik Member of the Elnes Formation. Another example of the exclusively planktic faunas are the Silurian successions in the black shale units of Thuringia, Germany (Schauer 1971). Silurian faunas of Arctic Canada are found in shale successions with abundant limestone concretions. Both benthic and planktic graptolites are common in these successions (e.g. Lenz & Kozłowska-Dawidziuk 2004).

Altogether, it may be stated that the diversity and composition of graptolite faunas is quite variable and open to interpretation. We only have a limited understanding of the factors controlling the bedding plane composition of graptolite faunas. Our interpretations are restricted by our poor knowledge of the paleoecology and distribution of graptolite faunas, as these might in most cases be explained as sorting by sedimentation effects (currents, storm events, etc.), seasonal patterns of diversity such as plankton blooms, or even by selective destruction through decomposition of the organic material of the tubaria.

Population Structure

Very little is known about the population structure of graptolites, as most bedding-plane faunas are post-mortem associations indicated by current orientation or various sorting effects on the colonies (e.g. Rigby 1993a). Cisne and Chandlee (1982) investigated the distribution of graptoloids in the Taconic Foreland Basin of eastern North America. They used bulk samples from shales and limestones to understand the distribution of benthic and planktic organisms. Due to the poor preservation of the graptolites in the succession, the authors only identified the taxa to the genus level. Interestingly, many of the assemblages were dominated, sometimes to the exclusion of any others, by a single taxon. A clear trend showed the restriction of individual taxa to certain lithologies. The authors were able to recognize a depth zonation, with *Orthograptus* in the shallowest, climacograptids in the intermediate and *Corynoides* in the deepest-water strata. It remained, however, unclear whether the assemblages represented life or death assemblages and the authors did not discuss the paleoecological and taphonomic problems with graptolite records in the succession. Rigby (1993a) investigated slabs with bedding-plane associations of graptoloids, interpreted as local populations and not time-averaged accumulations. She used length–frequency curves and a survivorship analysis to understand the population structure of the investigated taxa. The length–frequency of the tubaria indicated continuous growth through the whole life-span of the colonies. The populations died either from environmental stress or from problems with the unlimited growth, as indicated by the shape of the survivorship curves.

The Graptoloid Habitat

As planktic graptoloids lived unattached in the Paleozoic oceans, they were independent of the sediment types in which they are preserved. Most graptolite assemblages have to be interpreted as death assemblages and are most probably modified through current sorting or other taphonomic influences. Time averaging might have had a considerable influence on the diverse faunas found in anoxic sediments. Individual colonies settling on an anoxic sea floor would not be destroyed through organic decay and would accumulate for a considerable time interval until a sedimentary event

covered the fauna and preserved it for paleontological investigation. Briggs et al. (1995) did not notice any visible decay of the fusellum of *Rhabdopleura* during their experiments. After more than ten weeks the material was still intact, and this might hold true for other graptolites. The research indicates that the graptolite fusellum is not as easily destroyed by bacteria under oxic conditions as we might think, but might be even more durable in an anoxic environment. The encasing sediments, thus, provide very little clue to the living habitat of the graptoloids. The presence of graptoloids in sediments of any type may not indicate that they were living in the region, but just document that they were preservable under the existing conditions. Transport into a suitable environment and quick subsequent burial and protection from exposure to oxic conditions may have been important for the preservation of the organic tubaria of graptolite faunas.

All geological indications show that graptoloids lived under normal oxygenation and salinity and were not adapted to extreme environments, but the particulars of the preservation of their organic tubaria (see Chapter 5) may mask their original distribution. Graptolites have been discovered from very shallow, coastal sedimentary environments, but are not found in hypersaline strata or strata influenced by lacustrine or brackish conditions. A clear differentiation into faunas with a preference for warmer or colder waters can be noted through the biogeographical restriction of many faunal elements, as well as a depth differentiation into shallow water and deeper water (oceanic) faunal elements (see Biogeography). Fortey and Cocks (1986) used the distribution of the oceanic isograptid biofacies to track continental margins in the Middle Ordovician and support plate tectonic considerations with paleontological data (Figure 4.8).

Berry et al. (1987), based on the observation that graptolites are most commonly found in black shales, introduced the idea that graptolites inhabited a denitrified low oxygen zone of the Paleozoic oceans that was supposedly more widespread in the Early Paleozoic oceans (Berry & Wilde 1990). Graptolites could have migrated up and down in the water column for their daily feeding rhythm. The idea of an active vertical migration

Figure 4.8 Graptolite paleobiogeography of the Middle Ordovician. Isograptid biofacies in black; shallow-water graptolite biofacies in grey (based on Fortey & Cocks 1986, Fig. 3).

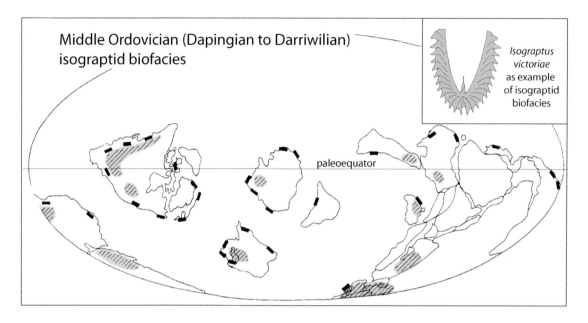

of the graptolites for feeding and predator avoidance, based on the lifestyle of modern euphausiids (Antezana 2009; Cooper et al. 2012), may be quite useful in the understanding of graptolite lifestyles, even though it is difficult to verify. The model suggests that graptoloids stayed in the surface waters during night times for feeding and descended to the oxygen minimum zone in the daytime. Thus they could avoid predators during the day and were able to filter-feed during ascent and descent.

Spatial Distribution

A distinct latitudinal and depth differentiation of graptolite faunas has been established, especially for Lower to Middle Ordovician faunas (Cooper et al. 1991), indicating considerable influence of oceanographic and climatic controls on graptolite distributions. The faunal differentiation was recognized through the difficulties of correlation of Lower to Middle Ordovician faunas between the British Isles and North America, and Skevington (1973, 1974) introduced the "Atlantic Province"

and "Pacific Province" of graptolite faunas for the time interval (see also Chapter 4). The reasons for the faunal differentiations have been hotly debated and a number of different causes suggested. Ecological controls like ocean currents or land barriers, climatic belts, water depth and water temperatures have been regarded as the main influences, but also biotic factors such as competition have also been considered. The two main emerging models include surface water temperature and depth stratification, and Goldman et al. (2013) discussed the various models that have been introduced into the literature and considered influences from both factors as important for the ecological (depth-related) and biogeographical distribution of graptolite faunas.

Depth Distribution

Graptolite faunas from shallow-water sediments are often low in diversity or may include monospecific assemblages, while the deeper-water faunas are often highly diverse. A typical example is the record of azygograptids and pendent

Figure 4.9 Transgressive succession in the Lower Ordovician of North Wales, showing early appearance of *Azygograptus* species in the most shallow facies (after Beckly & Maletz 1991, Fig. 1).

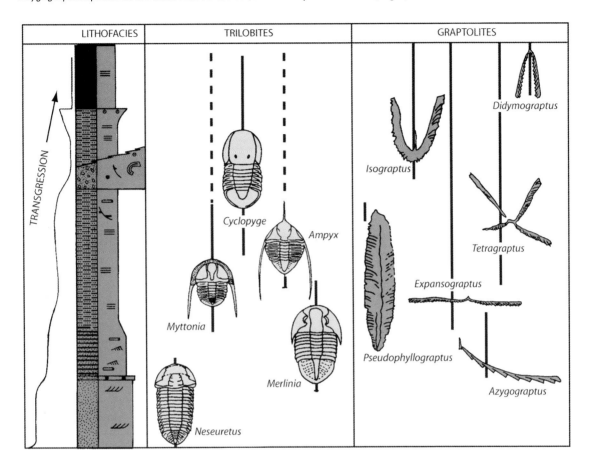

didymograptids (tuning-fork graptolites) in Middle Ordovician shallow-water successions of Britain. Beckly and Maletz (1991) discussed the facies association of *Azygograptus* species and stated that these are generally the first graptolites to occur in transgressive successions above the basal Arenig unconformity in North Wales (Figure 4.9). Other examples include the distribution of the Darriwilian *Didymograptus* species, but also the record of biserial axonophorans in the Upper Ordovician. Goldman et al. (2002) described the example of a monospecific association of *Amplexograptus perexcavatus* from the Lebanon Limestone of Tennessee, and referred to other monospecific graptolite assemblages of the cratonic regions of Laurentia.

Cooper et al. (1991) differentiated an inshore biotope (the "didymograptid biofacies"), a shallow epipelagic biofacies and the deeper mesopelagic biofacies (the "isograptid biofacies") (Figure 4.10) that was best for intercontinental correlations in their depth differentiation model. The inshore biofacies is characterized by the presence of an assemblage of entirely endemic faunal elements. The shallow epipelagic biofacies includes cosmopolitan faunal elements associated with some endemic taxa. The deeper-water biofacies is characterized entirely by the presence of pandemic faunal elements. The model also included as an additional complication the biogeographical differentiation of the faunas within the three biofacies.

Figure 4.10 Depth distribution of graptolites (based on Egenhoff & Maletz 2007, Fig. 11).

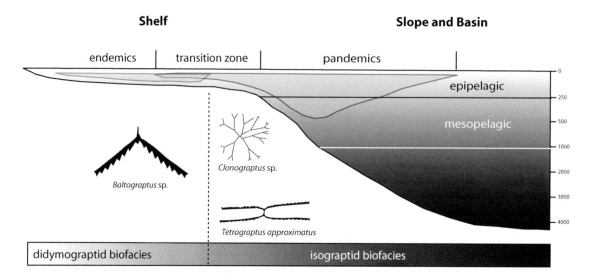

Finney and Berry (1997) compared the record of Ordovician graptolite faunas from the Vinini Formation of Nevada with that of plankton in the modern oceans. They observed that graptolites were common in waters above the shelf margins, but scarce in the open oceanic waters, as is most modern zooplankton. From this they interpreted upwelling zones with their high availability of nutrients and enhanced bioproductivity along the shelf regions as the favoured habitat of graptolites. Even without the development of upwelling, data indicate that graptolite faunas were most diverse along the shelf break and diversities diminished towards the deeper ocean basins (Chen et al. 2001; Finney & Berry 2003; Podhalańska 2013) (see Figure 4.6).

Egenhoff and Maletz (2007) used the spatial distribution of graptolite species in the Lower Ordovician succession of the Tøyen Shale of southern Sweden to trace sea-level changes in black shale successions. While the succession at Mt. Hunneberg is largely dominated by endemic graptolite faunas of the Atlantic Province, occasional layers with predominantly deep-water faunal elements occur and provide information on transgressive phases (Figure 4.11). The deeper-water faunas were able to invade the shallow-water region only when the sea-level rose and then replaced the shallow-water endemics or

mixed with these. Thus, the faunal differentiations made it possible to recognize maximum flooding surfaces (mfs) in a sequence stratigraphic concept through the invasion of pandemic, deeper-water graptolite faunas onto the shallow shelf areas.

Biogeography

Goldman et al. (2013) discussed in some detail the current ideas on the biogeographical differentiation of graptolite faunas. They concluded that the simple model of Skevington (1973), differentiating "Atlantic Province" and "Pacific Province" cold-water and warm-water faunas, needed several adjustments. The biogeography appeared to be influenced by a multitude of physical and chemical factors of the oceanic environment. They also acknowledged that the endemicity of graptolite faunal elements is not entirely based on a temperature gradient in the Paleozoic oceans. Certain faunal elements, such as *Geniculograptus typicalis* or *Paraorthograptus manitoulinensis*, are clearly endemic to the northern Appalachian Basin of Laurentia and not found on any other continents. A restriction to individual depositional basins or isolated oceanic circulation cells cannot be excluded (Goldman et al. 2013).

Figure 4.11 Diabasbrottet section, Västergötland, Sweden, GSSP section for the base of the Floian Stage, Ordovician System, graptolite biostratigraphy and recognition of maximum flooding surfaces (mfs) (based on Egenhoff & Maletz 2007, Fig. 3).

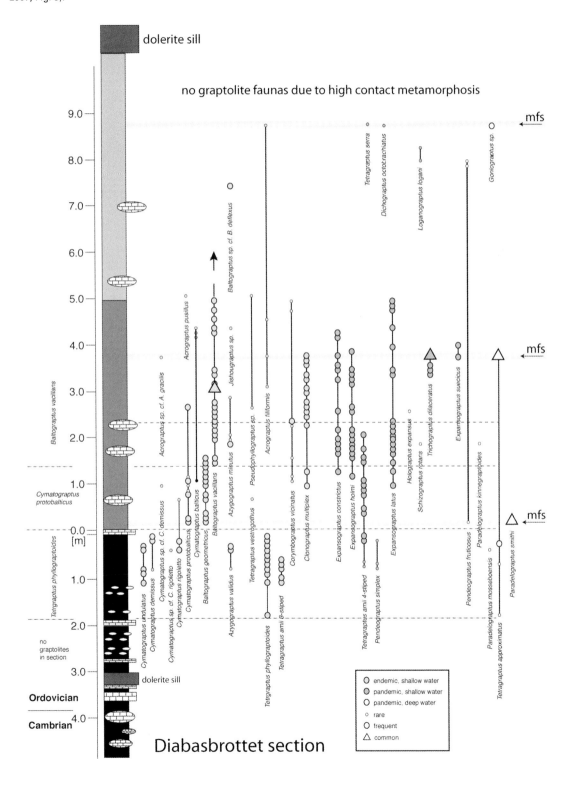

Diabasbrottet section

Figure 4.12 Paleocontinent reconstruction map for the Dapingian (Middle Ordovician) showing graptolite localities and distribution of graptolite biofacies. Black dots are localities that have graptolite successions during the Dapingian. Grey shaded ovals indicate oceanic (isograptid) biofacies and striped ovals indicate shelf (didymograptid) biofacies (based on Goldman et al. 2013).

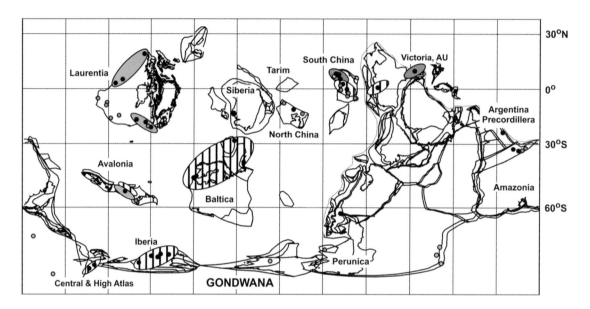

Goldman et al. (2013) considered the terms "Atlantic Province" (cold-water faunas) and "Pacific Province" (warm-water faunas) as misleading since they refer to modern geography, and preferred the neutral descriptions of high, medium and low paleolatitudes. They described the changing biogeography of the Ordovician from a number of time slices and provided biogeographical maps (Figure 4.12). During the Hirnantian (Uppermost Ordovician) the characteristic cold-water faunas dominated by the Diplograptina were replaced largely by the Neograptina that subsequently invaded the southern high latitudes and led to the evolution of the monograptid faunas in the Silurian (Goldman et al. 2011; Melchin et al. 2011).

The Silurian apparently showed a lesser degree of faunal provincialism, as can be shown by the worldwide use of a standard biozonation (Goldman et al. 2013). A considerable faunal differentiation was still present in the Sheinwoodian between Arctic Canada and the faunas of the Iapetus/Rheic Ocean region, but this diminished subsequently (Lenz et al. 2012). Štorch (1998a) provided a map

comparing the distribution of Llandovery graptolite faunas, in a concept comparable to the Ordovician faunal provinces. He differentiated cold-water and warm-water faunas as well as a peri-Gondwanan Europe (PGE) assemblage with a distinct depth differentiation. Similar differences are still present in the Ludlow–Pridoli time interval, but have not been investigated in detail. The Lower Devonian graptolite faunas are restricted in their distribution to equatorial regions and are of low biodiversity. Koren (1979) suggested that this biogeographical restriction may have been an important factor in the final extinction of the planktic graptolites.

Historical Biogeography

Historical biogeography aims at tracing the historical changes in spatial distribution of graptolite faunas, finding geographically definable spots or regions from where radiations started, and locating the geographical origins and the migration patterns of individual groups. Zhang and Chen (2007) used the biogeographical distribution and restriction of Middle Ordovician graptolite faunas on the Yangtze Platform as the main argument to predict the area of origination of the axonophoran graptolites, and concluded that they most likely derived from South China. The genus *Pseudisograptus*, a predecessor of the axonophorans (Mitchell et al. 1995; Fortey et al. 2005), was never found in the shallow-water region of the Yangtze Platform, but is present in the deeper-water region of the Jiangnan Slope and Zhujiang Basin regions. Thus it is a deep-water faunal element, and the descendants, the axonophorans, exemplified by the genus *Levisograptus* (*Undulograptus austrodentatus* group in their paper), progressively expanded their distribution area and invaded the shallow-water regions of the Yangtze Platform (Figure 4.13).

Goldman et al. (2013) used a parsimony-based analysis to trace the roots of the Axonophora and the main biserial groups. They took the latest cladistic analysis of the group by Mitchell et al. (2007a) and used a modified Fitch parsimony analysis to interpret the origin of the clades. Ancestral states were found by the evaluation of the overlap of geographical regions occupied by the taxa above the node in question (Goldman et al. 2013) (Figure 4.14). The data supported the origin of the Axonophora as suggested by Zhang and Chen (2007), but also provided the information that other groups may have originated in shallow-water regions and dispersed from there. Thus, migration from both shallow-water regions and deeper-water regions can be recognized during the life history of the graptoloids.

The origin of the expansograptids, the two-stiped horizontal to subhorizontal graptoloids, can be traced to the high latitudes, where they represented shallow-water endemics in early Floian times. They migrated into the deeper water of the oceans and were found worldwide in younger strata (Goldman et al. 2013). Quite a number of diversification events can be traced in this way to the high to mid paleo-latitudes, but the interpretations are based on a fairly small data-set as the authors acknowledged, and may have to be revised in the future. Still, they show that we can learn a lot more about the complex evolutionary and diversification patterns of graptolite faunas.

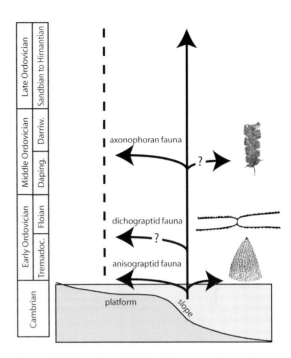

Figure 4.13 Example of historical biogeography and origination of graptoloid clades (based on Zhang & Chen 2007, Fig. 3).

Graptolite Life History

Some general ideas on graptolite lifestyle and life history can be gained through the investigation of extant pterobranchs, but most information is based on the tubaria, their mode of preservation in sediments, fossil accumulations, the taphonomy of the assemblages, and other aspects of the fossil record. Thus, it is based on theoretical considerations and not on actual fossil data. A comparison of colony growth patterns of fossil taxa with the established development of extant pterobranch colonies provides the estimations we use to interpret the life history of the extinct graptolite taxa.

Graptolite life span

It is difficult to estimate the life span of a fossil graptolite colony. Therefore, we have to look at the growth of the modern *Cephalodiscus* and *Rhabdopleura*. Dilly (1986) described the tube building of *Rhabdopleura* and stated that it took the zooids several hours to produce a single fusellus,

Figure 4.14 Area cladogram illustrating the geographical regions occupied at the nodes (see Goldman et al. 2013 for details).

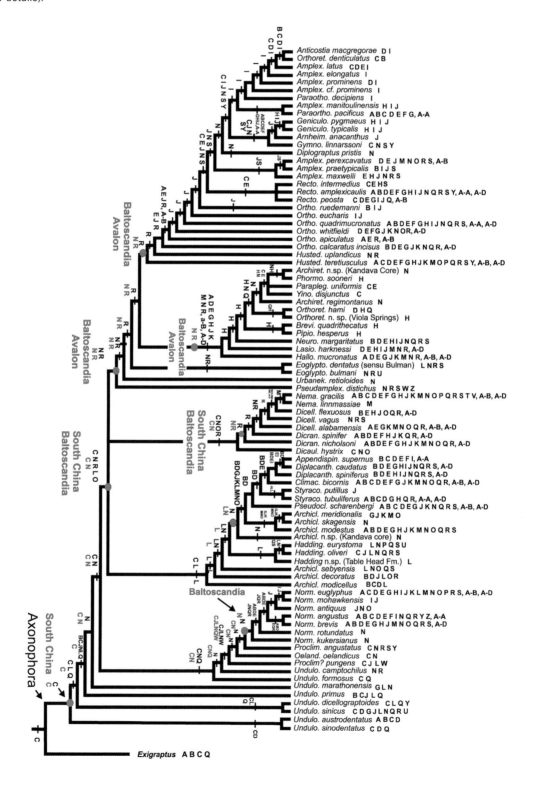

while Rigby and Dilly (1993) observed an estimate of eight hours for the secretion of a single growth increment, but considered the growth rates of *Cephalodiscus* and *Rhabdopleura* to be highly variable. Based on their estimations, the secretion of a single complete thecal tube of *Rhabdopleura* would take several days. The secretion of a *Rhabdopleura* colony with numerous zooidal tubes would then take weeks and months. However, the exact life expectance of the colonies is unknown and may only be estimated through the growth information provided by Rigby and Dilly (1993).

Several generations of *Rhabdopleura* zooids could successively inherit a single zooidal tube (Stebbing 1970a), extending the life expectancy of a colony considerably. Based on this estimate, Rigby and Dilly (1993) suggested that it took 6.6 days for a graptolite to produce a single theca. In their estimation, the secretion of the metasicula of *Amplexograptus maxwelli* alone would have taken about 20 days. If this is a correct estimate, then large graptolite colonies may have lived for many years. The life span of a 75 cm long specimen of *Monograptus priodon* could have been as long as 13 years, a considerable life span for a planktic organism. Rigby (1993a) used the length of *Orthograptus quarimucronatus micracanthus* tubaria to calculate the age of the colonies in days using the known growth of *Cephalodiscus* and *Rhabdopleura* specimens (Figure 4.7B). As the graptolite zooids were able to repair their tubaria and may even have recolonized individual thecae whose producers died, the life span of a graptolite colony may have been even longer. Possible changes in the metabolism and life history of the pterobranchs during the transition from a benthic to a planktic lifestyle might make these estimates highly speculative and not valuable for a comparison with the life expectancy of planktic graptolites.

Growth limitations
We can learn a lot from the fossil graptolite colonies, especially if we have many specimens of various growth aspects and can pull the data together and interpret them by comparing them with extant organisms. We can understand many aspects of their lifestyle, as we understand modern organisms. Even though we cannot observe them alive, we can use the fossil remains and ask the right questions to get a better picture.

It seems that the colony growth of graptolites was unlimited in many species, and constraints were only due to the limitation of the construction, the stability of the tubaria and the life expectancy of the zooids. Benthic colonies of large sizes are rarely found in the fossil record. Mostly, we have small fragments to judge from. These we can fit together and reconstruct the final dimensions of the colonies. Colony diameters have been estimated at 50 cm or more for some species (Bouček 1957) bearing thousands of zooids, but most specimens are much smaller. The only restriction appears to be ecological, through changes in the environmental conditions and predation or competition with other marine organisms.

Multiramous planktic graptolite colonies may have reached a diameter of more than 1 m in some cases, but a few centimetres to tens of centimetres in diameter were the more likely size. Straight monograptid colonies up to 1.45 m have been measured, but due to fragmentation, long specimens are rarely preserved complete. Compared with these extremes, most graptolites were most likely restricted in their growth, and in some cases we can even recognize this due to special features in the construction of their tubaria. In many retiolitid graptolites, a distal appendix (Figure 4.15D–E), a special theca, can be found at the tip of the tubarium. The appendix is differently shaped from the rest of the thecae. Other graptolites produce only a few thecae and are not found in larger colonies. The genera *Brevigraptus*, *Corynoides* or *Peiragraptus* have between three and eight thecae, but never more. The number is considered characteristic of individual species. *Corynoides* and its close relative *Corynites* bear a very small, coiled last theca (Figure 4.15C), but these are difficult to see in shale material. A different indication of limited growth in graptolite colonies may be seen in the common presence of nematularia, membranes at the end of the nema in biserial graptoloids. In the case of *Archiclimacograptus decoratus*, for example, the end of the nema bears a heart-shaped membrane and the colony stops growing when it reaches this feature (Figure 4.15A). In other species, the nematularia may be overgrown by the advancing colony, as in *Cystograptus*

Figure 4.15 Graptoloids showing growth limitations. (A, B) *Archiclimacograptus decoratus* (Harris & Thomas, 1935). (A) NMVP 31932, two specimens with heart-shaped nematularium (photo by A.H.M. VandenBerg). (B) *Rectograptus intermedius* (Elles & Wood, 1907), Viola Limestone, Oklahoma, showing nema inside and growing end. (C) *Corynites divnoviensis* (Kozłowski, 1953), glacial boulder, Germany (coll. Kühne) with reduced and coiled third theca. (D) *Holoretiolites erraticus* (Eisenack, 1951) with appendix, Sellin, Rügen, glacial boulder. (E) *Neogothograptus eximinassa* Maletz, 2008, Spandau, near Berlin, Germany, glacial boulder, showing appendix with strong reticulum. Scale indicated by 1 mm long bar in each photo.

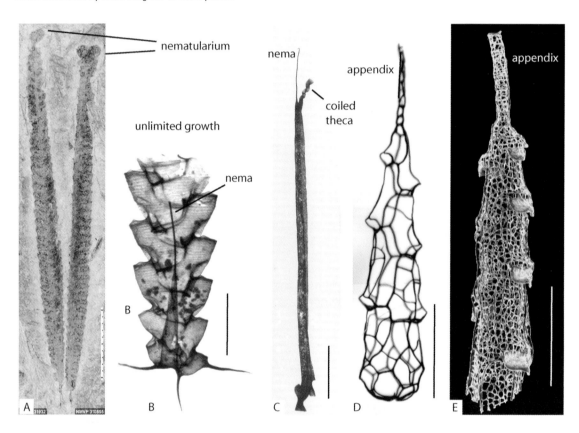

(Jones & Rickards 1967), but the nematularium may also be lengthened during the later growth of the colony and remain visible.

Graptolites and the Food Chain

The position of the graptolites in the Paleozoic food chain and in the benthic and planktic ecosystems has rarely been explored. Benthic graptolite faunas are seldom present in analyses of benthic communities, as these are usually interpreted from shelly faunas. Due to the organic material of the graptolite tubaria, these are not preserved in sediments in which shelly faunas are abundant. The examples of the Silurian benthic graptolites in Bohemia (Bouček 1957) and in North America (Erdtmann 1976) show that they may be associated with diverse communities of marine organisms like eurypterids, brachiopods, nautiloids, phyllocarids and other actively moving organisms. Associations with benthic colonial bryozoans have not been described, indicating that these preferred different environments.

Maletz and Steiner (2015) discussed new discoveries of Cambrian pterobranchs and the frequent misidentification of pterobranchs as algae or hydroids, aspects that might have a considerable impact on the interpretation of the Cambrian food chain. If these presumed fossil algae are actually pterobranchs, a considerable change in the interpretation of the Cambrian trophic structure and food chain would be the result. Filter-feeding pterobranchs need organically formed particles in the water column to feed upon, but will not serve as primary producers for other organisms. The lack of preserved algae as supposed primary producers could indicate differences in the trophic structure of the Cambrian in comparison with the modern situation. Alternatively, algae may just not have been preserved for environmental reasons during the Cambrian, even though the preservation of common algae occurs in even older strata (e.g. Steiner 1994).

Planktic graptolites are very dominant in certain sediments, but this is due to their preservation potential in anoxic environments and does not show a biological preference. Underwood (1993) explored the position of graptoloids in the Paleozoic ecosystem and concluded that the planktic graptolites filled a variety of ecological niches. They are often associated with a very insignificant benthos or are found exclusively preserved in the sediments, but micro-plankton in the form of radiolarians, acritachs and chitinozoans are often associated with planktic graptoloids. Thus, the indication is that the typical "graptolitic black shales" with their rich faunas may be an artefact of transport and preservation in these anoxic or dysoxic environments, and not an indication of the environment in which graptoloids lived. The association with this sediment type cannot be taken as an indication of an ecological connection.

Graptoloids can be regarded as primary consumers in the Lower Paleozoic food chain (Underwood 1998) and also may have provided a food source for organisms higher in the food chain. Therefore, indications of predation should be common, but the evidence is lacking. Very few fossil graptolites provide an indication of possible predation. Bull (1996) described various aspects of damage in *Dictyonema pentlandica*. She considered injuries from predation, disease and parasites to be among the reasons for these damages, but others appear to be related to environmental conditions. Underwood (1993) described remains of

Silurian monograptids as faecal packages and discussed the possible modes of predation on graptolites. He differentiated absorption, crunching and plucking as the three alternatives. Loydell et al. (1998) interpreted ovoid masses of fragments of *Mediograptus morleyae* as faecal remains and suggested that the predator was interested only in this species as other taxa were unaffected. No information on the identity of the predator was available, but the authors suggested a soft-bodied, probably nektic organism as the predator, which would not be recognizable in the fossil record. Zooid-plucking, as suggested by Underwood (1993), would probably leave no traces in the remaining graptolite tubaria, and thus is difficult to demonstrate from the fossil record. It could explain the high variability of thecal apertural types, with extreme constriction of the apertures in Silurian monograptids as a protection mechanism for the graptolites (Underwood 1993, p. 196). However, this remains speculative, as the gradational change of thecal styles along the stipes in many monograptids (e.g. "biform" monograptids: Hutt 1974b) may suggest a connection to the lifestyle and feeding mechanism of the graptolites. Lidgard (2008) discussed predation on bryozoans by tiny predators attacking individual bryozoan zooids, but the example may be difficult to compare to the situation in planktic graptolites as another planktic or nektic organism would have to be expected to prey on these.

Interestingly, in recent years a few suspension and filter feeding anomalocaridids have been described from Lower Paleozoic successions. These may have preyed on graptolites, but the record relies on few and often fragmentary specimens. Vinther et al. (2014) described *Tamisocaris borealis* from the Lower Cambrian Sirius Passet fauna, and interpreted this medium-sized anomalocarid as a nektonic suspension feeder based on the specialized frontal appendages. The presence of *Tamisocaris* may also indicate that planktic microorganisms were already common in the water column and did not evolve in the Upper Cambrian as suggested by Signor and Vermeij (1994). Thus, competition was already strong and graptolites did not invade an empty space. Filter-feeding anomalocaridids reached a considerable size in the Upper Tremadocian *Sagenograptus murrayi* Biozone, as can be seen in the 2 m long *Aegirocassis benmoulae* from the Lower Fezouata Formation of Morocco (van Roy et al. 2015). If correctly interpreted, these animals may have been potential predators on planktic graptolites.

Feeding Style

The extant pterobranchs *Rhabdopleura* and *Cephalodiscus* extract food from the water column with the help of their ciliated arms. Lester (1985) described the feeding of *Cephalodiscus*: the ciliated tentacles of *Cephalodiscus* move the water into the basket formed from the arms. The small food particles are trapped in the mucus on the tentacles and moved through ciliary action along the arm towards the mouth and are ingested here, while the water is ejected from the centre of the basket (Figure 4.16A). Information on the feeding of *Rhabdopleura* (Halanych 1993) indicates that food capture was accomplished by a local reversal beat of lateral cilia and mucus was not involved in the capture of particles.

We can assume that the feeding of the individual zooids in fossil graptolites was accomplished in a similar way as in modern pterobranchs. The main problem results in the interpretation of feeding currents in fossil graptolite colonies, for which no modern examples exist. Kirk (1990) suggested that an afferent flow is produced through the central opening of conical colonies and the efferent flow is through the sides of the net (Figure 4.16B). Melchin and Doucet (1996), however, recognized that this concept is not in accordance with observations on modern bryozoan colonies, and modelled the flow accordingly. In their interpretation the afferent flow is through the sides and the efferent flow goes through the centre of the colony (Figure 4.16C). Differently shaped colony forms (e.g. fan-shaped or complex shapes) would indicate a modified water flow and is influenced by the paleoenvironmental conditions under which the colonies flourished (Rickards et al. 1990).

Rigby and Rickards (1989) and Rigby (1991, 1992) used scale models of multiramous graptoloid colonies to understand their movement and feeding activities. Graptoloid colonies are postulated to have maintained a stable orientation in the water column, and were able to move vertically while

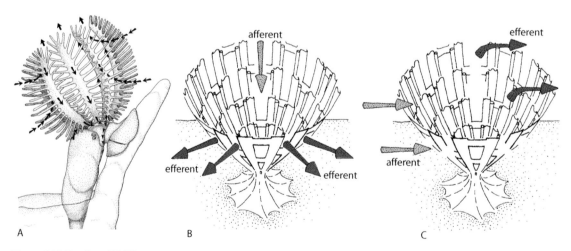

Figure 4.16 Feeding. (A) Ciliary currents in zooid of *Cephalodiscus*, rejection current in centre of arm basket, food particles moving downwards towards mouth (adapted from Lester 1985 with permission from Springer Science + Business Media). (B) Postulated water flow in dendroid graptolite colony (after Kirk 1990, Fig. 1a). (C) Modelled water flow in dendroid colony according to Melchin and Doucet (1996).

they rotated, to sample the maximum amount of the water column for food (see Figure 4.5).

The Diet

The dietary needs of the extant pterobranchs are difficult to investigate as these fragile organisms rarely survive the artificial environment of an aquarium for long periods. Rigby and Dilly (1993) kept *Cephalodiscus* colonies for a period of time to investigate the growth of the colonies, but did not comment on the food source of their specimens. Schepotieff (1906) mentioned remains of diatoms, radiolarians and crustacean larvae in the stomach of *Rhabdopleura*. Further information on the dietary needs of the extant pterobranchs is not available.

Some general estimation can be made for all graptolites based on the knowledge of their size and development. Graptolite zooids were just a few mm in size. Thus, they were able to digest only fairly small particles. If we assume that the food is completely digested internally, proceeding through the intestines, food particles had to be distinctly smaller than the zooids and the zooids would have been able to retract into the tubaria for digestion. Even simple thecal apertures are usually no more than

approximately 1 mm in diameter, and apertures are strongly restricted in many taxa. This would limit the size of the ingested particles even more. Thus, a diet of small phytoplankton would be reasonable (Bulman 1964), as even complete radiolarians and chitinozoans would have been much too large to be ingested by a pterobranch zooid.

Parasitism

A few parasites have been described from extant pterobranchs (Figure 4.17), but parasites are unknown in fossil pterobranchs. It is not known how common parasitic organisms infect modern pterobranchs, as the records are few and appear to be accidental. Parasitic copepods of the family Enterognathidae were found in the stomach of two *Cephalodiscus* species. *Zanclopus cephalodisci* has been recovered from the intestines of *Cephalodiscus gilchristi* (Calman, 1908) and the closely related species *Zanclopus antarcticus* is present in the stomach of *Cephalodiscus anderssoni* (Gravier, 1912). The small endoparasitic copepods of the family Enterognathidae appear to be host-specific to basal deuterostomes, to echinoderms and pterobranch hemichordates

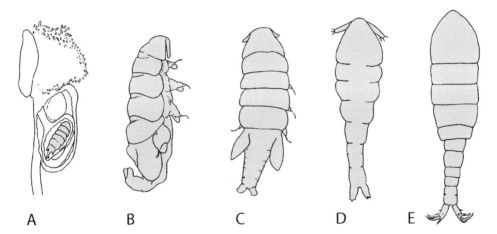

A   B   C   D   E

Figure 4.17 Parasites in *Cephalodiscus*. (A) *Cephalodiscus gilchristi* Ridewood, 1907, showing the position of the parasite *Zanclopus gilchristi* Calman 1908 in its stomach. (B–E) *Zanclopus gilchristi*, several specimens. (B–C) Female in two views. (E) Male in 5th copepodid stage, appendages omitted. (D) Earliest known larval stage (after Calman 1908, pl. 18, 19). Illustrations not to scale.

Figure 4.18 Possible epibionts on graptolites. (A) *Helicotubulus dextrogyra* (Kozłowski, 1967), holotype on fragment of *Mastigograptus* sp. (after Kozłowski 1967, Fig. 7). (B) *Clistrocystis graptolithophilius* Kozłowski, 1959 (after Kozłowski 1965, Fig. 1). (C) Tubotheca on *Acanthograptus* sp. (after Kozłowski 1970, Fig. 5). Illustrations not to scale.

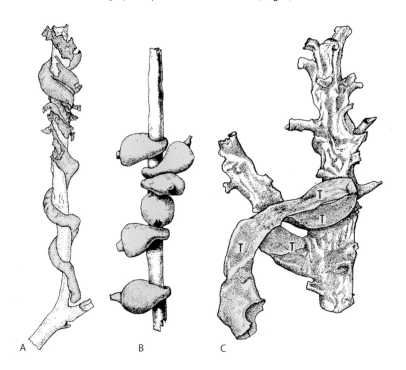

A   B   C

(Ohtsuka et al. 2010). They have a small, segmented body with strongly reduced appendages and attain a maximum size of only a few mm. A different parasitic species in *Cephalodiscus* is the protozoan *Neurosporidium cephalodisci*, which was found in the nervous system of *Cephalodiscus nigrescens* (Ridewood & Fantham, 1907).

Parasitic organisms on fossil graptolites have not been recognized unequivocally, as the zooids are not preserved in the fossil record and their anatomical details cannot be investigated. However, a number of unusual features in graptolite colonies may be regarded as either parasitic in origin or as epibionts, living on the surface of the graptolite tubaria. The tubothecae of Kozłowski (1970) may represent infestations of graptolites (Figure 4.18C), but could alternatively be harmless encrustations. They are formed as tubular growth structures, apparently made from the same organic material as the graptolite tubaria. They do not bear any fusellar structure, but appear to be secreted as a cortical structure by a graptolite zooid (Urbanek &

Mierzejewski 1982). These tubes may have been inhabited by a non-graptolitic organism, forcing the graptolite colony to produce an organic cover (e.g. Kozłowski 1970). Crowther (in Conway Morris 1981, 1982) suggested that the possible parasite attacked the graptolite colony and possibly attached itself to the stolon system, with the result that the adjacent zooids covered the organism with the layer of cortex. The single known specimen of *Helicotubulus dextrogyra* (Figure 4.18A) is another tubular remnant of an organism with a spiral shape around the stipes of *Mastigograptus* sp. (Mierzejewski & Kulicki 2003b), but its relationship to the tubothecae is unclear.

More common are short, open-ended tubes or closed blisters of various shapes on the surface of graptolite tubaria (Figure 4.19). Round or oval blister-like protuberances may be common in some graptolites but are difficult to see in shale material. Thus, they are generally found in chemically isolated graptolite specimens. These are regarded as epibionts, as they appear not to damage the

Figure 4.19 (A, B, E) *Anticostia lata* (Elles & Wood, 1907), specimens with round blisters on the surface of the tubarium and enlargement of blister (E). (C) *Hustedograptus teretiusculus* (Hisinger, 1837) with blister, relief specimen in shale. (D) *Geniculograptus typicalis* (Hall, 1865), proximal end with several short tubes on the surface (from Bates & Loydell 2000). (F) *Archiclimacograptus riddellensis* (Harris & Thomas, 1935), specimen with coiled structure. Scale indicated by 1 mm long bar in each photo.

graptolite colonies or modify their growth patterns and do not show a fusellar structure. Bates and Loydell (2000) described a spectacular colony of *Geniculograptus typicalis* with eight tubular outgrowths (Figure 4.19D). This development, even though apparently non-lethal, must have changed considerably the hydrodynamic behaviour of the colony. The walls of these tubes are cortical, thus produced by the graptolite colony and can be interpreted to have covered at least part of the epibiont.

Epibionts are also found on extant pterobranchs. Markham (1971) described and illustrated the tests of possible folliculinid ciliates on a tube of *Cephalodiscus* (*Orthoecus*) *densus*.

The strange *Clistrocystis graptolithophilus* (Figure 4.18B) has been found attached to *Mastigograptus* sp. fragments and is interpreted as representing possible eggs of cephalopods

(Kozłowski 1965). Radzevičius et al. (2013) described possible bacterial epibionts on Silurian graptoloids, but the more than perfect preservation of this material suggests that these coiled features possibly represent modern contamination.

Tubarium Repair

Damage is common in graptolites and may be related to various causes, from simple breakage to predatorial or ecological, and the real cause of a certain feature is often difficult to establish. The damage is recognized in the fossil record either through malformation of colonies on a larger scale or in the truncation of fusellar structure by new developments. Small-scale regeneration or rejuvenation can be seen in modern *Rhabdopleura* (Rigby 1994)

Figure 4.20 (A) *Rhabdinopora flabelliformis anglica* (Bulman, 1927a), showing regrown branches in the distal part of the colony (from Bull 1996, Fig. 1, reproduced with permission from The Palaeontological Association). (B) *Normalograptus scalaris* (Hisinger, 1837), isolated specimen from Dalarna, Sweden, distally lacking second stipe (based on Maletz 2003). (C) *Cardiograptus* sp. with one distorted stipe (from Han & Chen 1994, reproduced with permission from John Wiley & Sons). (D) *Dicaulograptus hystrix* (Bulman, 1932b), specimen with aborted second stipe (from Bulman 1932b, pl. 9, Fig. 9). (E) *Rectograptus gracilis*, regeneration of theca (after Kraft 1926, pl. 12). (F) *Bohemograptus bohemicus bohemicus* (Barrande, 1850) with misdirected first theca (from Urbanek 1970, Fig. 3, reproduced under the terms of the Creative Commons Attribution Licence 4.0, CC-BY-4.0). (G) *Slovinograptus balticus* (Teller, 1966) with partially biserial colony (from Urbanek 1997a, Fig. 10, reproduced with permission from Instytut Paleobiologii PAN). Illustrations not to scale.

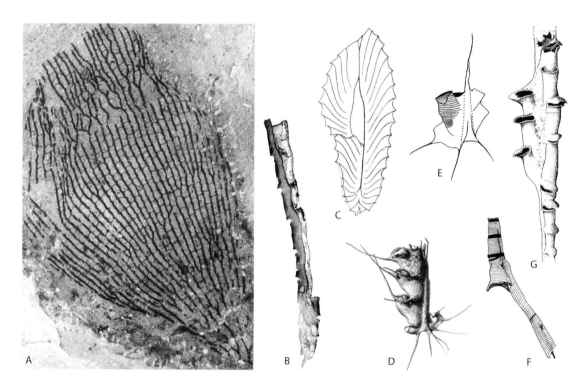

where discontinuities are interpreted as seasonal features, while others clearly indicate the repair of the tubes. Repair is common in Ordovician to Devonian planktic graptoloids (Bulman 1970a) and has most frequently been described from chemically isolated material. Kraft (1926, pl. 12) illustrated a specimen of *Rectograptus gracilis* with a repaired theca (Figure 4.20E).

The robustness of the tubarium design in graptolites can be seen in the ability of the colonies to survive even considerable malformation of the construction plan and the ability to resume the construction after considerable damage occurred.

Benthic graptolite colonies may abandon individual stipes after damage and fill in the gap with extra branchings or construct new tubarial features following the original design (Figure 4.20A).

Han and Chen (1994) described a colony of *Cardiograptus* in which one of the two scandent stipes was damaged but the colony was able to return to its normal development after a certain growth interval (Figure 4.20C). A second colony lost one stipe and was unable to redevelop it. Thus, it distally went on growing a single stipe, but this one was bent to grow as a central structure. Maletz (2003) described an unusual specimen of *Normalograptus scalaris* in which one stipe was apparently damaged and stopped growing after the development of the third theca, while the second stipe developed normally (Figure 4.20B). Interestingly, the specimen did not cover the dorsal wall of the single stipe with cortical tissue as it did with the lateral and ventral walls, but kept the wall thin and unprotected. Maletz, therefore, suggested that the secretion of the cortical material was at least in part genetically controlled and the zooids thus did not venture to the dorsal side that usually would be covered by the second stipe. Similarly, a specimen of *Dicaulograptus hystrix* (Bulman, 1932b), in which stipe two was abandoned after the growth of a single theca, was found in chemically isolated material from Öland by Holm (Bulman 1932b) (Figure 4.20D), indicating the ability of the graptolites to survive considerable malformation of their tubaria. Bates and Kirk (1991) even described a distally uniserial specimen of *Orthoretiolites hami* Whittington, 1954 and a second specimen of the same species forming a triserial colony.

Teller (1999) discussed the numerous developmental failures of Silurian to Devonian monograptids and referred to a number of papers illustrating these, but also stated that abnormalities seem to be less common in Ordovician times. Urbanek (1997a) described a spectacular colony of *Slovinograptus balticus* in which the monograptid stipe suddenly developed a second thecal series, but returned to its normal uniserial development after three paired thecae were formed (Figure 4.20G). Urbanek (1970) illustrated a specimen of *Bohemograptus bohemicus bohemicus* (Barrande, 1850) in which the first theca did not grow back on the sicula, but continued towards the sicular aperture, and then grew along the virgella (Figure 4.20F). These abnormalities appear to be rare in most species, but may be common in individual populations.

The single specimen of a *Climacograptus* from the Lower Devonian illustrated by Jaeger (1978, Fig. 2K) may represent a malformed monograptid and could easily be interpreted in comparison with the partially biserial colony of *Slovinograptus balticus* (Figure 4.20G) as additional material does not exist. Flügel et al. (1993) provided a photo of the single available specimen. Van Phuc (1998) named a new genus *Vietnamograptus* after a biserial graptolite found in the *Uncinatograptus hercynicus* Biozone of Vietnam, but did not figure the material.

Outlook

When early graptolite research centred on taxonomic and biostratigraphical purposes, these data led to a much better understanding of many ecological aspects of graptolite occurrences. The main problem for a long time, the lack of a connection of the fossil graptolites to extant organisms and only indirect interpretation of the lifestyle of the graptolites, obviously hindered the interpretation as living organisms. We can now plot graptolite distributions on paleogeographical maps and understand their ecological needs, follow the evolutionary origins and dispersal of various graptolite groups, and thus get a wonderful glimpse into the Paleozoic landscape, or rather, oceanic developments. With the combination of the paleontological data and the help of modern oceanographic investigations, we will certainly see more advances in the understanding of the paleoecology of the graptolites, and with this, of the ecology of the Paleozoic marine ecosystem.

5 Graptolites as Rock Components

Jörg Maletz

As for all other fossils, graptolites are part of the rocks in which they are preserved. Thus, searching for the rocks and trying to understand their development also helps us to understand the reasons for the presence and style of preservation of graptolites within them. Both cannot be seen independently, but we can sometimes artificially separate the fossil graptolite from the rock in which it was preserved by using acids and then investigate the fossil specimens. Even though the recognition of the graptolites during fieldwork is rarely difficult, it is often hard to understand why and how these delicate fossils are preserved and found in a particular rock type.

We see that organisms fall apart – often smelling quite badly – when decomposing, and very quickly nothing is left of their remains. Scavenging of other organisms will help the process, and of the many graptolites, only a very few fossilize. It is impossible to tell how many billions of specimens were ever around on the planet, populating the oceans, and never left any sign of their previous existence. It is not different in humans: most of us will never leave a trace, not even in our modern times when we think we are immortal. We are told that the internet will never forget anything and our traces are kept forever. Nothing can be farther from reality, as human history will tell us. Just a few hundred years from now, not much might be left from the digital world we know now and other storage media might be in use. Even a book might last longer as it is a physical existence, not just a glimpse into the storage of a computer system, from which it is easily eliminated, as we have all have seen by inadvertently deleting the photos of our last vacation.

Graptolites as physical remains of once-existing organisms are still there after more than 500 million years, and will be there as long as rocks from the

Graptolite Paleobiology, First Edition. Jörg Maletz.
© 2017 John Wiley & Sons Ltd. Published 2017 by John Wiley & Sons Ltd.

Lower Paleozoic time interval exist on our planet. They may remain as they were embedded in the sediment, or may have been modified considerably by compaction, by tectonic activities and diagenetic processes, or their casts filled with and replaced by minerals. All this we can see in the historical record of graptolite faunas when we carefully analyse their current status.

Graptolite Taphonomy and Preservation Potential

Taphonomic information from hemichordates, and especially from pterobranch or graptolite fossils, provide important clues as to the lifestyle, ecology, death and burial of these fossil organisms, but also for the subsequent geological history of the surrounding rocks. Therefore, all characteristics of the preservation need to be explored in some detail during the taphonomic analysis of pterobranch fossils. As the soft organic material of the zooids decomposes quickly after the death of the organism, skeletons and other tough material produced by the organism may be the only remains found in the fossil record. In the Pterobranchia, this is the tubarium (Figure 5.1), the colonial housing construction of the animals, and the zooids secreting this housing are not preserved.

Investigation of extant taxa shows a decomposition of the zooids within days after death (Briggs et al. 1995), leaving a very slim chance of preservation in the fossil record. Even though the preservation of fossil zooids has been claimed a number of times (Bjerreskov 1978, 1994; Rickards et al. 1991; Durman & Sennikov 1993; Loydell et al. 2004; Zalasiewicz et al. 2013), none shows convincing evidence of any anatomical detail. The tough organic walls of the pterobranch tubaria resist the effect of decomposition in an oxic environment,

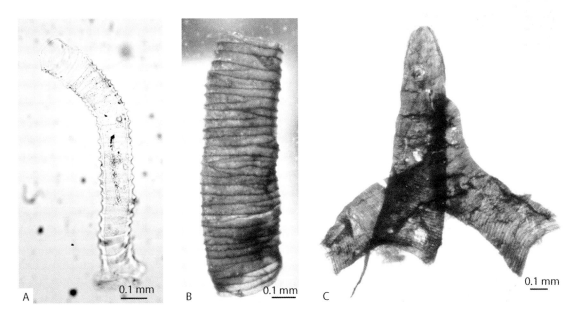

Figure 5.1 The preservable organic housing construction of the graptolites. (A) *Rhabdopleura normani* Allman in Norman (1869), single transparent tube of an extant graptolite. (B) *Rhabdopleura* sp., tube of a Middle Ordovician graptolite, Öland, Sweden. (C) *Xiphograptus primitivus* Maletz, 2010, proximal end of a planktic Ordovician graptolite. (B) and (C) are chemically isolated specimens showing the fossil preservation of the organic tubarium.

Figure 5.2 Graptolite preservation. (A) *Sigmagraptus praecursor* Ruedemann, 1904, NMVP 320445B, Victoria, Australia, multiramous sigmagraptine in typical black shale. (B) *Parisograptus caduceus* (Salter, in Bigsby 1853), NMVP 319348A, Victoria, Australia; darker, organically preserved graptolite in strongly weathered (bleached) shale. Photos provided by A.H.M. VandenBerg (Victoria, Australia).

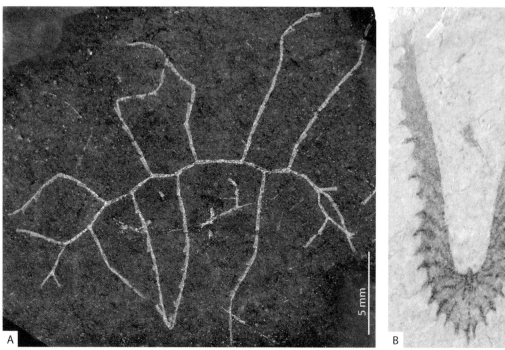
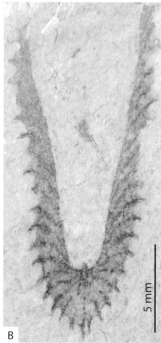

but are best preserved in dysoxic to anoxic sediments. Thus, graptolites are most commonly found in sediments identified as "graptolite shales" or "anoxic black shales" in the past (Figure 5.2A), which may be strongly weathered and in this case light coloured (Figure 5.2B). However, they may also be found in many other marine sediments, in sandstones, mudstones and limestones. Even volcanic ash beds or bentonites can yield important graptolite faunas, as the example of Mitchell et al. (1998) from the Precordillera of Argentina shows.

Death on the Sea Floor

Graptolites live either on the sea floor as benthic taxa or are freely swimming in the water column as planktic graptolites. This is already the first point at which a differentiation has to be made

concerning the potential for preservation as fossils. As shallow-water platformal regions are often destroyed through tectonic activities and due to major erosional events, the shallow-water benthic graptolite faunas are rarely found preserved *in situ* in their environment. Few records show benthic graptolites attached with their holdfasts to the sediment upon which they were growing. Erdtmann (1976) described the example of a diverse *in situ* benthic graptolite fauna from a Silurian inter-reef environment on the North American platform as an exceptional preservation, but most benthic graptolites are discovered as fragmented and transported specimens in deeper water settings (e.g. Hall 1865; Bouček 1957; Maletz 2006a). Thus, information on the original environment and ecological context is not available for this material.

The situation is even more complicated for the planktic graptolites. Even though it is clear that they were moving in the water column, their ecological requirements are less clear. Due to the active swimming habit and the water currents in the oceans, it is highly unlikely that the sediments in which they are found provide useful evidence for their original living environments. Dead colonies may have been transported for hundreds of kilometres in the oceans and been deposited far from the region in which they once flourished. After arriving on the sea floor, the graptolites may have been buried and preserved, although the process was not altogether straightforward and preservation depended on whether the sea floor was oxygenated or not. If the sea floor was oxygenated, graptolites might have been destroyed by organic decay or eaten by other organisms, but the organic material of the graptolite was quite resistant and probably also survived oxic environments for a certain time. In anoxic or dysoxic environments, in which organic decay was hindered, the potential for preservation of the graptolite tubaria, however, might have increased considerably. The graptolites, settling on such a sea floor, were slowly buried. Sedimentation rates on deep ocean floors may be less than 1 mm/century, and similar and lower rates may be inferred from radiometric age calculations of graptolite shale deposits (Cooper & Sadler 2010: graptolite zone durations). In terms of graptolite preservation, this suggests that the graptolites were exposed at the surface for quite some time before they were eventually covered by sediment. Even in the absence of a scavenging benthos, this suggests that the upper parts of these tubaria would suffer microbial degradation, but no evidence of such decay is typically found.

Interestingly, spiraliform graptolites like *Spirograptus turriculatus* in such settings (Plate 16B–C) typically do not have sediment infilling the cone-shaped spaces inside the whorls, as happens with such graptolites preserved within turbidite deposits. Rather, they were subsequently compactionally flattened on to a single two-dimensional plane, suggesting that clastic sediment was somehow excluded from their interior as they were being slowly buried, almost as if these graptolites had been wrapped in clingfilm (plastic wrap) prior to burial (Jones et al. 2002). An explanation for this common phenomenon

(Figure 5.3) is that the "clingfilm" took the form of enveloping organic matter in the form of microbial mats and marine snow, which subsequently decayed, leaving little trace except for some of the amorphous carbon that gives the black shale its hue. As well as excluding sediment, this would have helped to prevent surface degradation of the graptolite prior to "true" burial.

On an oxic sea floor, the carbon from the sediment is recycled back into the overlying water as dissolved CO_2. The deposits often retain evidence of bioturbation, mainly of *Chondrites* type (Figure 5.4A). Graptolites are usually rare in such deposits, although it is not entirely certain whether this is due to simple scavenging of the tubaria and aerobic microbial decay. However, many dark shales apparently are not truly anoxic, and recently bioturbation has been discovered to be common in black shales (Egenhoff & Fishman 2013), even associated with graptolites (Figure 5.4). The distribution of oxic and anoxic facies is generally thought to provide the primary control on graptolite occurrence in strata, but this might not be entirely true as the graptolite fusellum is surprisingly stable and preservable under various conditions. Thus, it probably was not the prevalence of anoxic seas in the early Paleozoic that has ensured that graptolites are a common and stratigraphically useful fossil group (cf. Berry & Wilde 1978; Wilde & Berry 1984; Berry et al. 1987).

The continuity of the graptolite record depended very much on paleoceanographic setting. Anoxia seemed generally, but not necessarily, a deeper-water phenomenon then (Berry & Wilde, 1978), in contrast with oceanic conditions today where an "oxygen-minimum" layer is commonly present in mid-waters, with deep water ventilated by strong contemporary thermohaline circulation. Thus, deeper-water settings in general generated more continuous graptolite records. For instance, in the Llandovery, the condensed, presumed deep ocean floor succession in the Polish Bardo Mountains (Porębska & Sawłowicz 1997) is continuously graptolitic, with no oxic interludes. In the more marginal though still likely oceanic setting of the Southern Uplands of Scotland (Toghill 1968) and Bohemia (e.g. Štorch et al. 1993), this interval is mostly anoxic, with subordinate oxic interludes.

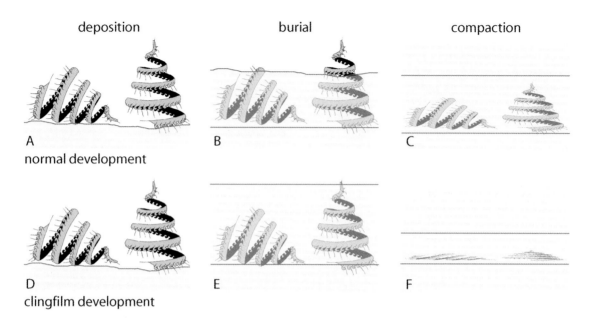

deposition   burial   compaction

A   B   C

normal development

D   E   F

clingfilm development

Figure 5.3 "Clingfilm" preservation of *Spirograptus turriculatus* (Barrande, 1850). (A–C) Preservation in turbidite depositional system. (D–F) Preservation in clingfilm mode (based on Jones et al. 2002, Fig. 4); see also Plate 16 for *Spirograptus turriculatus* preservation.

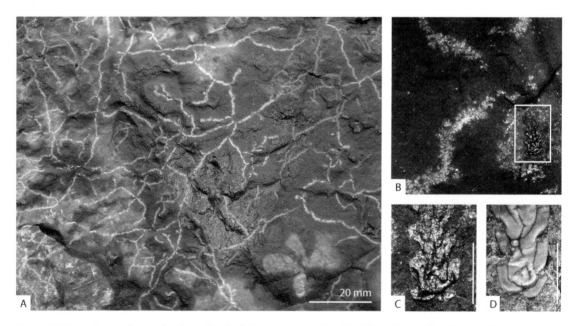

20 mm

Figure 5.4 Graptolites and trace fossils at Spudgels Cove, western Newfoundland, Darriwilian shales of American Tickle Formation with pyrite-stained trace fossils (*Chondrites* sp.) and *Archiclimacograptus* sp. (A) Field photo showing traces. (B) Photo of shale from same layer showing traces and graptolite (white box). (C) *Archiclimacograptus* sp., enlarged from (B), SP4.12.01b. (D) *Archiclimacograptus* sp., SP4.12.19, well-preserved specimen from same layer, coated with ammonium chlorite to show details. Scale is 1 mm in (C–D).

The water depth idea, however, might have to be disputed when looking at the black shale intervals in the Lower and Middle Ordovician of Scandinavia. The Upper Alum Shale Formation and the Tøyen Shale Formation are crowded with graptolites (e.g. Monsen 1925, 1937; Bulman 1954). New interpretations of these "black shale intervals" indicate a fairly shallow-water, shelf margin origin for the sediments (Egenhoff & Maletz 2012; Egenhoff & Fishman 2013).

In largely or wholly oxic shelf settings, graptolite occurrences are rare and often restricted to a few layers (e.g. Goldman et al. 2002). Either the graptolites were destroyed through the constant transport and redeposition of sedimentary particles in clastic settings, or biologically through organic degradation in carbonate settings, or through bioturbation. Thus, these factors might constitute the primary control on graptolite preservation in marine sediments, and the presence of anoxia or dysoxia represents a minor aspect only.

Beckly and Maletz (1991) recognized that the genus *Azygograptus* was the first graptolite appearing in transgressive successions in Britain, where it was usually found in monospecific assemblages. Increased faunal densities and changes in faunal associations were used by Egenhoff and Maletz (2007) to demonstrate transgressive phases in uniform black shale units of Scandinavia, but changes in sediment types were not noted. Graptolites might have been swept onto the shallow shelf region and deposited in and covered by the oxic sediments.

In Situ Death Assemblages

While sessile organisms die in the environment in which they live, and may or may not be preserved there as fossils, the story is much more complex when we consider that many fossils are fragmented and transported as dead organisms and not found in their original environment. The benthic taxa may be found more or less *in situ* in rare cases and sometimes associated with corals, sponges and other benthic organisms, and may be analysed in a similar fashion. Detached and transported fragments often figure prominently in planktic graptolite assemblages, and may also be accompanied by shelly fragments. In these cases, it is clear that the material is transported and not *in situ*.

Planktic graptolite taxa invariably show some indications of post-mortem transport. Like the planktic foraminifera, coccolithophores and other such fossils, they are never found in the environment they lived in, which was somewhere in the water column. They may just fall down through the water column and settle beneath the part of the water column they lived in, but this is not easily detectable. Usually, they can be interpreted as transported death assemblages, and to understand the nature of their fossil record, one must consider the steps in the path from their habitat in life to their final condition as fossils.

Transport of Graptolite Tubaria

Following the death of a planktic graptolite colony, the tubarium will sink to the sea floor, since it is no longer supported by the movement of the zooids. Sinking most likely did not happen immediately, because the decay of the soft organic tissues of the zooids may have produced some uplift and the decay gases would have acted to keep the colony afloat. Only with their dispersal could the colony sink to the sea floor and be preserved in the sediment. The delay in sinking would have allowed considerable transport of the colonies by water currents after death. Therefore, the descent of the dead colony had a lateral component, depending on local current patterns, and provided means of considerable post-mortem dispersal of the fossils. Deposition on the sea floor often shows the prevailing current direction through the orientation of the graptolite tubaria (Figure 5.5A). The current orientation of graptolite colonies can be used to infer water currents and basin geometries (e.g. Schleiger 1968; Moors 1970; Rigby 1993a). However, other accumulations may not show any current orientation, but specimens appear to be randomly oriented (Figure 5.5B). During the descent of the colony through the water column, the decay of the zooids occurred, but the tough organic tubarium most likely stayed undamaged. Destruction of the tubaria through scavenging is a possibility, but there is little evidence for predation on graptolites (Underwood 1993; Loydell et al. 1998). It is unlikely that the scavenging of abandoned graptolite tubaria was a significant item in the early Paleozoic food web, given the

Figure 5.5 (A) Current-aligned specimens of *Orthograptus apiculatus* (Elles & Wood, 1907) from Laggan Burn, Ayrshire, Scotland (photo by Denis Bates, Aberystwyth, Wales). (B) *Uncinatograptus uniformis* (Přibyl, 1940), Loitzsch quarry, Thuringia, Lower Devonian. Scale indicated by 1 cm long bar in each photo.

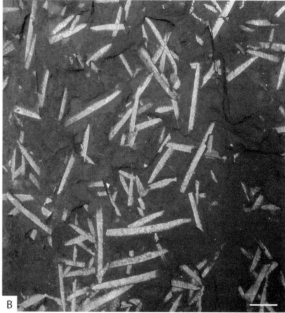

likely greater nutritional value of graptolite soft tissue than the fusellum. Physical destruction of the tubaria also seems unlikely except through a combination of wave action, biochemical processes and microbial decay. If such processes were common, we might expect to see a significant amount of fragmentary graptolite debris, equivalent to the "hash" of shelly deposits, on bedding planes of the typical graptolite shale facies, together with more complete graptoloid specimens. This is locally common in the Tremadocian part of the Scandinavian Alum Shale Formation (e.g. Bulman 1954: pls. 7, 8), but otherwise has rarely been noted.

Burial and Preservation

Burial of graptoloid tubaria was usually simply by covering the specimens with the next layer of sediment and separating them from the oxic environment in which they would decay. This could take place either through rapid entombment within a turbidite deposit, or by more or less slow burial on a sea floor where pelagic fallout dominated.

Anoxic conditions were not needed when graptolites were incorporated within turbidite units, as the surrounding sediment would be enough to isolate the specimens from the damaging oxygen contact. The graptolites were picked up by the current from regions upcurrent, where they originally dropped to the sea floor. They had to be transported into the deeper water region by turbidity currents before they were too much affected by decay on the sea floor.

Hydrodynamically, the graptolites seem to have behaved like fine to medium sand grains and, within a Bouma unit, are sometimes abundant in the current-rippled A- and C-horizons (Finney & Berry 1997; Cooper et al. 2012). They may be current-aligned (Figure 5.5A), either parallel or perpendicular to the current depending on current speed, and variably fragmented. The tubaria may still be well preserved in relief and filled with the surrounding sediment (Figure 5.6). The sediment may enter the thecae and common canal and prevents

Figure 5.6 Different views of graptolite colonies filled with sand in turbidite layer, *Levisograptus primus* (Legg, 1976), St. Pauls Inlet, western Newfoundland, Lower Head Sandstone, Darriwilian. (A) Apertural (scalariform) view, coated with ammonium chloride. (B) Oblique view, uncoated. (C) Reverse view, showing proximal development in partial relief. Scale indicated by 1 mm long bar in each photo.

flattening through compaction of the sediment layer. In examples where they are found at the base of the turbidite, they may leave distinctive brush-like trace fossils (Smith 1957; Palmer & Rickards 1991: pl. 136).

More commonly, graptolites can be found in the intervening fine-grained E-horizon of the Bouma sequence or in the intercalated hemipelagic black shales (Hills & Thomas 1954; Moors 1969, 1970; Cooper 1979a), where they may accumulate in large quantities. The most typical facies for graptoloids is within fine-grained, dark-coloured, carbon-rich, finely laminated mudrocks or "graptolite shales", where graptolites are the dominant and commonly the only macrofossils. These deposits represent the pelagic or hemipelagic sedimentation upon deep sea floors. This facies typically preserves its fine primary lamination because a burrowing and crawling benthos was excluded by persistent anoxia. In the prescient words of Marr (1925, p. 125) this facies represented conditions in "poisoned water below the 100-fathom line". According to Finney and Berry (1997, 1998) and Wilde et al. (1989), among others, the graptolites lived in oxygenated waters above, and after death fell into the anoxic realm below.

In the most condensed facies, where the deposits are essentially pelagic, they comprise mainly biogenic material that has settled from above. In the early Paleozoic, the main components were radiolarians, organic-walled planktic organisms such as acritarchs, and amorphous organic matter of indeterminate origin, transported downwards as faecal pellets and marine snow, but possibly also calcareous microorganisms (e.g. Munnecke & Servais 2008). This would have been accompanied by exceedingly fine-grained silicate components in the form of far-travelled aeolian dust (likely a significant component given the paucity of vegetation on the land surfaces of those days) and volcanic ash. In hemipelagic deposits, this pelagic material is finely interlaminated and "bulked up" with thin mud layers, which in effect represent highly dilute, slow-moving distal turbidites, the transported sediment being derived from such sources as winter floods and storm-affected shelf areas.

Diagenesis

Considerable changes happened to the organic material of the graptolite tubaria during the long periods within the thick piles of sediments in which they were preserved for the future. After compaction of the sediments, fluids passed through, modifying the chemical composition of the organic material of the graptolites, but more important were later developments during the metamorphosis and deformation history of the sediments. At the end, the graptolites were not even always recognizable as such.

Once the graptolites were buried in the sediment, no matter how thin this protecting sediment layer may have been, diagenetic processes began to modify the tubaria. It was these early phases of mineralization that enabled the tubaria to be preserved in a manner that could resist subsequent compaction, and even, to some extent, later tectonism and cleavage development. The most common means of achieving this is via pyritization (Figure 5.7), whereby pyrite can fill, partly or wholly, the interior of the tubarium, to yield a faithful internal cast of the graptolite that may retain imprints of details such as fusellar structure and even cortical bandages. The mode of this infill is astonishingly helpful to the paleontologist for understanding the precise development and formation of the graptolite colony, if it fills the entire interior of the colony. Every turn and branching of the thecal tubes can be followed in the internal cast of the specimens, which provides important information about the taxonomy and evolution of these organisms. The morphology of within-graptolite pyrite has been described by Bjerreskov (1991, 1994) and it includes such features as pyrite stalagmites and stalactites where pyrite has not infilled the graptolite completely but has lined the tubarium interior. In some cases the pyrite can be seen to form masses in front of the thecal apertures, or even cover whole colonies (Bjerreskov 1978). An outer compact cover of the tubaria with pyrite is rare and unwarranted for us as paleontologists, as it tends to obscure rather than reveal the colony form. Often, however, pyrite outside the colonies is finely dispersed in the sediment and does not form massive deposits (Figure 5.7C).

Pyritization *per se* has been well studied (Berner 1970, 1984; Raiswell & Berner 1985; Raiswell & Canfield 1998; Roychoudhury et al. 2003). The iron is derived from mineral matter, with Fe^{2+} being moderately soluble in reducing conditions. The sulphur is derived from sulphate, which is abundant in seawater, and from which the oxygen has been stripped, as an energy source, by sulphur-oxidizing bacteria, after the oxygen has been exhausted. But why does the pyrite precipitate so commonly, and so abundantly, in the graptolite tubaria rather than being disseminated through the mudrock? It is unlikely that this is due to local reducing conditions due to remnant fragments of zooid, as these have decayed away prior to burial. Rather, it may be due to simple physical effects, the graptolite tubaria simply representing a large, fluid-filled tube encased within less permeable mud, so allowing rapid diffusion and hence rapid supply of iron and sulphide ions to growing pyrite crystal nuclei. The phenomenon, although widespread, is irregularly dispersed on a small scale, even in the same bed, where there may be a mixture of wholly pyritized, partly pyritized and unpyritized graptolites, and these may be associated with various amounts of irregular pyrite nodules, that crystallized away from graptolites. These differences likely reflect subtle variations in local lithology, permeability and redox conditions within the near-surface deposits. For the paleontologist, perhaps the most revealing and "taxonomically friendly" form of preservation is as "part-relief" where pyritization was partial with subsequent flattening, or took place after partial compaction of the graptolite. Full three-dimensional pyritization (Figure 5.7E) may require more careful excavation of the tubarium to obtain a good profile view. Taxonomically useful features such as spines being outside the plane of flattening may easily be missed.

Other means of preservation can be within carbonate nodules, such as the Cape Phillips Formation of Arctic Canada (Melchin 1987; Lenz & Kozłowska-Dawidziuk 2004; Lenz et al. 2012) and the Kallholn Formation in Dalarna, central Sweden (Hutt et al. 1970; Loydell 1991a; Loydell & Maletz 2004, 2009). The concretions form at an early stage prior to the compaction of the sediment, and thus are able to preserve graptolites uncrushed and in three dimensions. These kinds of nodules tend to form below the sulphate-reducing zone, where the carbonate crystallizes both inside and outside the graptolite. More rare is the presence of graptolites in siliceous nodules or cherts (e.g. Kozłowski 1949). It provided unsurpassed information to Roman Kozłowski, who is the only person who

Figure 5.7 Diagenesis in graptolites. (A–B) *Amplexograptus perexcavatus* (Lapworth, 1876), JM 79, Lebanon Limestone, Murphreesboro, Tennessee (see Goldman et al. 2002), chemically isolated specimen showing internal and external growth of small pyrite crystals. (C) *Cochlograptus veles* (Richter, 1871), LO 1071t, showing colony surrounded by dispersed pyrite crystals. (D) *Cymatograptus bidextro* Toro & Maletz, 2008, specimen showing black fusellum and pink mineral fill (see also Pl. 8A). (E) *Rivagraptus bellulus* (Törnquist, 1890), LO 1128t, polished section showing pyrite fill of tubarium, fusellum shown as black outline and in the interthecal septae. Scale indicated by 1mm long bar in each photo.

has ever looked at Cambrian and Ordovician benthic graptolites in such detail. His material was largely derived from siliceous concretions of Tremadocian age, collected at Wysoczki in the Świętokrzyskie Mountains or Holy Cross Mountains of central Poland. Many of the taxa he described have never been found again. Maletz (2009) described the preservation of Middle Ordovician graptolites surrounded by silica in carbonates with silicified shelly faunas, largely ostracods. In this case, the early-formed silica prevented the tectonic distortion of the graptolites and kept the specimens intact during chemical dissolution of the carbonate. A silicification or replacement of the graptolites did not take place, however.

Metamorphism and Organic Maturation

Metamorphism dramatically modifies rocks and can completely destroy all incorporated remains and features like sedimentary structures, but also all evidence of the presence of fossils. Metamorphism includes heating of the rocks in various ways, from regional metamorphosis through to increasing basin fill, to local contact metamorphosis from magmatic sills and dykes. At a certain depth regional metamorphosis produces heating and recrystallization of minerals in the sediment through the overburden pressure and an increased temperature gradient. Mineral-rich fluids move through the rock succession and leave their traces in the form of new mineralizations.

The effect of the rising temperature on the graptolite fusellum is similar to that on more commonly investigated organic material – coal. Coalification happen to all organic material through increasing temperature gradients in the sediments, and leads to considerable changes in the optical properties of the organically preserved graptolites.

The earliest changes in the matter of the fusellum seem to involve subtle alterations in biochemistry, but obviously visible changes of the organic material does not occur. Actually, flattened graptolites freed from sediments with hydrofluoric acid may retain considerable flexibility. Specimens may be several cm long and you can literally wrap them around your finger without them breaking into pieces, as material from the Middle Ordovician Table Head Group (see Albani et al. 2001) indicates. Morphologically, the structure of the fusellum can be preserved almost to macromolecular levels of detail, with collagen-like banding on the smallest observed fibrils (Towe & Urbanek 1972; Bustin et al. 1989). Yet, collagen has not yet been detected by chemical analysis. Instead, the chemical composition of morphologically finely preserved fusellum is made up of an aliphatic polymer, likely formed by *in situ* molecular transformation at some early stage in burial (Gupta et al. 2006; Gupta 2014). Subsequent burial of the graptolite, until the phase of late diagenesis/metamorphism, largely involves maturation of the tubarium. The changes here parallel those in buried palynomorphs, with loss of volatiles and a diminution of the H, N, O content of the fusellum, which are essentially released as a contribution to petroleum generation. The matter that is left becomes progressively richer in carbon until all that remains is a carbon film. As these changes occur, the colour of the fusellum changes progressively from straw-yellow to orange to brown to black, a series that, as with palynomorphs and also conodonts, may be used as a geothermometer for recording maximum burial temperatures.

The coalification or maturation of the organic graptolite fusellum makes the material fragile and brittle. Often the fusellum shows fractures visible on the surface of the specimens, and the material is no longer chemically isolatable from the surrounding rocks. The material considerably changes in its optical appearance (Teichmüller 1978; Goodarzi 1990; Hoffknecht 1991). In recent years the maturation of graptolite fusellum has been investigated from several regions (Bertrand & Héroux 1987; Riediger et al. 1989; Malinconico 1992, 1993; Varol et al. 2006; Petersen et al. 2013). Graptolite reflectance data can be compared to determine the organic maturity of sediments, and compared to vitrinate reflectance and even to the colour alteration index (CAI) of conodonts (Goodarzi & Norford 1985, 1989; Hoffknecht 1991). It preserves the burial history of the sediments through the optical properties of the organic material. A transition from unaltered to highly coalified states can easily be recognized in graptolite fusellum along a temperature gradient. Unaltered graptolite fusellum is dark brown to black in colour, as is typical for the Cow Head Group of western Newfoundland or the Scandinavian Tøyen Shale at Skattungbyn, Dalarna (Figure 5.8A). As the temperature rises, the black fusellum changes its colour to a slight silvery shine of the dark material, as in some material from the Lerhamn drillcore of Scania, Sweden (Figure 5.8B). Eventually, a strong silvery shine appears, that can even be mistaken for pyritization of the material and is typical of many graptolites from the southern Scandinavian Tøyen Shale of Norway and Sweden (Figure 5.8C) and of many other regions (Figure 5.8D).

Figure 5.8 Coalification of graptolite fusellum. (A) *Pseudophyllograptus densus* (Törnquist, 1879), CN 2234, Skattungbyn, Dalarna, remains of black fusellum with some weathering. (B) *Baltograptus* sp., LO 10582t, Lerhamn drillcore, Scania, Sweden, low coalification. (C) *Expansograptus latus* (Hall, 1907), Diabasbrottet, Sweden, relief specimen, fusellum highly coalified through contact metamorphosis. (D) *Archiclimacograptus wilsoni* (Lapworth, 1876), SM A 19619, highly coalified, Dob's Linn, Scotland. Scale indicated by 1 mm long bar in each photo.

As the last step in the "burning off" of the graptolite fusellum, the organic material starts to disappear from the rock and the graptolite is completely eradicated. Sometimes a faint outline of the previously existing specimen may still be visible as a ghostly shadow, but eventually it also disappears. Only when the graptolite tubarium is filled with pyrite or other durable minerals might the specimen be recognizable as the remains of a graptolite for a while longer (Figure 5.9D).

The changes in the appearance of graptolites on a temperature transect can be used to understand regional metamorphism of rock successions, but similar effects are also visible in contact metamorphic rocks. The Mossebo and Diabasbrottet sections at Mt. Hunneberg in Vastergotland, south central Sweden, show a transition of the graptolite fusellum related to contact metamorphosis from an overlying Permo-Carboniferous dolerite sill. Here the transition can be seen over just a few metres in the succession (see Egenhoff & Maletz

2007). Graptolites about 10m from the dolerite sill are silvery shining (Figure 5.9A), showing moderate coalification. Only two metres higher up, the situation changes. The originally black shale changes to lighter colours and new minerals start growing in patches (Figure 5.9B). The graptolite fusellum is poorly preserved and thin, and starts to disappear, leaving faint outlines. Another two metres higher up the succession, the shale is light brown and hard. Graptolites have disappeared or are only faintly recognizable. A number of specimens appear to be better preserved, but a closer look shows that these were originally preserved as relief or partial relief specimens filled with pyrite (Figure 5.9C–E). Often the thecal apertures in which no pyrite was present have disappeared, and only partially preserved specimens remain (Figure 5.9D). The complete outlines are often not recognizable any more, and the specimens are not identifiable to species level. However, in some cases a latex cast may still show details of

Figure 5.9 Coalification and contact metamorphosis. (A) *Pendeograptus simplex* (Törnquist, 1904), CN 1801, *Tetragraptus phyllograptoides* Biozone, high coalification, silvery fusellum. (B) *Baltograptus geometricus* (Törnquist, 1901), LO 1585 T, *Cymatograptus protobalticus* Biozone, flattened specimen with patches of surrounding mineral growth partly obscuring and distorting the specimen. (C, E) *Baltograptus* sp., coated latex cast (C) and mould (E) of specimen, showing crystal growth. (D) *Baltograptus jacksoni* Rushton, 2011, *Baltograptus vacillans* Biozone, originally pyrite-filled parts preserved only. All specimens from Hunneberg, Sweden. Scale indicated by 1 mm long bar in each photo.

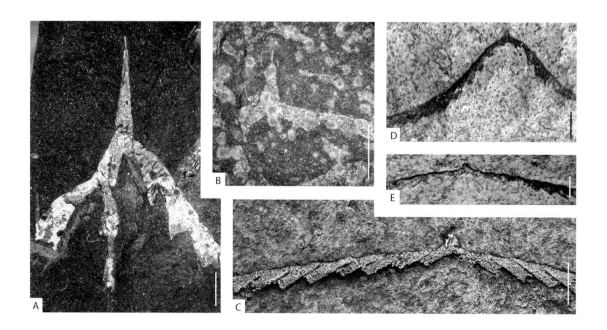

the proximal development, as in the illustrated specimen of *Expansograptus pusillus* (Törnquist 1901) (Figure 5.9C), where the imprints of contact metamorphic crystal growth can also be seen, even though the original specimen (Figure 5.9E) may not show much detail.

Mineral Replacement

Mineral replacement of the organic material of a graptolite tubarium is not easily achieved, if at all, during diagenesis or metamorphosis. The material is quite durable and not as readily replaced as mineralized shells of brachiopods, bivalves or trilobites. Underwood (1992) investigated the preservation of graptolites and stated that pyrite does not replace the graptolite fusellum. Even though the silvery pyritic internal casts of graptolites

may be mistaken as replacements, pyrite replacement has not been discovered, and in all cases the pyrite forms only a fill of the tubarium, as might be seen when the fusellum is still preserved (Figure 5.7E). A replacement may be found when the organic material has been destroyed through weathering and is replaced by the precipitation of minerals in the empty spaces (e.g. Maletz & Steiner 2015: *Sphenoecium wheelerensis*). Only the SEM-EDS investigation (elemental mapping and point analyses) indicated the lack of organic material and the replacement of the specimen by clay minerals.

The preservation of graptolites as replacements in silica was discussed in a number of German papers (Hundt 1934, 1946; Stürmer 1951, 1952; Greiling 1958), but has not been verified. Horstig (1952) described this preservation in more detail and stated that the graptolites are filled with silica,

Figure 5.10 (A, B) Stipe fragment of *Didymograptus* sp., showing pattern of sediment fill and chlorite pressure shadow distributions (based on Mitchell et al. 2008, Fig. 3). (C) Multiramous graptolite on cleaved black shale, largely showing preservation of pressure shadow minerals (light colour), JOS 23.1a, Sandia Region Peru (Maletz et al. 2010).

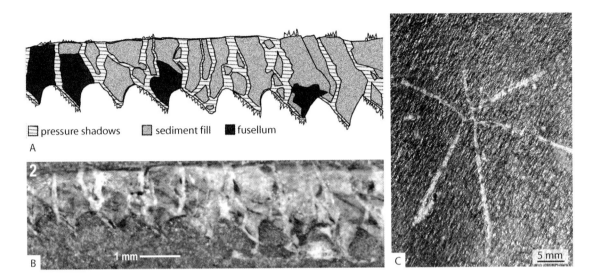

but that the fusellum is still present, thus the preservation does not represent a replacement.

A replacement may be seen of the internal fill of a graptolite tubarium when the original pyrite is replaced by other minerals, but this has not been investigated so far. Toro and Maletz (2008) described pink to white minerals filling Floian specimens of *Cymatograptus bidextro* Toro & Maletz, 2008 (Figure 5.7D; Plate 8A–B). They interpreted the minerals as an iron mineral, probably goethite, a possible weathering product of the oxidation of pyrite. Thus, the replacement is not to be understood as a metamorphic replacement of a mineral, but as a late-stage weathering effect.

Tectonic Deformation

The graptolite itself deforms as the rock around it is tectonically strained (Figure 5.10). This tectonic deformation often produces considerable changes in the original shape and dimensions of the specimens, and these are difficult or even impossible to determine without better-preserved material for comparison (Figure 5.11). Eisel (1908) discussed the tectonic deformation of the Silurian graptolites from Thuringia, Germany, in some detail and pointed out the problems of identification of deformed specimens. Maletz et al. (1998) demonstrated the extreme deformation of the graptolite *Pristiograptus* (?) *gerhardi* (Kühne, 1955) from the Silurian of Albania, where most graptolite fragments were impossible to relate to undeformed material. Simple retro-deformation of graptolites is possible using modern computer programs, and may yield impressive results as shown by strongly deformed specimens of *Arienigraptus zhejiangensis* from the Darriwilian of Peru (cf. Figure 5.11A– D). A number of papers deal with deformation and retro-deformation of graptolites to improve our understanding of the tectonic effects (e.g. Jenkins 1987; Cooper 1970, 1990; Williams 1990), but often tectonic deformation is not considered when discussing graptolite faunas and their dimensional characteristics. In some local tectonic situations, the maximum stress may be vertical, with increases in dimensions along both other axes, so the graptolite is then flattened to appear larger overall. This may explain the "giant graptolites" of the classic Wenlock locality at Goni, Sardinia, for example (Gortani 1922).

Figure 5.11 Tectonic deformation. *Arienigraptus zhejiangensis* Yu & Fang, 1981. (A) JOS 21.1, original specimen, tectonically deformed. (B) Retro-deformed specimen, vertically shortened by about 50%. (C) JOS 21.2, original specimen. (D) Retro-deformed specimen, horizontally shortened to about 75%. (E) Partial relief specimen, latex cast for comparison. *Didymograptus murchisoni* (Beck). (F) Proximal end, flattened, showing tectonic lineation, preserved largely as pressure shadow minerals. (G) Fragment, fusellum in dark red, strong tectonic lineation visible. (A–D, F–G) Sandia Region, southern Peru. (E) Lerhamn drillcore, Scania, Sweden. Scale indicated by 1 mm long bar in each photo.

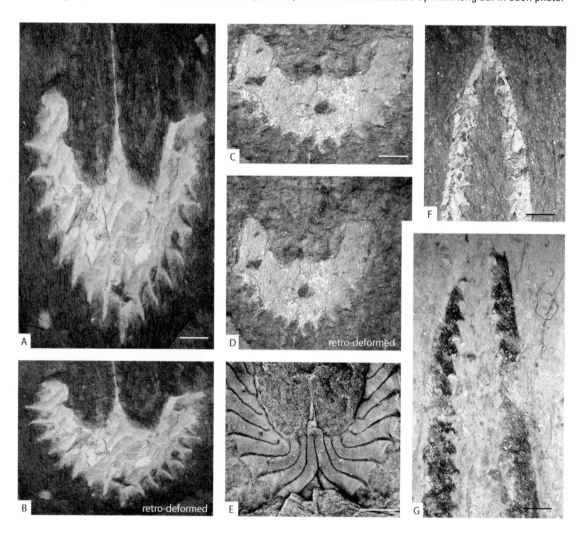

Any graptolite identification and interpretation from cleaved rocks must take this phenomenon into account, at the least by including the trace of the bedding/cleavage intersection on illustrations, and by noting differences in dimensions such as width and 2TRD (two thecae repeat distance: see Howe 1983) relative to this. Deformation may begin before there is obvious cleavage development and may only be made apparent by detailed examination of the specimens (e.g. lectotype of *Neolagarograptus tenuis* in Zalasiewicz et al. 2011, Fig. 4).

Perhaps the most characteristic type of graptolite preservation is as pale films on a dark shale background (Figures 5.10C, 5.11A–D, F). These specimens are clearly and easily outlined on the shale surfaces in stark contrast to the embedding medium, but it is now clear that they do not represent the true graptolites. When Palmer and Rickards (1991) used the term "Writing in the Rocks" for the title of their graptolite book, this is what they probably imagined as the best impression of a graptolite. At closer investigation it is certain that these light-coloured markings on shale surfaces are based on phyllosilicate minerals, formed during metamorphism and deformation of the surrounding sediment. The mineral is commonly referred to as chlorite, although part of the material has a significant potassium content and is not technically nor mineralogically a chlorite mineral. In fresh specimens, the mineral might be quite variable in colour. It can be whitish, light green or even bluish, and may weather to a yellow or reddish hue.

This phenomenon has generally been referred to as a "pressure shadow" effect, with the light-coloured mica mineral growing into the voids formed by a resistant object, either the organic material of the graptolite tubarium or the pyrite fill, as deformation proceeds (Underwood 1992). Both have different properties to the surrounding sediment and thus react in a different way to the tectonic deformation. The pressure shadow minerals are often identified in the earlier literature as gümbelite, and were considered as a replacement of the graptolites (e.g. Hundt 1924, p. 47; Chapman & Thomas 1936; Jaeger 1959). Richter (1853, p. 442; 1871) was probably the first to describe the pressure shadow mineral as a fibrous silvery white to greenish mineral in specimens from the Silurian of Thuringia, Germany, but Geinitz (1852) already mentioned the material and identified it as "Talk" (talcum). Gümbel (1868) recognized the mineral as pyrophyllite and compared it to similar material associated with many Carboniferous plant fossils. Kobell (1870) scientifically described the mineral as gümbelite, but this name has rarely been used in the mineralogical literature.

The development of pressure shadow minerals during tectonic deformation of the rocks appears to be quite complex. Where graptolites occur together with other fossils and with inorganic mineral bodies, these possess different types of enveloping minerals. In graptolites the phyllosilicates are chlorite-dominated, and around arthropod fragments illite dominates, while arthropod soft parts may have a kaolinite-dominated film (Page et al. 2008). The differences are interpreted as an indication of maturation of the organic material, formed as a late diagenetic or metamorphosis stage feature and not related to exceptional preservation, as has been done for the famous Burgess Shale biota in the past (Butterfield 1990; Butterfield et al. 2007). The variation in the mineralogical composition of the pressure shadow minerals suggests that there has been some kind of organic/inorganic reaction, in which the chemistry of the organic material catalyses the growth of specific phyllosilicates. This appears to start in the gas window, and then develops through low-grade metamorphism, with the phyllosilicates being oriented into cleavage-parallel fibres as the rocks are folded. Phyllosilicates can grow internally, too, in the space left where there has been pyritization, and here it has a random fabric of interlocking crystals (Butterfield et al. 2007).

In pyritized graptolites, the phyllosilicate can occupy a volume as great as, or even greater than, the pyrite internal cast itself (Figure 5.10). It surrounds the graptolite, although it is particularly thick in the "pressure shadow" regions with respect to slaty cleavage, and thinner or even absent at right angles to this, where the rock matrix has been pressed against the graptolite (see Underwood 1992). Pyritized graptolites resist deformation to some extent, in that they do not deform plastically, and details of 3D morphology can be preserved, even in strongly cleaved rocks. However, they can be subject to brittle fracture along their weakest or thinnest points, usually at the prothecal-metathecal contacts, as was shown by Sudbury (1958) for Silurian *Demirastrites* specimens from the Rheidol Gorge (see Figure 15.2B).

At a smaller scale, in non-pyritized graptolites, the fusellum may be stretched into generally square or rectangular fragments, between which there are usually fibrous syntectonic phyllosilicates (Mitchell et al. 2008). Closely-set parallel fractures can usually be seen in the fusellum, indicating the tectonic distortion and

Figure 5.12 Graptolites in weathered (yellow) shale from the Ordovician of China. (A) *Prorectograptus uniformis* (Chen, in Mu et al. 1979), flattened specimen with fusellum preserved as small carbon flakes. (B) *Undulograptus formosus* (Mu & Lee, 1958), relief specimen in obverse view, weathered pyritic cast with partial cover of black fusellum. (C) *Azygograptus* sp., flattened, flakes of fusellum perserved. (D) *Baltograptus geometricus* (Törnquist, 1901), weathered pyritic cast with cover of black fusellum. (E) *Arienigraptus* sp., mainly yellow-stained pressure shadow minerals. Scale indicated by 1 mm long bar in each photo.

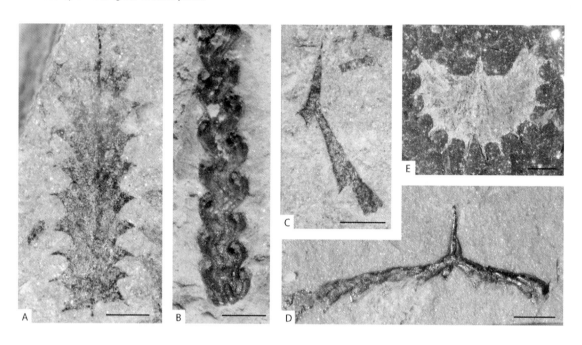

the direction of deformation (Figure 5.11E–F). The pattern of fracture generally seems to relate more to the pattern of tectonic forces than to the original structure of fusellum and cortex. Hence, its pattern may provide valuable information on tectonic history, especially as it develops prior to the discernable imposition of slaty cleavage and other obvious indicators of rock deformation. Graptolites may survive and remain identifiable even with levels of significant metamorphism, especially if protected in nodules or boudins, at least as far as greenschist facies (Dieni et al. 2005).

Weathering

Weathering of graptolites in surface exposures of the rock successions is present in all regions worldwide, but has rarely attracted the attention of paleontologists. The typical black shales in which graptolites are found easily weather to lighter coloured soft material on the surface, from which the graptolites often stand out in a darker colour (Figure 5.12). This is very typical for the successions of Australasia. The graptolite shales of Victoria, Australia, in general are light grey, yellowish or pinkish with the graptolites found in black, dark brown, but also greenish or reddish colours (see photos in Rickards & Chapman 1991). Often the specimens are surrounded by light-coloured halos on less weathered black shale surfaces. The material of the Lower Tremadocian La 1 of Victoria shows the graptolites in white or silvery colours on a black shale surface (Cooper & Stewart 1979), indicating relatively unweathered material. The weathering colours obviously are strongly influenced by the presence of organic material and the pressure shadow minerals. The colour can vary also through later weathering and the effects of fluids moving through the sediments. Fresh, unweathered graptolite material of the *Didymograptus murchisoni* Biozone from Abereiddy Bay, Wales, shows the light green colour of the pressure shadow minerals (Underwood 1992).

Weathering of the pyrite fill of graptolites may lead to yellow or reddish iron mineral colours of the internal casts or staining of the surrounding rocks. Relief specimens may appear dark brown to black in the weathered rocks, even though the original pyrite is gone and little organic material remains (Figure 5.12). The pink and white colour (Plate 8A–B) of the graptolite *Cymatograptus bidextro* (Toro & Maletz, 2008) may be formed through weathering of iron minerals (pyrite?) and it seems that the colour of the mineral is originally pink, but weathers quickly to white when exposed. This preservation has only been found at a few localities in Argentina so far.

Outlook

Even though we have learned a lot about the preservational potential and preservational aspects of fossils, and in this case of the graptolites, the interpretation of individual specimens and faunas remains difficult. Sometimes it is quite easy to say "this is a real specimen", but more often a detailed investigation of the material reveals that this is not the case. Many graptolites have been altered considerably from the original state in which they were embedded in the sediment. They are flattened, tectonically distorted, coalified or in extreme cases even replaced with minerals. This part of their history is often neglected, but potentially provides a wealth of geologically important information. It is valuable for taxonomic purposes, for the exploration of the petroleum industries, and also for paleoenvironmental interpretations, including the estimation of sedimentation rates and time-averaging of fossil assemblages. So we need to be much more careful in analysing the fossils on our working tables.

6 Graptolites and Stratigraphy

Jörg Maletz

One of the key aims of geology is to build up a history of the Earth – not least because this is a prerequisite to working out how the Earth functions. This history must be one that encompasses the whole Earth, and not just one part of it. In practice, this means building up partial, fragmentary histories of many places, through the study of local successions of rock sequences – in short, stratigraphy. Then we have to piece together these local histories into regional and finally into global histories. We have to place events into a framework of time and space, for only then do we have a chance of establishing cause and effect between, say, massive volcanic eruptions in one part of the world, and evidence of climate change or mass extinctions of species elsewhere.

The key we use to piece together the geological history of our planet is the correlation, the establishment of time lines, between geographically separated successions. This is the way to find out whether events in one part of the world took place at the same time as, or earlier or later than, events in other parts of the world. One of the oldest ways is the use of fossils, and a precise and highly reliable succession of fossil faunas has been recognized over the last two centuries that reflects the evolution of our planet and the living organisms on its surface. This is what geoscientists call biostratigraphy. Graptolites, or more precisely, planktic graptolites, are one of the best groups to establish biostratigraphical successions in the Lower Paleozoic, and thus they have become one of the exemplary groups of fossils for dating rock sequences and correlating them across the planet. Graptolites come in numerous forms and change considerably over short time spans. Thus, they are great indicators of time equivalence and provide the optimum for biostratigraphical work.

Graptolite Paleobiology, First Edition. Jörg Maletz.

Graptolite research does not stop with the work on biostratigraphy, but goes far beyond it in its application in the geological sciences. As one of the geological features available from many rocks, graptolites can provide valuable information for a number of purposes, from understanding the geography of the past, to explore plate distributions and movements through millennia. They can be used to recognize and trace natural resources like oil, gas and mineral deposits. Thus, it is a hike into geological application that makes graptolites so important even for geologists not interested in fossil collecting and paleontology.

Biostratigraphy and Graptolites

The most robust and effective way to establish correlations in rock successions, at least from the Cambrian onward, is through the use of fossils – that is, through exploiting the succession of species that have appeared and then died out, as biological communities have evolved and changed on this planet. Biostratigraphy produces the basic framework for all other, more detailed or specialized methods of stratigraphic correlation. Thus, from the Phanerozoic onwards, the geological history can be worked out in much greater detail than is possible for the long Precambrian time interval, where fossils as time indicators are largely missing.

Certain fossils are restricted to certain rock intervals, and paleontologists have worked out a general succession of faunas since the 19th century. The concept of using fossils for dating rocks and correlating them was well established even before Darwin (1859) published his ideas on evolution and with this produced the scientific mechanism behind the concept of biostratigraphy. This method has the beauty of being based on a succession of unique and unrepeated events representing a directed history. Each biological species has only a single origin in evolving from some precursor species, an interval of time during which it lives, and a point in time when it becomes extinct. Putting together a timescale based on the origins and extinctions of many species therefore provides a clear, unambiguous record of the passage of geological time.

Biostratigraphy can be called a method of relative dating, as it does not produce any numerical ages – a number representing an age in millions of years as can be done by absolute dating with radiometric methods. Geologists successfully practised this method, even though they did not know whether the successions that they studied were millions of years old or maybe just a few thousand years. With the advent of radiometric dating, the biostratigraphical and other stratigraphical units can be calibrated with numerical dates, so that the absolute age and duration of events can be worked out. Radiometric dating, however, does not make relative dating superfluous. It merely adds another dimension or more information to the methodology of dating rock sequences and understanding geological time.

Among fossil groups, some are better time indicators than others for some obvious reasons. Fossils that are large, rare and restricted to particular regions or facies, such as dinosaurs, are, in practice, poor indicators of geological time. The kinds of fossils that are useful in practical biostratigraphy are, by contrast, small, common, widespread, relatively independent of facies and rapidly evolving. They should also be easily identified to serve as index species for biostratigraphy and correlations. Graptolites score highly in most of these respects (Figure 6.1). Their planktic mode of life allowed them to spread widely through the oceans, as far as their ecological tolerances allowed.

Unlike many microfossils, graptolites are rarely reworked by erosion from one stratigraphic unit into another, as sedimentary reworking would destroy the specimens, except when preserved in larger clasts. Reworking of graptolites can be seen in the common occurrence of graptolites in Quaternary moraines bearing glacial erratic boulders in Central Europe (e.g. Kozłowski 1949; Urbanek 1958; Eisenack 1951). The Darriwilian (Middle Ordovician) graptolites of the Daniels

Figure 6.1 Fossil correlation. The presence of *Nicholsonograptus fasciculatus* can be used to correlate Middle Ordovician sections in Norway and western Newfoundland, the basis for biostratigraphical correlation of lithological successions on two different continents (based on data in Maletz et al. 2011).

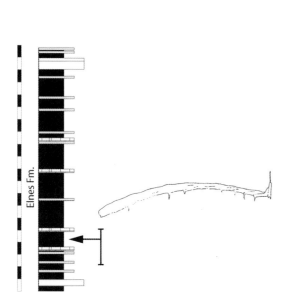

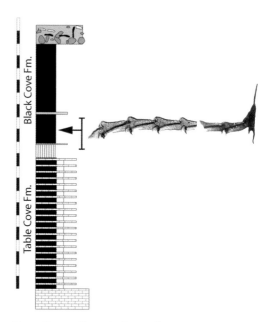

Slemmestad, Norway

Newfoundland, Canada

Harbour Conglomerate in western Newfoundland, in which clasts with well-preserved graptolites from the older Table Cove Formation are incorporated (Whittington & Rickards 1969; Stenzel et al. 1990; Albani et al. 2001), can easily be recognized to be "out of sequence" when compared with the normal succession of faunas of the region. Thus, graptolites in general show valuable stratigraphic reliability. Graptolites from Lower Ordovician olistoliths in Middle Ordovician olistostromes were found in the Hamburg Klippe of Pennsylvania (Ganis et al. 2001; Ganis 2005), showing larger-scale reworking of graptolite faunas and helping to unravel the sedimentary and tectonic history of a Paleozoic foreland basin on the eastern side of North America.

Only planktic graptolites are used in biostratigraphy, as these are widespread and independent of control by the sediment type, but in rare cases benthic graptolites have been used as biostratigraphical index species locally when planktic species are not available (see *Acanthograptus sinensis*

Biozone in the Upper Tremadocian of South China: Zhang et al. 2010). Graptolites are not completely independent of facies, and are most common, diverse and widely used in deeper water settings where they are often virtually the only macrofossils present (Chapter 4). They are typically confined to strata laid down on anoxic or dysoxic sea floor environments, excluding burrowing, crawling or scavenging benthic animals as well as oxygen. In this way, graptolite tubaria were protected from the effects of both scavenging and aerobic oxidation processes until they were buried under a protective layer of sediment (see Chapter 5).

Graptolites are often rare or absent in shallow-water, shelf or inshore settings. Where present, they generally comprise assemblages that are less diverse than contemporaneous deep-water communities, or are represented even by monospecific assemblages. Lateral variations of graptolite faunas along depth or temperature gradients in larger ocean basins often make it

Figure 6.2 Correlation of graptolite faunas across the Iapetus Ocean using biofacies overlap and transitional faunas (based on Maletz et al. 2011).

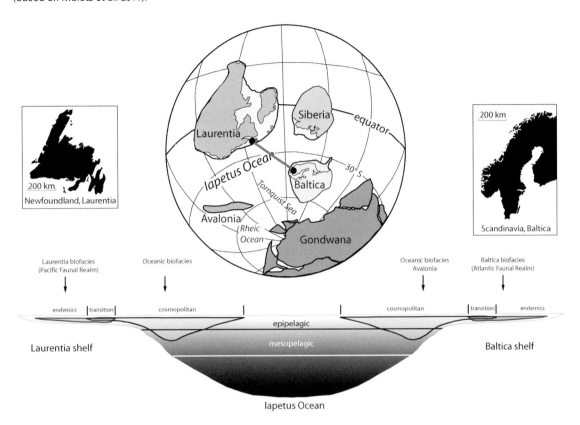

difficult to correlate these faunas, and different local biozonations may be used. Maletz et al. (2011) provided a good example from the two sides of the Paleozoic Iapetus Ocean (Figure 6.2), where faunal differences between the Laurentian margin (tropical region) and the Baltica margin (cold-water region) appeared to be pronounced. However, the detailed study of Maletz et al. (2011) showed that the two regions have quite a number of faunal elements in common and a precise biostratigraphical correlation of the faunas was possible. A key part of global stratigraphy thus lies in establishing ties between regional biozonations, mainly on the basis of species in common between them, but also using other stratigraphic evidence. Products of such correlation attempts include global "standard biostratigraphies" in which the units are broader and therefore with poorer time-resolu-tion, but more widely traceable (cf. Silurian graptolite zonal sequence of Koren et al. 1996; or the Ordovician time slices: Webby et al. 2004b).

William Smith (1769–1839), often dubbed the "father of British geology", recognized the faunal succession in rock sequences as a basic concept for stratigraphic practice and thus established biostratigraphy as a tool to unravel geological history and understand deep time. From here it was not far to the idea of using graptolites as a stratigraphic tool. It was James Hall (1850) who recognized the value of graptolites to identify certain strata, and thus demonstrated the useful-ness of graptolites for geological purposes and to develop a graptolite biostratigraphy. At the same time, Barrande (1850) also provided information on the biostratigraphical distribution of the grap-tolite faunas in Bohemia. Nicholson (1868a) demonstrated the vertical distribution of the graptolites in the British Ordovician and Silurian, but did not provide a biostratigraphical framework or biozonation. Linnarsson (1876) added biostratigraphical data from the graptolite successions of Scandinavia, but it was Lapworth (1878) who, in his important paper on "The Moffat Series", established the zonal concept for graptolites. He demonstrated the importance of biostratigraphical investigations for the interpretation of the complexly tectonized, extremely thick greywacke succession of the Ordovician and Silurian of southern Scotland. The graptolite biostratigraphy established by Lapworth for the region is still valid and useful after nearly 150 years (Fortey 1993).

A key early debate leading to the concept of biostratigraphy was between Lapworth and Barrande, who had been identifying similar, distinctive assemblages of graptolites in Ordovician to Silurian strata. Barrande, a classical supporter of Cuvier's "theory of catastrophism", described the faunas from tectonic blocks as "colonies", faunas interpreted as originating from a number of extinctions and creations (see Barrande references, 1859–1881). He thought that these faunas possessed environmental rather than time significance. Lapworth, however, interpreted the assemblages he observed as representing a succession of species through time. He used these assemblages to solve major geological problems (Lapworth 1878). Lapworth's interpretation proved to be essentially correct, and the succession of zones he established became the nucleus of modern graptolite biozonation. Gertrude Elles and Ethel Wood subsequently produced over a couple of decades in the early 20th century the monumental *Monograph of British Graptolites* (Elles and Wood, 1901–1918) that established a biozonation of Ordovician and Silurian graptolites for Britain, and this work became something of a global standard.

Graptolites are the main key for the biostratigraphical subdivision of Ordovician and Silurian strata, and in the 1960s it was recognized that even the Lower Devonian time interval bears graptolite faunas useful for biostratigraphical correlations (Jaeger 1959, 1978, 1988). The lower Devonian graptolite faunas are of fairly low diversity and monotonous monograptids prevail, but up to eight graptolite zones can be differentiated in the Lochkovian to lower Emsian time interval (Figure 6.3). The youngest planktic graptolites occur in the basal Emsian *kitabicus* conodont zone (Yolkin et al. 1997, Fig. 3; Becker et al. 2012), but the exact correlation of the Pragian/Emsian boundary with the graptolite record is still debated.

Zonal names are generally based on certain – mostly common – index species. Lapworth (1880b) used index species to define a biozonation for the Ordovician and Silurian in Britain and numbered the zones from 1 to 20 for easy access. Zones 1–9 belong to the Ordovician and zones 10–20 are of Silurian age. Elles and Wood (1914) revised the succession, based on the data from their previous taxonomic descriptions, in a table documenting the zonal distribution of the British Ordovician and Silurian graptolite taxa and numbered their zones from 1 to 36. The two numbering systems denominate different zones with the same number, and thus are not comparable. Unfortunately, the numbering systems of Lapworth (1880b) and Elles and Wood (1914) became standards in Germany, where for a long time the zonal affiliations were noted only through the numbering system (e.g. Eisel 1899; Münch 1952). Index species were added to the numbered succession (Eisel 1903), but the zones were usually only identified by their numbers. Unfortunately, both numbering systems were also used in parallel to each other for a long time. Jaeger (1991) provided the latest version and also indicated the incompleteness of the Elles and Wood (1914) succession by adding un-numbered intervals, before the numbering system was finally abandoned (Maletz 2001a).

Types of Graptolite Zones

Due to the sporadic nature of graptolite distribution – and of fossils in general – biozones are a theoretical concept and are defined locally, in lithological sections, but are not expected to represent exactly correlatable units. To establish graptolite biozones, thorough, systematic collection is

Figure 6.3 Devonian graptolite biozonation, showing the most important taxa and their actual ranges (compiled from data in Jaeger 1988; Loydell 2012; Lenz 2013).

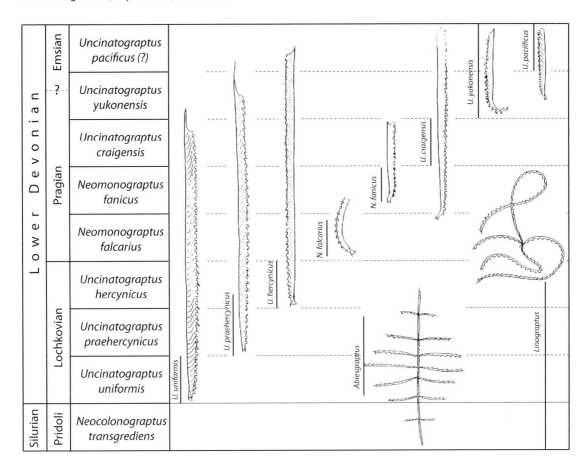

needed in key reference sections. One such is the well-exposed, graptolite-rich succession at Dob's Linn, in the Southern Uplands of Scotland, which spans some 60 metres and approximately 20 million years. Here, Charles Lapworth in the 19th century (and many paleontologists subsequently) logged the stratigraphic distribution of many graptolite taxa through the rocks, and recognized a succession of graptolite assemblages characterized by particular combinations of more or less distinctive species. In this succession, also the base of the Silurian System has been defined by a "golden spike" at the FAD (first appearance datum) of the species *Akidograptus ascensus* (Figure 6.4) (Holland 1985; Williams 1983, 1988; Melchin 2003).

Each graptolite biozone is named after at least one key taxon. Usually, but not always, this is a taxon that appears at the base of the biozone, and then ranges through part or all of the zone. In biostratigraphy, we often use the concept of the FAD (first appearance datum) and LAD (last appearance datum) to define certain biostratigraphic levels in a specific faunal succession (see Figure 6.5). The FAD defines the earliest record (lowest occurrence in section) of a species in a stratigraphical succession, and the LAD recognizes the last record (highest occurrence in section) of that species. These levels are locally defined in a particular section and may not be correlatable exactly with other sections, as fossil occurrence often is fortuitous and recovery of important specimens may be by chance.

Figure 6.4 The Ordovician/Silurian boundary interval at Dob's Linn, Scotland, showing ranges of important graptolite taxa (based on https://engineering.purdue.edu/Stratigraphy/gssp/ordsil.htm).

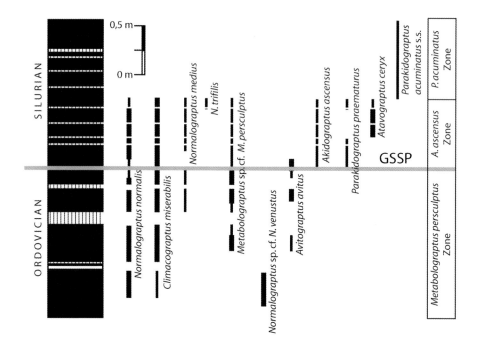

Figure 6.5 Examples of biozone types as discussed in the text. FAD (first appearance datum), LAD (last appearance datum).

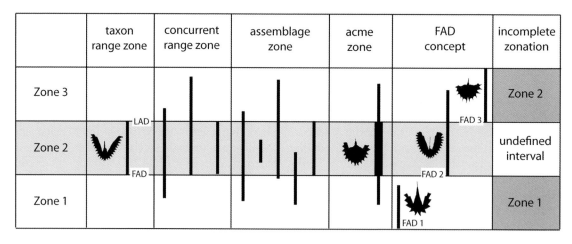

Biozones are often based upon reference sections, but they are not unambiguous indicators of time. Rather, they are proxies for the passage of geological time. Their boundaries, in detail, will always cut across time planes to some extent, because graptolite species often appeared and disappeared at different times in different parts of the world, or their record may be incompletely sampled in the section. Unlike the case with chronostratigraphical units, biozones do not have type sections, nor is their base fixed by a "golden spike" placed at such a section.

Graptolite biozones, like all fossil zones, are assemblage zones, units characterized by a certain combination of species found in this interval, ideally with a restricted vertical range and a wide horizontal distribution. This zone is usually named after a single, characteristic species found in the interval. The name-giving species may be restricted to the biozone, with its first appearance being used to define the zonal base, but other faunal elements may also be present and even useful for the identification of the zone. The index species may, equally, range outside its biozone, either above its upper boundary or rarely below it (in an acme zone). Quite a number of different types of biozones have been described in the literature, and most of them are difficult to compare in detail (Figure 6.5). In the past, the most commonly used zone is a taxon range zone, showing the distribution of a single species from its first to last occurrence in a section. A concurrent range zone shows the interval in which two or more species co-occur, but individual species may be present above or below the interval. It is similar to an assemblage zone, which is characterized by the co-occurrence of a number of species. However, individual species may be lacking in the zone and the interval is still recognizable. An acme zone is based on the common occurrence of a certain species, but the species might be present also below and above the zone, representing an interval in which the species is most common. Thus, for biostratigraphical purposes, the total assemblage and the incoming species are important. If the base of a biozone is defined by the new appearance of more than one species in a section, it may be due to a smaller or larger gap in the lithological succession, as these species may not appear exactly synchronously in time and earlier occurrences may not be shown. The main problem by using different concepts is that biozonations from different regions are not easily correlated or compared.

The nature of the appearance and disappearance of fossils, in this case of graptolites, also needs to be considered in biostratigraphy. Each graptolite species must have evolved from a pre-existing species. Sometimes, evidence has been found for a more or less gradational passage from one form to another in one place or region. Such instances of observed micro-evolution are sporadic, but enough examples exist that present us with some important information on evolutionary changes and provide a key for precise biostratigraphical zonation. The succession of the isograptids (Cooper 1973) provides a great tool for a precise biostratigraphy of the Dapingian to lower Darriwilian time interval (see also Chapter 10). A number of successively evolving species appear during this time and are commonly used with great precision for biostratigraphy. However, the differentiation of the individual species and subspecies may be difficult due to overlapping in the natural variation of the taxonomically useful characteristics in the populations of these taxa. Also, the evolutionary changes within the Silurian monograptid genus *Spirograptus* (Figure 6.6) is proven to provide important information for the biostratigraphical differentiation within the Telychian (Loydell et al. 1993), and allows a glimpse into the evolutionary changes within the genus.

More commonly, a species will have evolved from some local small ancestral population and then, at some point in time, it will have escaped and migrated across the oceans, to become much more widespread. Thus, the appearance of new species reflects the timing of immigration, and not of evolutionary origin. This might have an advantage, as the sudden and seemingly synchronous appearance of a morphologically distinct form over a wide area may be used as a time line for biostratigraphical purposes, since it is instantaneous in relation to the available stratigraphic resolution.

In recent years the FAD and LAD concept has been introduced into graptolite research and especially to graptolite biostratigraphy (see graphic correlation below). The FAD or first appearance datum represents the stratigraphically earliest occurrence of a certain species and is used to define the base of a certain graptolite zone (Figure 6.5). The top of the zone is not defined, but is represented by the defined base of the overlying graptolite zone in the succession. Thus, each zone is defined at a single level in the stratigraphical succession, and it is not necessary to define base and top of a biostratigraphical unit. Defining only the base of a unit, and recognizing the base of the overlying unit to represent also the top of the lower unit, avoids the creation of undefined intervals.

Figure 6.6 Biostratigraphy of the Aeronian/Telychian boundary, showing the evolution and biostratigraphy of the genus *Spirograptus*, length of zones and subzones not to scale (based on Loydell 1992; Loydell et al. 1993).

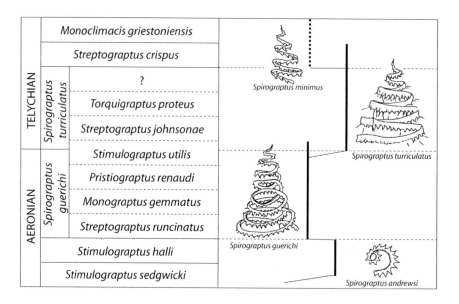

Graptolites and Chronostratigraphy

Our modern geological time scale (Gradstein et al. 2012) consists of formally defined units, termed geochronological units (e.g. the Silurian Period, the Llandovery Epoch). Parallel to these units based on theoretical concepts, physical stratal "time-rock" units of chronostratigraphy (e.g. the Silurian System, the Llandovery Series) are defined using multiple criteria (e.g. fossils, lithostratigraphical units, event beds, bentonites, chemostratigraphy, paleomagnetism). These units are based on physical successions and precise correlations of units in marked sections, based on the definition of the base of the units, the Global Stratotype Section and Point (GSSP) concept with the golden spike (Figure 6.7) as the visible physical documentation in a defined stratotype section. Graptolites provide the greater part of the underpinning of the geological time framework in the Lower Paleozoic, and define 13 out of 15 stage boundaries between the bases of the Ordovician and Devonian Systems (Figure 6.7). Only two boundaries are defined by other fossil groups: conodonts and brachiopods, respectively. Detailed information on the chronostratigraphy can be found online on the website of the International Commission on Stratigraphy (www.stratigraphy.org).

When Lapworth (1879) introduced the Ordovician System and the concept of a tripartite Paleozoic to our stratigraphic framework, he strongly favoured the differentiation based on fossil faunas, and Lapworth (1880b) discussed the succession of graptolite faunas in the Paleozoic in some detail from his investigations. But it took a long time until the Ordovician and Silurian Systems were finally internationally accepted by the International Geological Congress in Copenhagen in 1960 (Gradstein et al. 2012).

The Cambrian–Ordovician boundary is now formally defined by the incoming of the conodont species *Iapetognathus fluctivagus* at the 101.8 m level in the measured and illustrated section at Green Point, western Newfoundland (Cooper et al. 2001). However, the level at which this was recorded – in the GSSP section at Green Point – is very close to the level at which the earliest plankic graptolites are found (Cooper et al. 1998, 2001). Thus, the appearance of these graptolites in rocks around the world provides a useful practical guide to the recognition of the basal Ordovician strata and a close approximation for the base of

Figure 6.7 Lower Paleozoic GSSPs based on graptolites. The right side shows the "golden spike" for the base of the Floian at Diabasbrottet, Västergötland, Sweden, and a specimen of *Tetragraptus approximatus*, the graptolite species defining the boundary.

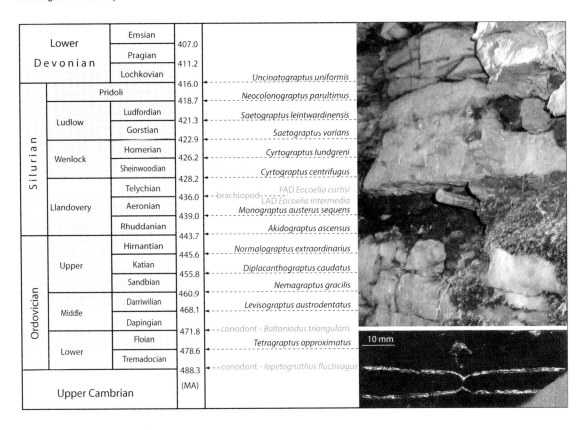

Lower Devonian		Emsian	407.0
		Pragian	411.2
		Lochkovian	416.0
Silurian	Pridoli		418.7
	Ludlow	Ludfordian	421.3
		Gorstian	422.9
	Wenlock	Homerian	426.2
		Sheinwoodian	428.2
	Llandovery	Telychian	436.0
		Aeronian	439.0
		Rhuddanian	443.7
Ordovician	Upper	Hirnantian	445.6
		Katian	455.8
		Sandbian	460.9
	Middle	Darriwilian	468.1
		Dapingian	471.8
	Lower	Floian	478.6
		Tremadocian	488.3
Upper Cambrian			(MA)

Uncinatograptus uniformis
Neocolonograptus parultimus
Saetograptus leintwardinensis
Saetograptus varians
Cyrtograptus lundgreni
Cyrtograptus centrifugus
FAD *Eocoelia curtisi*
brachiopod — LAD *Eocoelia intermedia*
Monograptus austerus sequens
Akidograptus ascensus
Normalograptus extraordinarius
Diplacanthograptus caudatus
Nemagraptus gracilis
Levisograptus austrodentatus
conodont - *Baltoniodus triangularis*
Tetragraptus approximatus
conodont - *Iapetognathus fluctivagus*

10 mm

the Ordovician System itself. The bases of the succeeding Silurian and Devonian Systems are both defined primarily by the incoming of particular graptolite taxa: *Akidograptus ascensus* for the base of the Silurian at Dob's Linn in Scotland (Holland 1985) and *Uncinatograptus uniformis* for the base of the Devonian at Klonk in the Czech Republic (Chlupác & Kukal 1977).

Graptolites and Absolute Ages

Graptolite biozones do not represent absolute ages and have to be seen in comparison with older and younger strata or graptolite biozones. Thus, it is not possible to provide ages in millions of years by using graptolite biostratigraphy alone. However, these relative dates of the graptolite biozonations can be combined with radiometric estimates of time, producing a general idea of the average duration of graptolite zones (see ages in Figure 6.7). This means an indirect dating of graptolite biozones through radiometrically datable lithostratigraphical units like bentonite beds, layers of ancient volcanic ash (e.g. Obst et al. 2002; Sell et al. 2011). These have frequently been used as markers to correlate Paleozoic successions (Kolata et al. 1986, 1996; Mitchell et al. 2004). However, a direct radiometric dating of bentonite beds associated with graptolite faunas was not done or was not possible.

The latest compilations of biostratigraphical and radiometric data (Sadler et al. 2009, 2011; Cooper & Sadler 2012) suggest that the graptolite biozone durations in the Paleozoic vary greatly. There is in general poorer time resolution for the Ordovician than for the Silurian. Thus, the

Ordovician graptolite biozones average about 1 million years in the Early and Mid Ordovician, and nearer to 2 million years in the Late Ordovician. In the Silurian, Llandovery graptolite biozones average 0.8 million years, while for the Wenlock, Ludlow and Pridoli, graptolite biozones average 0.5, 0.3 and 0.35 million years respectively. Devonian graptolite zones again show lower resolution, in excess of a million years. However, a number of graptolite biozones can be split locally into several subzones, increasing the biozonal resolution considerably. Loydell (1992) for instance, has shown that the combined *Spirograptus guerichi* and *Spirograptus turriculatus* biozone interval of the British Isles can be split into up to seven subzones. Each of these refined units represent time intervals approximately 100,000 years in length (Figure 6.6).

Graptolite Biozonations

The use of graptolites to subdivide rock strata can be as simple as recognizing their presence and so immediately dating the enclosing rock succession to the Paleozoic – much as one uses the presence of trilobites for the same purpose, or ammonites to establish a Mesozoic age. If the specimen is a graptoloid (e.g. a planktic graptolite, and not a benthic one), the age interval can be narrowed down to the Ordovician to lower Devonian. Graptolite faunas evolved considerably over time, and distinct groups of graptolites can be regarded as typical of certain time intervals, thus allowing fairly precise dating of successions even for the non-specialist. The graptolite biostratigraphy is closely connected to evolutionary trends in the graptolite faunas, and more details on the identification of graptolites from the various time intervals will be provided in the taxonomic chapters.

A great number of biostratigraphical tables have been published in the literature, and it is often difficult to search through the wealth of data as compilations are rare and most of the tables are based on local data. Loydell (2012) provided the most recent update of the biostratigraphy of the graptolites and should be consulted for a precise correlation across and between continents. Many local biostratigraphies have been established independently from local or regional investigations and are still difficult to correlate due to the presence of endemic or local faunal elements (see also remarks on paleogeography in Chapter 4).

During the early Ordovician, multiramous graptoloids were the most important graptolites, providing most of the index fossils. The succession started with the earliest planktic graptolite of the genus *Rhabdinopora* (Figure 6.8A), often used to define the base of the Ordovician System in the past and long discussed as a possible index species for chronostratigraphical purposes (see Cooper et al. 1998, 2001). From here it was not difficult to derive the multiramous *Anisograptus*, *Adelograptus* (Figure 6.8B) and *Clonograptus* (Figure 6.8C) species, among others, that are common in the interval from the Lower Tremadocian to the Lower Darriwilian in many regions. These are differentiated through their proximal developments and are often difficult to identify. The Floian to Dapingian interval is dominated by pauciramous taxa like those of the genera *Didymograptus*, *Expansograptus* (Figure 6.8D), *Tetragraptus* (Figure 6.8E) and related others. In the Dapingian to Lower Darriwilian the genera *Isograptus* (Figure 6.8F), *Arienigraptus* (Figure 6.8G) and *Pseudisograptus* were the best for biostratigraphy, as their shapes make them easy to recognize. In detail, the taxonomy of the isograptids is, however, more complicated, and a number of genera are differentiated through their special proximal developments (see Chapter 10).

These were succeeded by the biserial graptolites, as the dominant group throughout the rest of the Ordovician. They originated approximately at the base of the Darriwilian Stage and quickly became the most common group of graptoloids, interestingly developing secondarily multiramous taxa later on (e.g. *Nemagraptus*). The dicranograptids, including the V-shaped *Dicellograptus* (Figure 6.8L) and Y-shaped *Dicranograptus*, and the curved, multiramous nemagraptids (Figure 6.8I), were also important and distinctive elements at various times in the mid to late Ordovician. Details of their development and taxonomy are found in Chapter 11.

Following near-extinction during the latest Ordovician glaciation (Melchin et al. 2011), a few

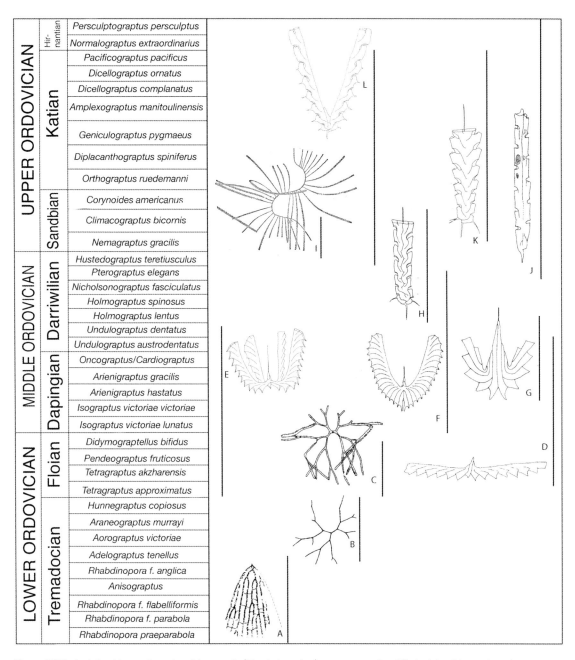

UPPER ORDOVICIAN	Katian	Hir-nantian	Persculptograptus persculptus
			Normalograptus extraordinarius
			Pacificograptus pacificus
			Dicellograptus ornatus
			Dicellograptus complanatus
			Amplexograptus manitoulinensis
			Geniculograptus pygmaeus
			Diplacanthograptus spiniferus
			Orthograptus ruedemanni
	Sandbian		Corynoides americanus
			Climacograptus bicornis
			Nemagraptus gracilis
MIDDLE ORDOVICIAN	Darriwilian		Hustedograptus teretiusculus
			Pterograptus elegans
			Nicholsonograptus fasciculatus
			Holmograptus spinosus
			Holmograptus lentus
			Undulograptus dentatus
			Undulograptus austrodentatus
	Dapingian		Oncograptus/Cardiograptus
			Arienigraptus gracilis
			Arienigraptus hastatus
			Isograptus victoriae victoriae
			Isograptus victoriae lunatus
LOWER ORDOVICIAN	Floian		Didymograptellus bifidus
			Pendeograptus fruticosus
			Tetragraptus akzharensis
			Tetragraptus approximatus
	Tremadocian		Hunnegraptus copiosus
			Araneograptus murrayi
			Aorograptus victoriae
			Adelograptus tenellus
			Rhabdinopora f. anglica
			Anisograptus
			Rhabdinopora f. flabelliformis
			Rhabdinopora f. parabola
			Rhabdinopora praeparabola

Figure 6.8 Ordovician biostratigraphy of Laurentia (North America) as an example of Ordovician biostratigraphy. A number of typical graptolite genera for the Ordovician are shown. (A) *Rhabdinopora* (Lower Tremadocian). (B) *Adelograptus* (Upper Tremadoian). (C) *Clonograptus* (Floian). (D) *Expansograptus* (Floian to Lower Darriwilian). (E) *Tetragraptus*, reclined (Floian to Lower Darriwilian). (F) *Isograptus* (Dapingian to Lower Darriwilian). (G) *Arienigraptus* (Upper Dapingian to Lower Darriwilian). (H) *Archiclimacograptus* (Darriwilian to Katian). (I) *Nemagraptus gracilis* (Sandbian). (J) *Normalograptus* (upper Darriwilian to Silurian). (K) *Amplexograptus* (Sandbian to Katian). (L) *Dicellograptus* (Uppermost Darriwilian to Katian). Graptolite illustrations from various sources, not to scale.

Figure 6.9 Silurian graptolite biostratigraphy (based on Koren et al. 1996; Loydell 2012), showing some of the characteristic Silurian graptolites. (A) *Akidograptus ascensus*. (B) *Petalolithus*. (C) *Dimorphograptus*. (D) *Demirastrites*. (E) *Rastrites*. (F) *Stimulograptus*. (G) *Spirograptus*. (H) *Streptograptus*. (I) *Monoclimacis*. (J) *Oktavites*. (K) *Retiolites*. (L) *Cyrtograptus*. (M) *Pristiograptus*. (N) *Heisograptus*. (O) *Saetograptus*. (P) *Bohemograptus*. Graptolite illustrations from various sources, not to scale.

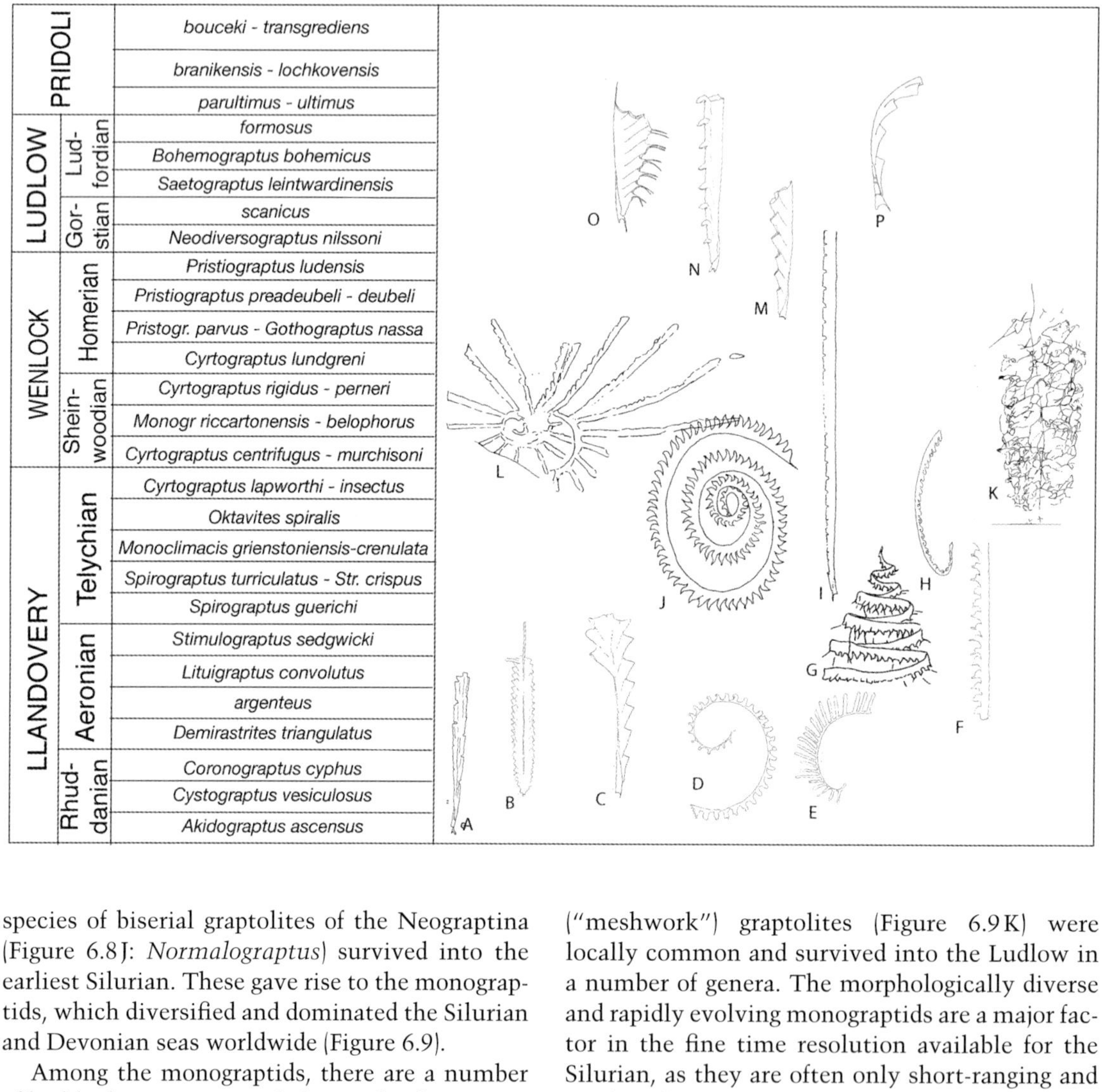

PRIDOLI			*bouceki - transgrediens*
			branikensis - lochkovensis
			parultimus - ultimus
LUDLOW	Lud-fordian		*formosus*
			Bohemograptus bohemicus
			Saetograptus leintwardinensis
	Gor-stian		*scanicus*
			Neodiversograptus nilssoni
WENLOCK	Homerian		*Pristiograptus ludensis*
			Pristiograptus preadeubeli - deubeli
			Pristogr. parvus - Gothograptus nassa
			Cyrtograptus lundgreni
	Shein-woodian		*Cyrtograptus rigidus - perneri*
			Monogr riccartonensis - belophorus
			Cyrtograptus centrifugus - murchisoni
LLANDOVERY	Telychian		*Cyrtograptus lapworthi - insectus*
			Oktavites spiralis
			Monoclimacis grienstoniensis-crenulata
			Spirograptus turriculatus - Str. crispus
			Spirograptus guerichi
	Aeronian		*Stimulograptus sedgwicki*
			Lituigraptus convolutus
			argenteus
			Demirastrites triangulatus
	Rhud-danian		*Coronograptus cyphus*
			Cystograptus vesiculosus
			Akidograptus ascensus

species of biserial graptolites of the Neograptina (Figure 6.8J: *Normalograptus*) survived into the earliest Silurian. These gave rise to the monograptids, which diversified and dominated the Silurian and Devonian seas worldwide (Figure 6.9).

Among the monograptids, there are a number of highly distinctive morphologies that have overall time significance. These include the triangulate monograptids of the genera *Demirastrites* (Figure 6.9D) and *Rastrites* (Figure 6.9E), the genus *Streptograptus* (Figure 6.9H), and the cyrtograptids (Figure 6.9L). Biserial graptolites persisted for a short while into the Silurian, while retiolitid

("meshwork") graptolites (Figure 6.9K) were locally common and survived into the Ludlow in a number of genera. The morphologically diverse and rapidly evolving monograptids are a major factor in the fine time resolution available for the Silurian, as they are often only short-ranging and easily differentiated based on the details of their thecal morphology.

The Devonian graptolites are more strongly reduced in their diversity, and only about 20 species may be differentiated during this time interval. Many more species have been described in the literature, but often can be recognized as synonyms of other species (Lenz 2013). The Devonian faunas consist largely of straight or proximally curved, planar monograptids. The secondarily multiramous genera *Abiesgraptus* and *Linograptus* with their simple thecal style and low thecal overlap look quite different and represent the last bloom of the linograptids, appearing first in the Ludfordian, Upper Silurian (Urbanek 1997b), but disappearing before the end of the Lochkovian (see Figure 6.3).

The graptolite biozonation continues to be developed, and revisions are common. Taxonomic uncertainties continue to be the most severe constraints upon the use of graptolites in biostratigraphy. Some species remain poorly understood or inadequately described, but are still used for biostratigraphical purposes. This is especially common in the Silurian, where the proximal end development and even thecal details are basically unknown for some of the common monograptid taxa. The type material of many classical species is too poorly preserved, or poorly described, to serve as a reference by modern standards. In many instances, too, several distinct taxa have been described under the same name, or for other reasons were not differentiated in the past. Better definition of taxa, for instance, by the re-description of type material, will in the future lead to further refinement in graptolite biostratigraphy. Certain poorly documented time intervals and faunas have only been described recently in more detail. The uppermost Silurian, late Ludlow to Pridoli monograptids may serve as an example here. Many of these faunal elements have been described only since the 1970s (Tsegelniuk, 1976; Koren & Suyarkova 1997; Urbanek 1997a) and their distribution is still not explored in detail (see Chapter 13).

Graphic Correlation

Graphic correlation is a method estimating first and last appearance datum (FAD and LAD) of taxa in a composite section based on all available evidence from as many sections as possible. Shaw (1964) and Miller (1977) described the method in detail. Cooper and Lindholm (1990) introduced the graphic correlation method to graptolite biostratigraphy

and produced a composite range chart based on data from 14 sections worldwide for the Lower to Middle Ordovician (Figure 6.10). Sadler (2004) improved the method and established the CONOP (constrained optimization) method for a computerized graphic correlation of the Ordovician and Silurian graptolite successions, producing a composite graptolite biostatigraphy based on nearly 2100 appearances of individual graptolite taxa (Cooper et al. 2014) and even incorporating radiometric dates (Sadler et al. 2009; Cooper & Sadler 2012; Melchin et al. 2012). The method can also be used to understand diversity changes and to develop ideas on evolutionary patterns like extinction and origination events (Sadler et al. 2011).

Graptolites and Exploration

Biostratigraphy, and with it graptolite biostratigraphy, is not a scientific method in itself and done in an "ivory tower" scientific environment, but needs to be seen in a geological context. Scientists may have to answer the questions as to why they are doing their work – establishing a graptolite biostratigraphy or working on taxonomic or evolutionary problems. One of the simplest answers obviously is the geological dating of rock sequences as we have discussed before, but graptolite biostratigraphy does not end there. Lapworth (1878) established graptolite biostratigraphy as an important tool for geological mapping, for unravelling complexly folded and faulted lithological sequences, based on the distribution of graptolite faunas. Thus, his work in the Moffat Series of Britain was a milestone in graptolite research, especially as it was the first time graptolites were used to solve a geological problem. At nearly the same time, the graptolite work of T.S. Hall (1895) and later of Robert Keble in a number of publications laid the ground for understanding the Ordovician graptolite succession of Australasia (VandenBerg & Cooper 1992). At the time, the strata-bound gold deposits of the Bendigo and Castlemaine goldfields generated an interest in the graptolite successions of Victoria, Australia (Phillips & Hughes 1996; Hegarty et al. 2003). Thus, graptolite biostratigraphy was used to understand the structural geology

Figure 6.10 Graphic correlation of Lower Ordovician graptolite sequences, composite standard sequence of FA (first appearance) events plotted against Australasian sequence (modified from Cooper & Lindholm 1990, Fig. 2).

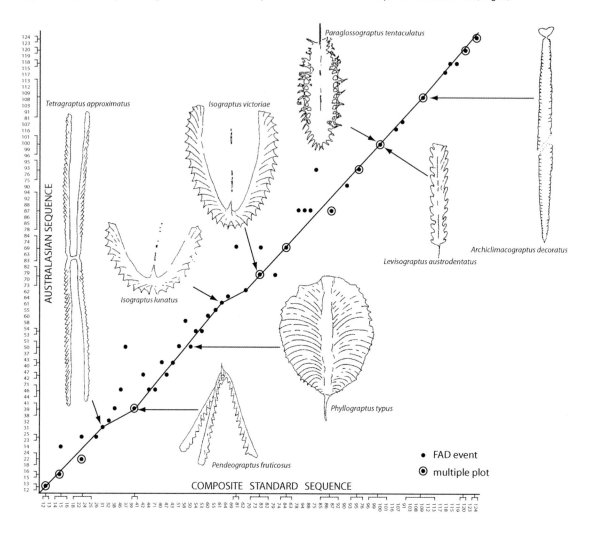

Oil and Gas

Graptolite shales have been and will be understood as oil and gas source rocks, as graptolites are made from organic material. Thus, graptolites are an important potential source of hydrocarbons, as Podhalańska (2013) stated in her research on the of the region and – more importantly – use its results for the economic exploration of one of the largest gold deposits in the world.

black shale facies of the Silurian Baltic Basin. She recognized that graptolites provide a useful tool to allow the identification of potential occurrences of unconventional hydrocarbon deposits, including shale gas. In her opinion, graptolites are one of the instruments that enable exploration of potential source rocks for hydrocarbons, including shale gas, without the need for complicated and expensive analytical studies, and equally important to the investigation of TOC (total organic carbon) contents.

The reflectance of the graptolite fusellum can be used to determine the organic maturity of sediments, as its behaviour is similar to that of other organic materials like plants, chitinozoans and scolecodonts (Teichmüller 1982; Bertrand & Heroux 1987; Bertrand 1990; Petersen et al. 2013). Thus it can be compared to vitrinite reflectance (Goodarzi & Norford 1985, 1989). Graptolite reflectance data have been compared to the conodont colour alteration index (Hoffknecht 1991), providing information on maximum overburden through the known depth–temperature gradient (Bergström 1980). Thus, the graptolite fusellum preserves information on the burial history of the entombing sediments through changes in the optical properties of the organic material. Through paleotemperature evaluation and thermal alteration, the metamorphism and overburden of the sedimentary succession can be evaluated. Most graptolite specimens appear to be dark brown to black in unaltered sediments and sediments showing only minimal burial and heating histories. Higher temperatures during burial or contact and regional metamorphosis result in lighter, silvery films of material forming the graptolite remains (see Chapter 5).

In Scandinavia and the Baltic States, early Paleozoic successions were long known to provide important economic resources (Althausen 1992). In particular the Cambrian–Ordovician Alum Shale Formation, including the early Tremadocian former "*Dictyonema* Shale" (Veski & Palu 2003), and the Middle Ordovician kukersite deposits (Bauert & Kattai 1997; Dyni 2005; Kann et al. 2013) of Estonia, were the focus of intense exploration. The richest beds of the kukersite of Estonia can reach an organic content of 40-45 wt% (Bauert 1994). Schovsbo et al. (2014) discussed the potential of the Alum Shale Formation (including the early Tremadocian "*Dictyonema* Shale" interval) as an unconventional gas play in Denmark and provided the geological model that underlies this assessment. They showed the prospective areas for gas in the Norwegian-Danish Basin.

Exploration for oil and gas also led to the recognition of a thick succession of Middle Ordovician rocks below northeastern Germany and especially the island of Rügen, and further north into the southern Baltic Sea, of which no surface exposure exists (e.g. Franke et al. 1994; Hoffmann et al. 1998). Biostratigraphical investigation of the encountered graptolite faunas (Maletz 1998a, 2001b) documented considerable tectonic distortion and a stacking of the succession – an interpretation that would have been impossible without the paleontological research. Even though the main goal of the project was not reached, the scientific community gained a lot of insight into the geology of the subsurface of the southern Baltic Sea.

Economic factors are also the driving force behind the investigation of Silurian "hot shales" in northern Africa and Arabia in recent years (Lüning et al. 2000, 2005; Loydell et al. 2013a, b). These shales, an interval biostratigraphically restricted to the Rhuddanian (*Parakidograptus acuminatus* to *Atavograptus atavus* biozones, or possibly *Coronograptus cyphus* Biozone), are important hydrocarbon source rocks. They are the origin of 80–90% of the hydrocarbons of Paleozoic age in North Africa and Arabia. The organic-rich shales were formed in laterally discontinuous basins on the northern rim of Gondwana, where their thickness was controlled by an early Silurian paleorelief shaped by glacial processes related to the Upper Ordovician glaciation event. Extensional and compressional regional tectonics also influenced the distribution of the shales. Thus, precise biostratigraphical correlation of the succession over the large area involved is important to understand the source potential of this deposit for economic exploration (Lüning et al. 2000).

Uranium Exploration

Althausen (1992) discussed prospecting for uranium in the "*Dictyonema* Shales" of Estonia after the Second World War. Similarly, uranium prospecting was carried out in the graptolitiferous Silurian successions of Thuringia, Germany, by the SAG/SDAG Wismut, producing a total of 220,000 tonnes of uranium ore between 1947 and 1990. During this time interval, East Germany (German Democratic Republic; DDR) was one of the largest producers of uranium in the world (Kämpf et al. 1995; Czega et al. 2006). The Ronneburg mining area was at one time the largest uranium mine in Europe, covering an area of 60 km^2 and extending to a depth of 940 m (Paul et al. 2002), but paleontologists will remember mainly the invaluable monograph on the Silurian graptolite fauna of Thuringia by Schauer (1971), in part collected from a number of sections in the Ronneburg area. Remediation of the Ronneburg area after the end of the uranium exploration was an enormous undertaking and led to a better understanding of uranium toxicity and environmental implications (e.g. Kistinger 1999; Merkel & Hasche-Berger 2006).

Outlook

Graptolites have provided important relative ages for rock successions since the invention of biostratigraphy. The graptolite biostratigraphy, thus developed, changed considerably through time, but has always been used for applied geological purposes. Initially described as curious artefacts, graptolites are now used for precise dating of rock sequences, help to understand complex tectonic situations in mountain building processes, and find economic mineral deposits. Thus, there is potential for future geological exploration and exploitation of graptolites, and they remain one of the most successful fossil groups in geological research. If the future in petroleum research looks for organic-rich "hot" shales in the Lower Paleozoic, the graptolites are there!

7

Taxonomy and Evolution

Jörg Maletz

Taxonomy and evolution have been controversial issues in recent years, and we are far from a resolution. It is assumed that paleontology can provide direct information of evolutionary changes from the fossil record, while modern biology is concerned only with the results of millions of years of organismic evolution. The starting point, however, is the question of how to deal with taxonomic issues, before we can talk about evolution. Linnean taxonomy organizes groups of organisms into higher-level taxa (e.g. families, orders, etc.), while nomenclature, the closely related naming convention of organisms, determines what name is suitable for a certain group. Both together form the basis for evolutionary studies, and nomenclature and taxonomy influence how we interpret the evolution of organisms. Today, cladistic methods of taxonomy and interpretation of evolution are often used in the scientific world, and some scientists have considered modifying or even eradicating the traditional hierarchical Linnean System of nomenclature and replacing it with a non-ranked system of phylogenetic nomenclature ("The PhyloCode").

Obviously, the problems are complex, leading to fierce discussions on the correct use of taxonomy and nomenclature and the associated evolutionary interpretations. But why do we need all these discussions? We need an established taxonomic system for further interpretation of our fossils. A genealogical interpretation of fossil organisms is something comparable to a family tree in humans. Without the understanding of our ancestry, we are unable to follow our family history, find relatives and understand why we are here. So if we research the ancestry and phylogenetic relationships of the graptolites, we can get answers to a number of questions. When and where did graptolites originate? What did the ancestors look like? Why are graptolites so useful for dating rock successions?

Graptolite Paleobiology, First Edition. Jörg Maletz.
© 2017 John Wiley & Sons Ltd. Published 2017 by John Wiley & Sons Ltd.

Graptolites and Taxonomy

Graptolite taxonomy leads us back to the oldest descriptions of graptolites, but these are obviously simple and often not very precise due to lack of knowledge. Therefore, they do not indicate a comprehensive understanding of the strange fossils we call graptolites. Linnæus (1735, 1768) first used the name *Graptolithus* to describe inorganic markings on rocks. This term became the origin for the currently accepted general name of our fossil group, even though it is not used any more as a genus name. It now provides the name for a whole group of organisms in the form of the Graptolithina, established by Bronn (1849) about a hundred years later. Graptolite research started seriously with the monographs of Barrande (1850) and Hall (1865), where a single family, the Graptolitidae, was enough to include all graptolites known at the time. Linnæus established the binominal naming system we use today in describing our fossils, and it is his 10th edition of the *Systema naturæ* (Linnæus 1758) that has been selected as the starting point of modern taxonomy and remains the basis to which all taxonomists have to refer.

In the last decades a concept called cladistics has entered the scene, and non-cladistic approaches to taxonomy and interpretation of evolution are often considered to be less valuable or precise. Hennig (1950, 1965, 1989) introduced the cladistic method for the taxonomy of modern insects, but it is now applied for the interpretation of various groups of extant and extinct organisms, including graptolites (e.g. Fortey & Cooper 1986; Melchin et al. 2011).

The main goal behind any taxonomy is the idea of establishing taxa with a phylogenetic meaning, such as clades or groups that show the evolutionary relationships of the included taxa. This is achieved through the use of certain characteristics in the definition called synapomorphies. Synapomorphies are characteristics derived only in the group defined by them, also called shared derived characteristics. In the past, a trial-and-error method has often been used in taxonomy. Characteristics to define taxa were based on inference, not on knowledge of their applicability. Experience showed the way to mod-

ern taxonomic ideas, but misinterpretations were common. However, the flexibility of the Linnean System helped to develop ideas and has stood the test of time.

The main problem of any taxonomic concept in paleontology is the interpretation of morphological characteristics, the only characteristics available in the fossil record. In the biology of modern, extant organisms, DNA analysis provides another source of information not available from fossils. This information can be used to interpret phylogenetic relationships based on the similarity of the DNA as an independent measure from a morphological data-set. Additional information may be gained from biological aspects of the organisms, as there is information from biogeography, ecology, and so on.

Morphological characteristics are easily misinterpreted and may misdirect us to a wrong conclusion, as it invariably depends on the precise definition or interpretation of the characteristic by the scientist. The virgellar spine of the planktic graptolites (Figure 7.1) is a good example of misinterpretation in evolutionary studies of graptolites (Maletz 2010a). Initially, it was defined as a simple spine originating from the aperture of the sicula. This interpretation was used to infer that all graptolites bearing a virgellar spine can be included in a single taxonomic group, the Virgellina (Fortey & Cooper 1986). Thus, the Virgellina were introduced as a monophyletic taxonomic unit within the graptolites. The main problem that arose from the concept was the difficulty in understanding the constructional relationships of the tubaria of the phyllograptids (Family Tetragraptidae: Chapter 10) and the Axonophora (Figure 7.1C–F), since intermediate taxa were not available (Mitchell 1990). Intermediates are also missing between the two-stiped xiphograptids (Figure 7.1B) and the axonophorans, and a phylogenetic connection was hard to establish.

The phyllograptids possess a tubarium with four stipes united back to back, forming a cross-section (Holm 1895; Cooper & Fortey 1982). The proximal development is simple, with a central sicula and an isograptid proximal development like in their ancestors, the reclined tetragraptids (Chapter 10). The early Axonophora, however, had a sicula with a low metasicular origin of the first

Figure 7.1 The virgellar spine. (A) Sigmagraptine indet., virgellar spine not present. (B) *Xiphograptus lofuensis* (Lee, 1961), dorsal virgellar spine. (C) *Archiclimacograptus* sp., juvenile with ventral virgellar spine. (D) *Archiclimacograptus* sp., proximal end with two complete thecae. (E) *Archiclimacograptus* sp., longer specimen. (F) *Saetograptus leintwardinensis* (Lapworth, 1880a), specimen showing ventral virgellar spine and branched dorsal apertural tongue. V = virgellar spine, N = nema. Scale indicated by 1 mm long bar close to each specimen.

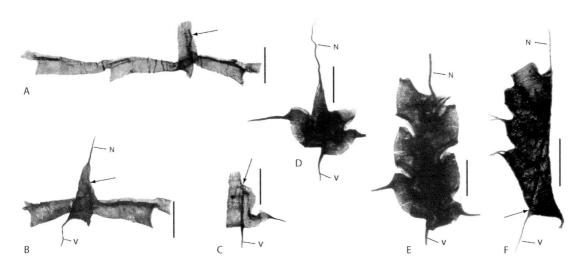

theca and a complex proximal development style, initially of a pattern U astogeny (Mitchell 1994). There were no intermediate taxa known from the fossil record to support this relationship. Therefore, a new investigation and interpretation of the virgellar spine was necessary. Numerous chemically isolated graptolites from the Lower to Middle Ordovician and a morphological analysis of many tubarium characteristics provided a surprising result: the virgellar spine originated a number of times independently in the Graptoloidea (Maletz 2010a). It was even necessary to differentiate a dorsal virgellar spine (Pterograptidae, Tetragraptidae) (Figure 7.1B) from a ventral virgellar spine (Axonophora) (Figure 7.1C–F), depending on the position of the spine on the dorsal or ventral side of the sicular aperture.

Nomenclature

A naming convention or nomenclature is quite important in taxonomy as a means of communication, but should not be confused with the taxonomy itself. The Linnean System provides a number

of ranks for the differentiation of taxa and the inclusion of species in successively larger units: the genus, family, order and phylum (Table 7.1). The individual taxa are included based on morphological characteristics of supposed monophyletic origin. Even though the term "monophyletic" is a modern invention, it is obviously the goal of even the earliest taxonomic system to sort organisms according to evolutionary novelties, showing phylogenetic relationships. At first, the approach was used by inference, but with time and accumulation of knowledge, the system improved in its reliability. Even our modern computer-aided cladistic analyses are based on the concepts developed through the work of numerous scientists trying to understand evolutionary patterns in all available groups of modern and fossil organisms.

The more recently developed system of phylogenetic nomenclature (PN) uses phylogenetic definitions based explicitly on cladistic diagrams, not on types and diagnoses of taxa as the Linnean system does. PN intends to get rid of the Linnean system of ranked taxa (see Cantino et al. 1999; Cantino 2004; de Queiroz 2006, 2007) and replace it with a system of rankless units. Some recent

Table 7.1 The taxonomy of the Pterobranchia (from Maletz 2014a).

Phylum **Hemichordata** Bateson, 1885, p. 111
  Class **Enteropneusta** Gegenbaur, 1870, p. 158
  ?Class **Planctosphaeroidea** van der Horst, 1936, p. 612
  Class **Pterobranchia** Lankester, 1877, p. 448
    Subclass **Cephalodiscida** Fowler, 1892, p. 297
        Family **Cephalodiscidae** Harmer, 1905, p. 5
    Subclass **Graptolithina** Bronn, 1849, p. 149
        Incertae sedis Family **Rhabdopleuridae** Harmer, 1905, p. 5
        Incertae sedis Family **Cysticamaridae** Bulman, 1955, p. 42
        Incertae sedis Family **Wimanicrustidae** Bulman, 1970, p. 52
        Incertae sedis Family **Dithecodendridae** Obut, 1964, p. 295
        Incertae sedis Family **Cyclograptidae** Bulman, 1938, p. 22
      Order **Dendroidea** Nicholson, 1872b, p. 101
        Family **Dendrograptidae** Roemer in Frech, 1897, p. 568
        Family **Acanthograptidae** Bulman, 1938, p. 20
        Family **Mastigograptidae** Bates & Urbanek, 2002, p. 458
      Order **Graptoloidea** Lapworth, 1875, in Hopkinson & Lapworth 1875, p. 633
        Suborder **Graptodendroidina** Mu & Lin in Lin, 1981, p. 244
          Family **Anisograptidae** Bulman, 1950, 79
        Suborder **Sinograpta** Maletz et al., 2009, p. 11
          Family **Sigmagraptidae** Cooper & Fortey, 1982, p. 257
          Family **Sinograptidae** Mu, 1957, p. 387
          Family **Abrograptidae** Mu, 1958, p. 261
        Suborder **Dichograptina** Lapworth, 1873, table 1, facing p. 555
          Family **Dichograptidae** Lapworth, 1873, p. 555
          Family **Didymograptidae** Mu, 1950, p. 180
          Family **Pterograptidae** Mu, 1950, p. 180
          Family **Tetragraptidae** Frech, 1897, p. 593
        Suborder **Glossograptina** Jaanusson, 1960, p. 319
          Family **Isograptidae** Harris, 1933, p. 85
          Family **Glossograptidae** Lapworth, 1873b, table 1 facing p. 555
        Suborder **Axonophora** Frech, 1897, p. 607
        Infraorder **Diplograptina** Lapworth, 1880e, p. 191
          Family **Diplograptidae** Lapworth, 1873b, table facing p. 555
            Subfamily **Diplograptinae** Lapworth, 1873b, table facing p. 555
            Subfamily Orthograptinae Mitchell, 1987, p. 380
          Family **Lasiograptidae** Lapworth, 1880e, p. 188
          Family **Climacograptidae** Frech, 1897, p. 607
          Family **Dicranograptidae** Lapworth, 1873b, table facing p. 555
            Subfamily **Dicranograptinae** Lapworth, 1873b, table facing p. 555
            Subfamily **Nemagraptinae** Lapworth, 1873, p. 556
        Infraorder **Neograptina** Štorch et al., 2011, p. 368
          Family **Normalograptidae** Štorch & Serpagli, 1993, p. 14
          Family **Neodiplograptidae** Melchin et al., 2011, p. 298
         Superfamily **Retiolitoidea** Lapworth, 1873b, table 1 facing p. 555
          Family **Retiolitidae** Lapworth, 1873b, table 1 facing p. 555

Table 7.1 (Continued)

Subfamily **Petalolithinae** Bulman, 1955, p. 87
Subfamily **Retiolitinae** Lapworth, 1873, table 1 facing p. 555
Superfamily **Monograptoidea** Lapworth, 1873, table facing p. 555
Family **Dimorphograptidae** Elles & Wood, 1908, p. 347
Family **Monograptidae** Lapworth, 1873b, table 1 facing p. 555
possibly several subfamilies

graptolite workers combined cladistics with a Linnean system of ranked taxa, an approach that led to a considerable increase in used taxonomic levels (e.g. Mitchell et al. 2007a; Maletz et al. 2009; Melchin et al. 2011), a praxis commonly used in systematics. Phylogenetic nomenclature uses the results of cladistic analyses and depicts the results in diagrams in which characteristics are sorted one by one in the succession of their first appearance, which represents an inherent ranking system. The differences between these two seemingly diametrically different concepts are thus more philosophical in aspect than of real practical importance (Nixon et al. 2003). Both cladistics and Linnean taxonomy rely on homologous characteristics to differentiate or sort taxa, in short, homologies, to reliably represent phylogenetic relationships. Each rank in a Linnean taxonomy is based on an interpretation by a specialist, and thus is subjective and influenced by personal opinions (see discussion in Benton 2007), as is any unranked node in a cladistic analysis.

Monophyly in Graptolite Taxonomy

The idea of naming only clades (monophyletic groups) and not grades (paraphyletic or polyphyletic groups) is implemented and strongly promoted by cladistics and in phylogenetic nomenclature. However, it is not a new idea, and it has been the underlying, even though rarely explicitly stated, aim of every taxonomic approach and every evolutionary interpretation since the introduction of the Linnean system (e.g. Haeckel 1866, 1868; Gegenbaur 1870, p. 78–81). At the beginning of taxonomic and evolutionary research on graptolites, the knowledge and understanding of synapomorphic characteristics (an unknown

term at the time of Linnæus) was just starting to emerge. Taxonomy was a "trial-and-error" system, using characteristics that appeared important and meaningful. This is easily visible in early graptolite work (e.g. Lapworth 1873b; Tullberg 1883; Gürich 1908), when number of stipes and uniseriality or biseriality of the tubaria were used as main characteristics for taxonomic differentiation. Many of the graptolite genera described in these taxonomies were identified as polyphyletic early on, as shown by the statement of Nicholson and Marr (1895, p. 538) that "the single genus *Monograptus* may contain descendants of more than one family". It was a basic fact to Ruedemann (1904, p. 478), who stated that "Their results point also to a polyphyletic origin of the large genera of this family and especially of *Tetragraptus* and *Didymograptus*" among others. Every specialist on these graptolite taxa would have to agree with the statement of Ruedemann. Jaeger (1978) discussed the trends ("Entwicklungszüge") in the evolution of graptolites following similar ideas, but clearly stated that the trends are descriptional and that identical patterns appeared often independently in various groups. Thus, he did not emphasize a phylogenetic meaning of these trends.

Graptolite Cladistics

The taxonomy and evolution of the Graptolithina has been hotly debated in recent decades, since cladistic analyses became increasingly popular and a number of groups within the Graptolithina have been investigated using cladistics (Fortey & Cooper 1986; Mitchell 1987; Bates et al. 2005; Mitchell et al. 2007a; Maletz et al. 2009; Melchin et al. 2011; Štorch et al. 2011). Cladistics helped us to understand the general and evolutionary relationships of a number of graptolite clades, but a complete analysis of all graptolite taxa is far from being even attempted. Important results from the available cladistic analyses include the recognition of the Anisograptidae as the ancestors of all planktic graptolites, and their inclusion in the Graptoloidea (Fortey & Cooper 1986) and the evolution of the axonophorans (Maletz et al. 2009). Through recognition of the proximal development types of the axonophoran graptolites by Mitchell (1987) and Melchin (1998), we reached a better understanding of part of the large group of the biserial graptolites, their taxonomy and phylogeny. The most recent improvement is the recognition of *Rhabdopleura* as an extant graptolite (Mitchell et al. 2013) (Figure 7.2), preceded by a similar suggestion by Beklemishev (1951a, b [various later editions in Russian, English and German]) who included the pterobranchs in the class Graptolithoidea. This result includes the notion

that graptolites are alive and not entirely extinct, an important aspect for the interpretation of fossil graptolites in a biological context.

More than 600 graptolite genera have been described in the past (Maletz 2014a), but relatively few are known in enough detail to be useful for any modern phylogenetic analysis. Therefore cladistic approaches are limited to groups like the Silurian retiolitids (Lenz & Melchin 1997; Bates et al. 2005) and the Ordovician to Lower Silurian axonophorans (Mitchell 1987; Mitchell et al. 2007a; Melchin et al. 2011). Enough material of these groups is available, yielding the morphological details necessary for a reasonable analysis.

In the analysis of modern taxa, the cladistic diagram provides information on hypothetical ancestors through the sister–group relationships, but what happens if we add another dimension, the fossil record, to this analysis? If we can pinpoint the ancestor directly from the fossil record? With the overwhelming influence of cladistics in modern taxonomy, focusing on sister–group relationships, little has been done in many years on the understanding of ancestor–descendant relationships (Dayrat 2005), but recently the issue has gained more interest (see Gavryushkina et al. 2014; Tsai & Fordyce 2015). Graptolite specialists commonly looked into the evolutionary origin of taxa to a defined ancestor, available from the fossil record (e.g. Sudbury 1958; Rickards et al. 1977; and many more). Ancestor–descendant relationships have been shown besides cladistic sister–group relationships as phylogenetic interpretations, highlighting the difference between both concepts. Lenz and Melchin (1997, Fig. 4), for example, showed a phylogenetic tree of the Retiolitidae based on a cladistic analysis (Lenz & Melchin, 1997, Fig. 3). Urbanek et al. (2012, Fig. 1) discussed the complex evolution of the *Pristiograptus dubius* group and provided a phylogenetic tree showing the iterative evolution of new taxa from a stem lineage (Figure 7.3). In this case, however, a cladistic analysis was not undertaken, but the interpretation was supported by a careful morphological analysis of the individual taxa.

Care should be taken in naming clades and nodes based on cladistic results and any additional knowledge of the taxa involved. A cladistic tree

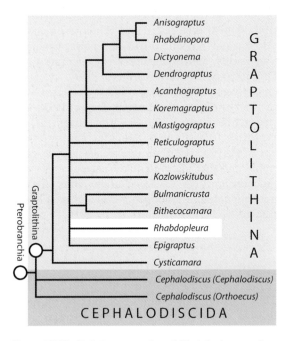

Figure 7.2 Cladistic interpretation of *Rhabdopleura*, strict consensus of 12 equally parsimonious trees obtained from the full 17 taxon set, showing the position of extant *Rhabdopleura* inside the clade of the Graptolithina (based on Mitchell et al. 2013).

Figure 7.3 The evolution of the *Pristiograptus dubius* lineage (based on Urbanek et al. 2012). Illustrated specimen is *Pristiograptus dubius frequens*.

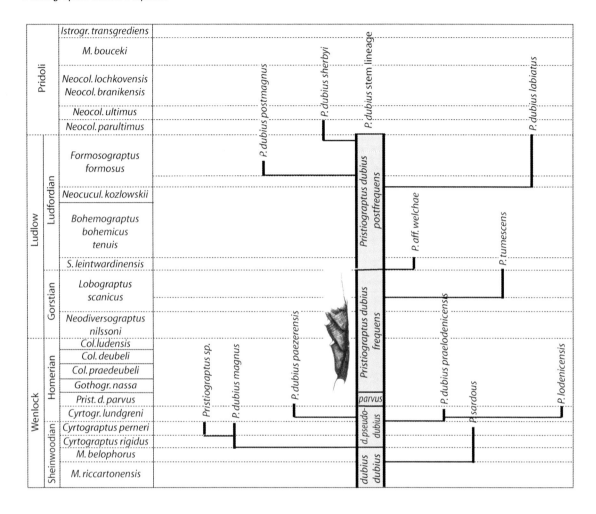

includes numerous nodes at which taxa potentially can be defined, and an excessive naming convention (as done in Maletz et al. 2009) could flood us with unnecessary new names (Maletz 2014a). It is not necessary to name all clades or nodes in a cladistic diagram, even though these nodes may provide important additional taxonomic information. Therefore, Melchin et al. (2011) preferred to use a system with fewer taxonomic levels. The taxonomy of the Graptolithina is certainly not settled and agreed with all graptolite taxonomists, and further discussion is necessary. Taxon names for more primitive and more derived taxa following Maletz et al. (2009) may be useful to describe inclusive groups such as the Axonophora and Pan-Axonophora in an evolutionary phylogeny. The cladistic concepts can here help to understand phylogenetic relationships quite precisely. Then, however, the naming convention may produce strange results, at least for the conservative taxonomist. For example, Mitchell (1987) included Middle Ordovician biserials in his order Monograptina, a decision not easily followed by all graptolite specialists. Thus, Štorch et al. (2011) decided to rename the group as the Neograptina to avoid misunderstanding and confusion.

Relationships of the Major Graptolite Groups

Graptolites have long been investigated and certain phylogenetic relationships have been worked out. Thus, the general patterns of graptolite evolution are known, but many details of the macro-evolution of the graptolites still need investigation or confirmation. A comprehensive cladistic analysis of the whole graptolite clade has never been attempted, and may be difficult due to the heterogeneous quality of the paleontological data and the research centred upon a few important groups. Thus, the cladistic analyses so far cover only a patchy landscape of graptolite diversity and are not able to produce a complete and coherent picture of the whole clade. Mitchell et al. (2013) analysed the benthic graptolites and included the modern pterobranchs, with the result that the authors were able to identify the extant *Rhabdopleura* as an extant graptolite. Unfortunately, the internal resolution of the clade of the benthic graptolites is low due to the lack of morphological information for many taxa, and the formerly established orders (e.g. Kozłowski 1949; Bulman 1970a; Bates & Urbanek 2002) have not been verified. The results indicate some differentiation, but also show that many of the high-level taxonomic units (orders) of Kozłowski (1938, 1949) may be unnecessary and the benthic taxa are in dire need of a modern taxonomic analysis.

Fortey and Cooper (1986) were the first to attempt a cladistic analysis of the graptoloids, suggesting a monophyletic origin of the planktic graptoloids, but also probably a polyphyletic differentiation of the bithecate Anisograptidae into the non-bithecate groups. Maletz et al. (2009) revised the Lower to Middle Ordovician multiramous to pauciramous graptoloids, but did not cover the origin and diversification of the Anisograptidae in detail, as the authors were interested in the non-bithecate taxa only. Some general trends were apparent from the analysis, the most important one being the differentiation of the Sinograpta and the Dichograptina.

Detailed analyses of the axonophorans graptolites (Mitchell 1987; Melchin 1998; Melchin et al. 2011) provided the basis for the understanding of the biserial graptolites and their proximal end construction. The analysis of Melchin et al. (2011) also provided additional insight into the evolution of the axonophorans and especially into the origin of the monograptids (Family Monograptidae) through a dimorphograptid ancestor. The evolutionary patterns of the Monograptidae have never been investigated in detail, but results are presented by following individual monograptid lineages (compilation in Rickards et al. 1977), and only a single, unpublished cladistic analysis of lower Silurian monograptids exists (Muir 1999).

Extinction Events and Radiations

Graptolite diversity varies considerably through the Paleozoic time interval (Sadler et al. 2011; Cooper et al. 2014) and quite a number of extinction events have been established (Figure 7.4). The number may, however, vary depending on the stratigraphic resolution used for the investigations (see Cooper et al. 2014, Fig. 2), and many of the extinction events are based on the investigation of local successions. The largest of these extinction events in the Ordovician was the Hirnantian or *Pacificus* Event (Koren 1979, 1991a; Melchin & Mitchell 1991; Chen et al. 2003, 2005), when very few species barely survived a major glaciation. Bapst et al. (2012) showed that during this mass extinction event, graptoloid diversity became decoupled from changes in disparity. As a result, the numerous species evolving during this interval are extremely difficult to identify (see Štorch et al. 2011). The Hirnantian event was one of the five great mass extinctions, and graptolites were not the only group of organisms affected so severely (see Webby et al. 2004a). All warmer water or tropical graptolite faunas went extinct and only a few cold-water generalists survived. Other Ordovician extinction events of the graptolites are of a lesser magnitude or have not been investigated in enough detail. A possible extinction event at the base of the Upper Tremadocian (see Sadler et al. 2011, Fig. 13) needs further verification, and the "Darriwilian (Dw3) depletion" of Sadler et al. (2011) is pronounced, but also was never investigated in detail. Štorch (1995)

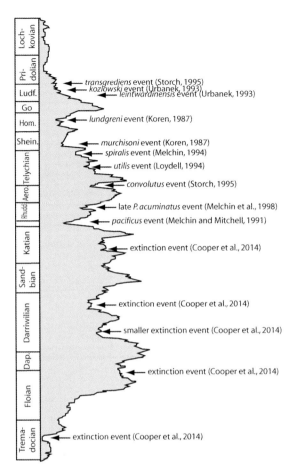

Figure 7.4 Extinction events in graptolite history, based on Cooper et al. (2014).

During the Lower Devonian, graptolite faunas were reduced in diversity and disparity. According to Koren (1979), the Devonian graptolite faunas show a substantial reduction in geographical distribution, and a gradual but not uniform mode of extinction, sometimes associated with a great abundance of specimens, but the taxonomic and morphological diversity was considerably reduced.

One of the surprising results of the work of Cooper et al. (2014) is the difference in the tempo of graptolite evolution between the Ordovician and Silurian. Graptolite species durations in the Silurian (0.69 Ma) appear to be much shorter than durations in the Ordovician (ca. 1.27 Ma) and led to higher turnover rates. These differences are also reflected in the length and spacing of graptolite biozones in the Ordovician and Silurian. It is simple to see on correlation charts (e.g. Loydell 2012) that the Silurian is much more densely zoned with graptolites than the Ordovician. The reason for these differences is not yet explored and needs some consideration in the future.

Evolutionary Lineages

Numerous evolutionary studies have been made of the graptolites to document both micro- and macro-evolutionary patterns. These follow the idea that a densely sampled fossil record through time traces the evolutionary changes of the taxa and provides an evolutionary lineage of a fossil group, a concept termed stratophenetics (e.g. Gingerich 1979, 1990). Elles' (1922) study of graptolite evolution based on the British faunas is one of the most comprehensive early overviews, and only one further study exists that covers the Silurian monograptids in detail. Rickards et al. (1977) supported a few of the lineages of Elles (1922), but largely came to different conclusions and illustrated quite a number of evolutionary lineages based on biostratigraphical ranges and constructional analyses of the tubarium development of the involved species. The work of Sudbury (1958) shows the evolutionary patterns established from a species-to-species comparison (Figure 7.5), and demonstrates a good example of evolutionary studies of Silurian grap-

compiled the extinction history of the Silurian monograptids and recognized eight extinction events in the Prague basin. The size of these events is still only in part explored, and many of these may either be artefacts of preservation in certain regions or represent minor events in graptolite history. The *Lundgreni* Event or the "Great Crisis" of Jaeger (1991), however, may be one of the major events and driving the final chapter of graptolite evolutionary history. At this level, most of the monograptids and retiolitids of the Wenlock disappeared, and through a short interval, very few surviving taxa can be found, before a new diversification took place.

Figure 7.5 The evolution of the *Demirastrites triangulatus* group, modified from Sudbury (1958) and Urbanek (1960).

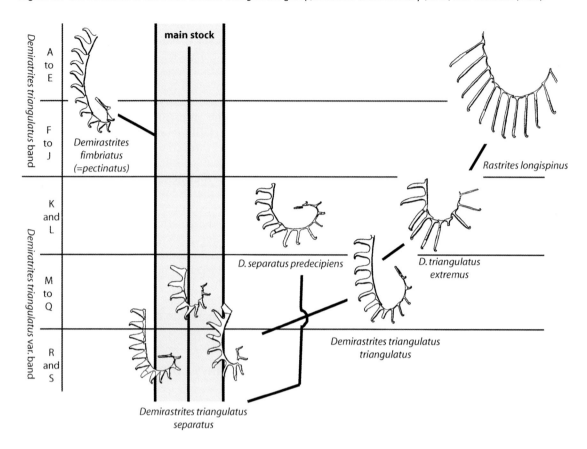

Demiratrites triangulatus band

A to E

F to J

K and L

Demiratrites triangulatus var. band

M to Q

R and S

main stock

Demirastrites fimbriatus (=pectinatus)

Rastrites longispinus

D. separatus predecipiens

D. triangulatus extremus

Demirastrites triangulatus triangulatus

Demirastrites triangulatus separatus

tolites. Urbanek (1960) attempted a detailed analysis of graptolite colonies and used a number of Silurian examples to show morphophysiological gradients in the astogenetic succession of thecae. Štorch and Loydell (1992) used the biostratigraphical ranges of the rastritids of the *Rastrites linnaei* group to document the origin of the group from *Rastrites hybridus* and the subsequent evolutionary diversification of this clade in the Llandovery.

A complex history of the evolution of the Upper Silurian to Lower Devonian monograptids can be seen through the micro-and macro-evolutionary investigations of the Cucullograptinae, Neocucullograptinae, Linograptinae and Neocolonograptinae by Urbanek (1963, 1966, 1995, 1997a, b). Urbanek described a number of subfamilies in his papers and derived them from

a *Pristiograptus* stock ancestor at various times in the Upper Silurian (e.g. Urbanek 1997a, Fig. 6), leading to distinct radiations and diversification events.

Convergent Evolution

If we look at the early understanding of graptolites, their taxonomy and evolution, we see a common theme: the misunderstanding of certain common characteristics. It starts with the number of stipes and ends with the style of thecae, but early on researchers understood that behind this apparent problem was a complex evolutionary history (Nicholson & Marr 1895), and that many of the characteristics used initially for graptolite

Figure 7.6 Convergent evolution of uniserial, single-stiped Ordovician and Silurian graptoloids. (A) *Jishougraptus novus* Beckly & Maletz, 1991. (B) *Azygograptus suecicus* Moberg, 1892. (C) *Nicholsonograptus fasciculatus* (Nicholson, 1869). (D) *Pseudazygograptus incurvus* (Ekström, 1937). (E) *Atavograptus ceryx* (Rickards & Hutt, 1970). Bars show the ranges of genera, not the species shown as examples. Reconstructions not to scale.

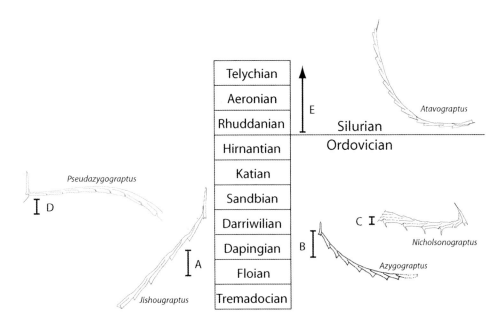

taxonomy were not homologous, but evolved independently more than once. This problem of recognizing homologies in graptolites is a good indicator of the common presence of convergent evolution as a basic theme in graptolite evolution. In this respect, it is surprising that convergent evolution has not been considered a typical expression in graptolite evolution and has rarely been demonstrated conclusively. Lenz and Melchin (2008) showed a remarkable example of convergent evolution in monograptids, comparing the construction of the planispirally coiled tubaria of *Cochlograptus veles* and *Testograptus testis* with their typical abrupt deflection of the post-sicular region of the tubarium, and the presence of a pseudovirgula.

Is the evolution of the uniserial tubarium, the monograptid condition, the indication of a mono-phyletic event, or is it an expression of convergent evolution? We now know that single-stiped, uni-serial colonies originate several times independently in the planktic Graptolithina (Figure 7.6). They first appear in the Floian (Lower Ordovician)

to Dapingian (Middle Ordovician) with the sigma-graptine genera *Azygograptus* and *Jishougraptus*. In the Darriwilian (Middle Ordovician) we have the sinograptid *Nicholsonograptus*, and in the uppermost Darriwilian to Sandbian (Upper Ordovician) the dicranograptid *Pseudazygograptus* appears, but the truly successful Monograptidae only appear in the basal Silurian. When we look at the construction of all these groups in detail, we recognize clearly that they develop independently and from different ancestors – a picture book example of convergent evolution (Figure 7.6).

The recognition of homologous characteristics in graptolites is so difficult because the mode of production of the originally tubular thecae, from fusellar halfrings through the zooids, strongly restricts the variation of the available features for the tubarium. These constructional limitations cannot be overcome, and thus many features evolved again and again independently during the evolutionary history of the graptolites. This is not only true for the construction of the virgellar spine (Maletz 2010a), but also for many other features,

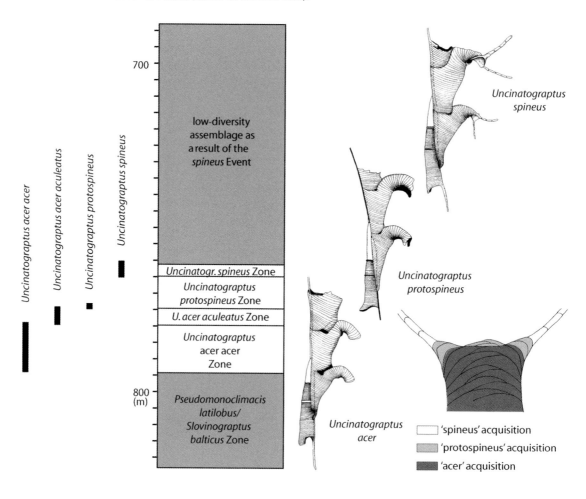

Figure 7.7 The micro-evolutionary changes of *Uncinatograptus* in the Ludfordian, Upper Ludlow, Silurian, from the Mielnik-1 borehole section of Poland (based on Urbanek 1995).

including the numerous apertural modifications of the monograptids. Urbanek (1995) documented the transition from the non-spinose *Uncinatograptus acer* to the spinose *Uncinatograptus spineus* through the intermediate *Uncinatograptus protospineus* by the successive addition of lateral lobes and eventually of paired lateral, apertural spines along a biostratigraphical gradient as a typical micro-evolutionary transition in late Ludfordian monograptid graptolites (Figure 7.7). The documented transition shows without doubt that the hooded, spinose thecal apertures in *Uncinatograptus* originated separately and independently from a constructionally identical aper-

tural development in the older *Monograptus priodon-flemmingi* lineage. Thus, a precise and detailed constructional and evolutionary analysis is necessary to recognize homologous characteristics in graptolite colonies and to understand the micro- and macro-evolution of the graptolites.

Outlook

Taxonomy and evolution are scientific concepts that are constantly in motion. Taxonomy changes with the addition of each new species and new information on anatomy or genetic aspects of an organism. It also changes with scientific paradigm changes, general concepts on taxonomy and nomenclature, and thus it is difficult to say which way graptolite taxonomy will go in the future. Certainly, cladistics will have a considerable influence on the further development of taxonomic ideas for graptolites and their evolution. The lack of interest of the specialists in certain graptolite groups (e.g. the benthic taxa), but also and maybe more importantly the lack of suitable material, will probably prevent a complete cladistic analysis of the graptolites, even though it is sorely needed. In other biostratigraphically relevant groups, analyses will certainly develop our taxonomic concepts for the future. Thus, a divergence in the understanding of the various clades of graptolites will increase, unfortunately, and with this, certain aspects in graptolite evolution may never be explored. The focus will be on the graptolite groups with a decent fossil record, and these will become picture-book examples for phylogenetic analyses.

8 Bound to the Sea Floor: The Benthic Graptolites

Jörg Maletz

The sea floor: a vast region of the modern oceans and home to numerous organisms, plants and animals, actively moving around or passively enjoying the bottom of the sea and waiting for food to be served. This was the home of the benthic graptolites in the early Paleozoic and it is the home of its few survivors. Here they originated and here they evolved into the first group of diverse colonial organisms the world has seen. Prior to the evolution of the corals and modern reef systems, the graptolites colonized the shallow sea bottoms worldwide and enjoyed the wide expanses of their chosen environment. Somewhere here also the exploration of the water column of the oceans and the conquering of the planktic realm started, but this was still far in the future, somewhere in latest Cambrian times, when the world changed again for the graptolites.

Exploration of the Paleozoic seas led the graptolites to their first bloom, a still strongly underestimated expansion of their diversity, poorly documented in the sediments of the time. Only a few scientists have been able to have a glimpse into this forgotten world of the Cambrian sea floor. It is due to the work of Roman Kozłowski (1899–1977) that this important episode in graptolite evolution is not completely overlooked and lost. So little is known, but the Tremadocian rocks chemically dissolved by Roman Kozłowski provided evidence of complex life and a diversity of benthic graptolites that has not been recognized in any later periods of geological history. Whether this is due to chance – preservation of faunas in unusual environments – or showing a "normal" situation, an undiscovered extinction of the early benthic graptolites, is unclear and the answer to this question may have to wait for future research.

Graptolite Paleobiology, First Edition. Jörg Maletz.
© 2017 John Wiley & Sons Ltd. Published 2017 by John Wiley & Sons Ltd.

"Rooting" the Graptolite Colony

All benthic graptolites are firmly attached to the sea floor, either with the whole weight of their colonies as encrusting organisms spreading on any hard substrate, or with an attachment disc, a small patch of tubarial material with which they anchor themselves to the sediment or rock surface. The general colony organization of the benthic, encrusting and erect-growing graptolites is similar to that of bryozoans, and the terminology for sessile colonial organisms as discussed by Jackson (1979) can easily be adopted. Jackson differentiated runners, sheets, mounts, plates, vines and trees as the six types for bryozoan colonies, of which only runner, sheets, vines and trees can be identified in benthic graptolites, indicating that these represent a group of organisms with different requirements and constructional limitations.

Extant rhabdopleurids are attached to corals, hydroids, rock surfaces or shells. They form a creeping main stem from which erect-growing, unbranched thecal tubes originate. We differentiate the thigmophilic ("touch-loving") (Stebbing 1970b, p. 212) or sheet-type *Rhabdopleura compacta*, densely covering small areas, and the runner-type *Rhabdopleura normani* loosely covering much larger areas. In a similar way, Cambrian benthic graptolites began to cover the sea floor with their encrusting colonies and later started to invade the water column, forming erect, bushy to tree-like shapes (Plate 1B–C).

The origin of the colonial lifestyle of the graptolite is uncertain, but probably their ancestors were similar to the extant *Cephalodiscus*, a benthic non-graptolitic pterobranch. Modern cephalodiscids already show many of the tubarium shapes employed by their colonial relatives. Their tubarium shapes range from dense mats of individual tubes to erect-growing, branched and bush-like shapes (see Chapter 1; Figure 1.2), but their zooids are not interconnected at the mature stage as they are in *Rhabdopleura*, and therefore they are not called graptolites (Mitchell et al. 2013).

The oldest colonially interconnected zooids of a graptolite are found in the Middle Cambrian (Figure 8.1), and comparable tubaria can be recognized in many regions of the world (Maletz &

Steiner 2015). A number of species of the colonial *Sphenoecium* were present in shelf regions of the continents of Australia, Baltica and Laurentia, indicating a possibly worldwide distribution even during this time period. The records from these widely separated continents indicate that graptolites even then were successfully occupying and exploring the world's oceans. They were able to move around and most likely populate the shelf regions of all Cambrian continents through their swimming larval stages, as do the extant rhabdopleurids. Either the evolutionary differentiation of early graptolites was extremely fast, or they experienced a long period of evolutionary changes not documented in the fossil record.

These early benthic encrusting graptolites were widely distributed in the Cambrian Series 3 (Figure 8.1), but have rarely been identified as graptolites. Often they were described as hydroids (e.g. Chapman 1917, 1919; Chapman & Thomas 1936), since information on fusellar structure was not available. One of the oldest species is *Sphenoecium obuti* (Durman & Sennikov, 1993) from the Middle Cambrian (Drumian Stage) of Siberia, initially described under the name *Rhabdopleura obuti* as the oldest rhabdopleurid and regarded as an argument to call the genus *Rhabdopleura* a "living fossil". The specimens show beautifully preserved fuselli, indicating their graptolitic nature. Maletz et al. (2005b) also demonstrated fusellar construction for other Middle Cambrian specimens from the Czech Republic and from North America. Even though specimens of *Sphenoecium wheelerensis* from the Spence Shale of North America (Plate 1B) are often poorly preserved and weathered, replacement minerals and imprints of the fusellum can demonstrate the fusellar nature of the erect thecal tubes (Figure 8.2).

Benthic graptolites need a hard surface for attachment, as they have to be permanently fixed to the substrate. The surface can be everything from a firm sediment or rock surface to parts of organisms, shells, bones or other skeletal remains. Benthic encrusting taxa will use the basal sides of their thecal tubes for attachment and do not need any special developments. The tubes can be anchored with a bit of cortical tissue to any surface

Figure 8.1 Cambrian Series 3, biostratigraphy and pterobranch faunas. (A) *Sphenoecium robustus* (Maletz et al., 2005b), Luh, Czech Republic. (B) *Yuknessia simplex* Walcott, 1919, Burgess Shale, holotype. (C) *Sphenoecium wheelerensis* Maletz & Steiner, 2015, Spence Shale, Utah. (D) *Sphenoecium mesocambricus* (Öpik, 1933), Alum Shale, Sweden (from Bengtson & Urbanek 1986). (E) *Sphenoecium discoidalis* (Chapman & Thomas, 1936), Heathcote fauna, Tasmania. (F) *Sphenoecium wheelerensis* Maletz & Steiner, 2015, Wheeler Shale, Utah. (G) *Sphenoecium robustus* (Maletz et al., 2005b), Konicek, Czech Republic. (H) *Sphenoecium wheelerensis* Maletz & Steiner, 2015, Marjum Fm., Utah. (I) *Sphenoecium mesocambricus* (Öpik, 1933), Krekling, Norway. Illustration of specimens not to scale.

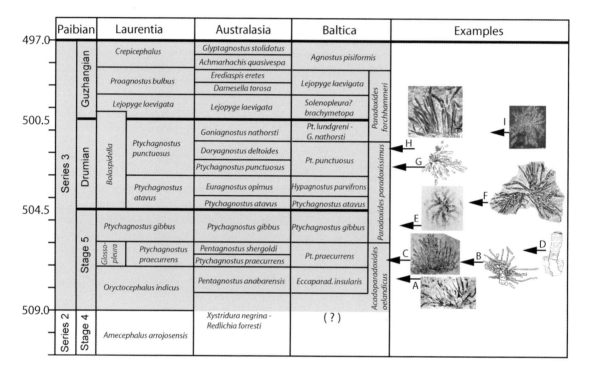

they choose. Later during the growth of the colony, additional cortical material can strengthen the attachment if necessary.

Erect-growing colonies have a stronger need for firm attachment, as the water currents may pull at them and a reliable attachment is crucial for their long-term survival. They initially attach the developing colony by the base of the erect tubular sicula, a small surface area that may not be enough to support a larger graptolite colony. Thus, a distinct attachment disc is often formed from cortical material secreted by the zooids to anchor the colony safely to the ground. Still, this attachment disc may be ripped off from the rock and is preserved in some fossilized specimens (Figure 8.3). Details of the attachment disc of bushy dendroid forms are only known from a few species, as the preserved fossil material is generally very fragmentary. Specimens preserved in life orientation possess considerable cortical overgrowth covering the initial attachment site, and do not show many details of the development of the attachment structures. The attachment structure can be described as a flat-bottomed, dendritic holdfast structure for most erect-growing dendroids (e.g. Wiman 1896b: *Dictyonema cavernosum*). A "root system" extending into the sediment has never been described for benthic graptolites and, if the pterobranch model is correct, it is unlikely that it existed. As the attachment is invariably produced as a secretion of cortical tissue by the zooids of the colony and not as a growing root like in plants, an extension into the sediment is not possible.

Figure 8.2 The Middle Cambrian graptolite *Sphenoecium wheelerensis* Maletz & Steiner, 2015. (A–C) KUMIP 204381, Spence Shale, Utah, USA, complete specimen and details showing impressions of fusellar construction. (D–E) WHE-001, Wheeler Shale, Utah, USA (see Maletz et al. 2005b). B–E are SEM-BSE photos showing the fusellum in black and providing evidence of the fusellar construction.

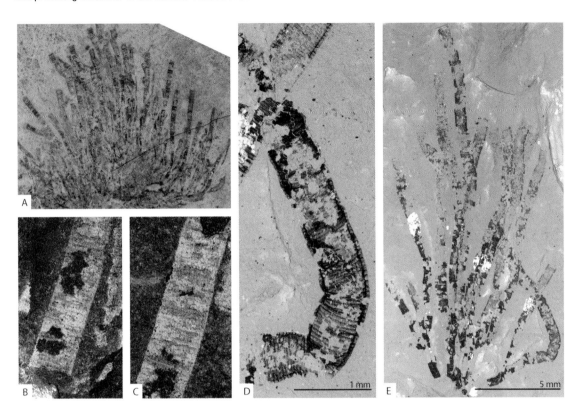

The Graptolithina

When we talk about the Graptolithina, we include all taxa, both benthic and the derived planktic graptolites. Mitchell et al. (2013) used the characteristic of "serial budding from an interconnected stolon system" to identify the Graptolithina and to separate them from the non-colonial Cephalodiscida. Interestingly, the benthic graptolites already include a high diversity of genus-level taxa in the Lower Paleozoic, but are poorly known and mostly ignored by taxonomists. Their taxonomy and phylogenetic relationships are not well known. Maletz (2014a) recognized five groups of largely encrusting graptolite groups as *incertae sedis* and did not assign them to a particular order within the subclass Graptolithina.

Family Rhabdopleuridae

The Rhabdopleuridae (Harmer 1905) includes the best-known benthic graptolites and the only living members of the Graptolithina. Close relatives are already present in the Middle Cambrian, but have recently been referred to the genus *Sphenoecium* (Maletz & Steiner 2015). A number of species are found on several paleocontinents, from Laurentia to Baltica, Australia and possibly Asia. They all share a creeping part from which erect, unbranched tubes grow (Figures 8.2A, 8.4). These erect tubes are interpreted as the actual thecal tubes in which the zooids live. Maletz et al. (2005b) documented the fusellar construction in a single specimen from the Middle Cambrian of Utah, identified at the time as

Figure 8.3 Colony shape and attachment in erect, benthic graptolites. (A) *Dictyonema* sp., conical colony without stem. (B) *Dictyonema* sp. without stem. (C) *Dendrograptus* sp., colony with robust stem, thecae not visible. (D) *Dictyonema* sp., specimen with long, thecate stem. (E) *Dendrograptus* sp., bushy colony without stem. (F) *Dictyonema cavernosum* Wiman, specimen with irregular attachment disc. Reconstructions based on Bulman (1928) and Chapman et al. (1996), not to scale.

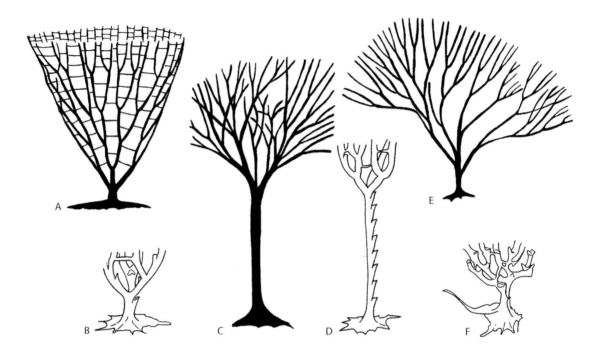

?*Cephalodiscus* sp. (Figure 8.2D–E), but now identified as *Sphenoecium wheelerensis*. Runner-type rhabdopleurids are present at Monegeeta, Victoria, Australia, and have been identified as *Archaeolafoea* in the past (Chapman 1919). The material shows a clearly encrusting taxon with meandering creeping main tubes and possibly isolated lateral tubes (Figure 8.4). Their colony shape is identical to that of the modern *Rhabdopleura normani* type (Figure 3.3A).

Camaroids and Crustoids

Kozłowski (1949, 1962) erected the orders Camaroidea and Crustoidea for these poorly known taxa. Mitchell et al. (2013) did not use the upper level taxonomy of Kozłowski, as accepted by Bulman (1970a), and Maletz (2014a) included the taxa in the family units Cysticamaridae Bulman 1955 and Wimanicrustidae Bulman

1970a. Both groups of essentially encrusting graptolites share the development of an inflated camara bearing an erect neck with a variously elaborated apertural part. While the Cysticamaridae (Figure 8.5A) mainly include thigmophylic species, the Wimanicrustidae (Figure 8.5B–D) are represented by runner-type colonies forming an irregular meshwork of linearly oriented thecae. The stolon system of the camaroids shows diad budding, but the crustoids typically have a triad budding system.

As the species of the crustoid and camaroid graptolites are mostly known from small fragments, the exact shapes of their colonies are uncertain. This is clearly evident in the numerous fragments of camaroids (Kozłowski 1949) for which the initial zooidal tube, the sicula, is unknown. The crustoids (Kozłowski 1962) are similarly poorly preserved, and few pieces of colonies with more than two or three thecae exist. Kozłowski (1962) described *Bulmanicrusta* fragments with

Figure 8.4 Middle Cambrian (Series 3) rhabdopleurids from Monegeeta, Australia. (A) *Archaeolafoea longicornis* Chapman, 1919, NMVP 13112, holotype. (B) Growing end of same. (C) NMVP 13114, holotype of *Archaeocryptolaria skeatsi* Chapman, 1919, growing end. (D) Central part of another colony from same slab. Scale indicated by 1 mm long bar in each photo.

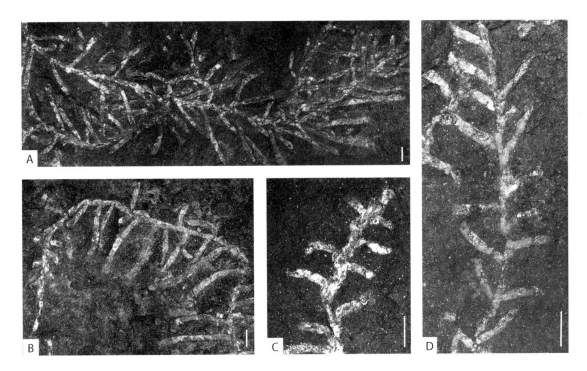

four thecae in a single row, and another fragment that shows four thecae and a branching point. Mitchell et al. (1993) identified and illustrated a member of the crustoids as ?*Bulmanicrusta*, showing a runner-type colony with fairly close branching intervals and the presence of graptoblasts. All crustoids and camaroids appear to possess well-defined zigzag sutures on the dorsal side of the camara and on both sides of the thecal necks (Figure 8.5), comparable to the dorsal zigzag sutures in the creeping part of *Rhabdopleura*.

The crustoids develop a very unusual structure, the graptoblast, as a normal and routinely formed feature. Graptoblasts (Figure 8.5D) are specialized stolothecae, modified into inflated bodies, interpreted as resting stages. Kozłowski (1949) described the graptoblasts initially as a new fossil group, the Graptoblasti, and established two genera with a number of species. Mierzejewski (2000) discussed the graptoblasts in some detail and

noted Kozłowski's (1971) suggestions that graptoblasts were formed by camaroids and crustoids, supported by Mitchell et al. (1993), who discovered a colony of *Bulmanicrusta*? sp. with preserved graptoblasts.

The Cyclograptidae

The general shape of the tubarium is known from a few Cyclograptidae (Figure 8.6) based on larger specimens, but many details are not available. The descriptions of most species are based on very little information from isolated thecal fragments. Maletz (2014a) diagnosed the group as essentially encrusting, but also developing short erect stipes, often with serially arranged tubular thecae. He included quite a number of genera in the family, but many of them are too poorly understood for a valuable interpretation. Kozłowski (1949) differentiated the

Figure 8.5 Wimanicrustidae and Cysticamaridae. (A) *Bithecocamara*, thigmophylic, sheet-like colony, one theca highlighted, reconstruction based on Bulman (1970a), Fig. 31, with permission from the Paleontological Institute. (B) *Bulmanicrusta*? sp. runner-type colony, specimen with irregular basal sheet, stolon and single partly preserved autotheca. (C) *Bulmanicrusta*, reconstruction, showing thecal style and graptoblast (after Urbanek, unpublished). (D) *Bulmanicrusta*? sp., graptoblast. (B, D) adapted from Mitchell et al. (1993) with permission from Cambridge University Press. Illustrations not to scale.

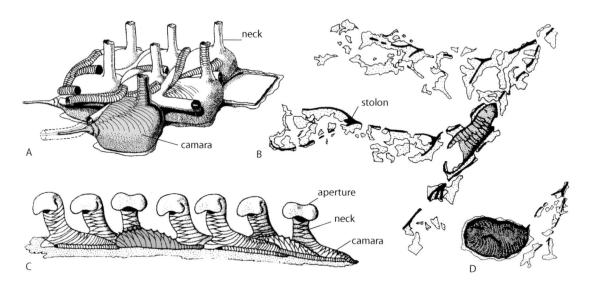

Tubidendridae with their characteristically coiled thecae from the simple Idiotubidae, a separation not used in Maletz (2014a). The thecal style of the members of the Cyclograptidae are quite variable, especially the thecal length and the development of dorsal and ventral rutelli. The coiled initial parts of the thecae are known from only few specimens (Figure 8.6C, F) and may not be present in all members. The erect stipes include tubular thecae with complex overlapping and may show infrequent branching.

The Dithecodendridae

The Dithecodendridae are characterized as erect colonies with slender but robust stipes and delicate, cone-shaped, completely isolated thecae. They reach a considerable size, but usually only small fragments of their tubaria can be seen in the fossil record, making identification difficult. Obut (1964) initially erected the family Dithecodendridae for erect-growing taxa with a slender stem and irregularly placed, elongated tube-shaped thecae. Maletz (2014a) included a number of genera in the family, based on the outline of the colony, and provided an emended diagnosis, but was unable to provide clear evidence for the presence of fusellar construction in most taxa. Fusellar construction can be detected only in a few members (e.g. *Tarnagraptus*: Sdzuy 1974). Thus many specimens could easily be included in the hydroids or other poorly known groups of fossils, if only the general outline of their colonies is seen.

The basal attachment is unknown for the Dithecodendridae. The slender but robust stem is apparently thickened by considerable amounts of cortical tissue, while the long thecal tubes show thin walls that probably do not include large amounts of cortex. This can be deduced through the preservation of *Tarnagraptus* and *Ovetograptus* from the Middle Cambrian of the Cantabrian Mountains of northern Spain (Sdzuy 1974; Maletz et al. 2005b; Maletz & Steiner 2015). The black organic material is thick on the stem, but the

Figure 8.6 The Cyclograptidae. (A) *Galeograptus nicholasi* Bulman & Rickards, 1966, dorsal view. Adapted from Bulman and Rickards (1966). (B) *Discograptus*, dorsal view. (C) *Tubidendrum bulmani* Kozłowski, 1949, fragment showing coiled thecae. (D) *Kozlowskitubus erraticus* (Kozłowski,1963), proximal end with erect stipes. (E) *Galeograptus*, lateral view. (F) *Dendrotubus wimani* Kozłowski, 1949, initial coiled part of thecal tube. (B), (E) and (F) adapted from Bulman (1970a) with permission from Paleontological Institute, and from Bulman (1970b) with permission from Société Belge de Géologie et de Paléontologie et d'Hydrologie. Illustrations not to scale.

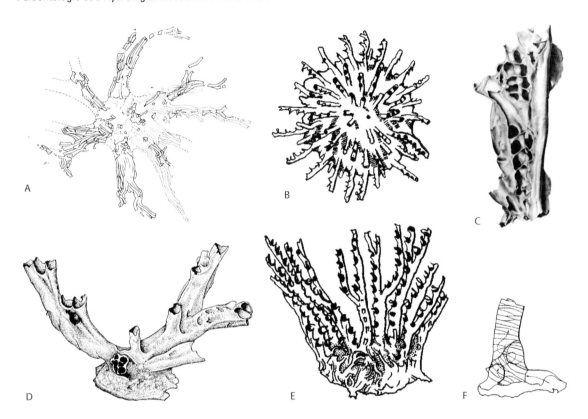

thecae are preserved as thin layers of fragmentary organic material (Figure 8.7).

The thecae originate irregularly on the erect stems in the Dithecodendridae, and sometimes may also be present in whorls or pairs, but as all specimens are flattened films of organic material, the details cannot be deciphered and information on the thecal origins is not available.

The Dendroidea

The order Dendroidea includes all benthic erect-growing graptolites in which the thecal development is based on a triad budding system and the differentiation of autothecae and bithecae (Maletz 2014a). The origin of these features is still uncertain, and the family Mastigograptidae with a slender stem and completely isolated metathecae, clearly showing a triad budding (Bates & Urbanek 2002), is unusual and difficult to relate morphologically to the remaining dendroid groups.

The Mastigograptidae

The Mastigograptidae (Figure 8.8) differ from all other Dendroidea through the possession of a slender stem and completely isolated thecal tubes. A general similarity to the Dithecodendridae can

Figure 8.7 Dithecodendridae. (A) *Ovetograptus gracilis* Sdzuy, 1974, SMF 30028, holotype. (B) *Tarnagraptus cristatus* Sdzuy, 1974, SMF 30021, holotype. (C–D) *Tarnagraptus palma* Sdzuy, 1974, SMF 30000, holotype. All specimens from the Cantabrian Mountains, Spain (see Sdzuy 1974).

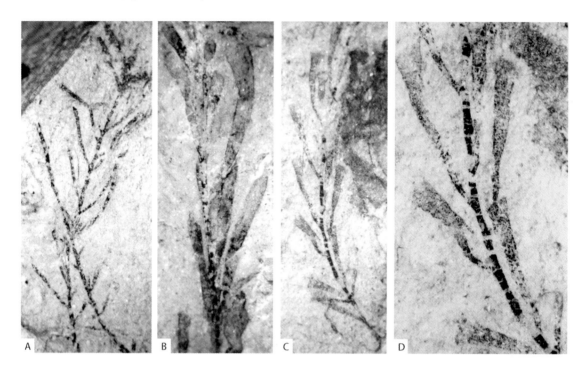

be noted in the colony shape. The genus *Mastigograptus* Ruedemann, 1908 has a complex history and was often misidentified or misinterpreted. The species *Dendrograptus tenuiramosus* Walcott, 1883 was originally defined as a bushy, multiramous colony with slender stipes, and Ruedemann (1908) was the first to recognize the delicate autothecae of this taxon. Eisenack (1934) and Andres (1961, 1977, 1980) described isolated fragments of *Mastigograptus* from glacial erratic boulders of northern Germany, and added considerably to our understanding of the genus. Andres (1961) recognized the fusellar development of the thecae and demonstrated the triad budding system, relating *Mastigograptus* to the derived graptolites. Bates and Urbanek (2002) investigated *Mastigograptus* in great detail and described the ultrastructure of the tubarium walls and the stolon system. They demonstrated the presence of an erect sicula and an attachment disc, anchoring the tubarium to rock surfaces and shells. A clear difference to all other graptolites is the stem, which

shows a strongly inflated stolon, filling the whole stolotheca. The thecal tubes are completely isolated and form slowly widening delicate and thin-walled tubes with irregular sutures of the fuselli. Andres (1961) differentiated autothecae and bithecae through their size differences, and identified the bithecae as the smaller thecal type.

The general shape of the two genera included in the Mastigograptidae, *Mastigograptus* Ruedemann, 1908 and *Micrograptus* Eisenack, 1934, is very similar to many of the specimens referred to the Dithecodendridae. However, in the Dithecodendridae, a differentiation of autothecae and bithecae as well as a triad budding system has never been demonstrated.

The Dendrograptidae

The Dendrograptidae (Plate 3C–D) include largely erect-growing colonies with bushy (Figure 8.9A) to fan-shaped and conical forms (Figure 8.9B). All

Figure 8.8 Examples of Mastigograptidae. (A–B) *Mastigograptus tenuiramosus* (Walcott, 1883), large fragment and detail showing thecae. (C–E) *Mastigograptus* sp., chemically isolated thecae showing triad budding. (A–D) adapted from Bulman (1970a) with permission from Paleontological Institute, and from Bulman (1970b) with permission from Société Belge de Géologie de Paléontologie et d'Hydrologie. (E) adapted from Andres (1977) with permission from Springer. Illustrations not to scale.

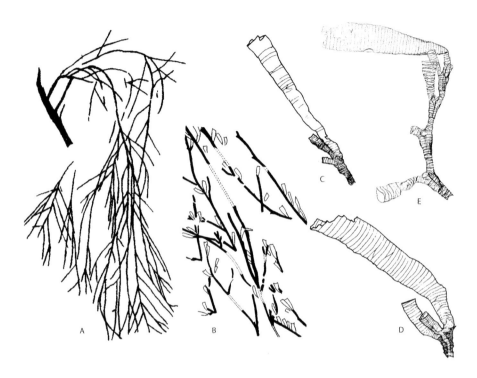

Dendrograptidae bear autothecae and bithecae. The thecae are arranged in a serial fashion with all apertures arranged in a single row and opening in the same direction. Dissepiments often connect the stipes in fan-shaped and conical colonies (Figure 8.9C). Maletz (2014a) discussed the Dendrograptidae and used a fairly inclusive approach for the family. Dendroid graptolites are long known, and numerous genera and species have been described. However, it is surprising how little is known about the Middle and Upper Cambrian members of the Dendroidea. Rickards and Durman (2006) compiled the Cambrian pterobranch records and included also the "dendroid graptolites". In their range chart (Rickards & Durman 2006, Fig. 3) they indicated the presence of the earliest dendroids of the genera *Callograptus*, *Dendrograptus*, *Desmograptus* and *Dictyonema* from the late Middle Cambrian *Paradoxides*

forchhammeri trilobite zone onwards, but presented little evidence of Upper Cambrian faunas. Berry and Norford (1976) described one of the oldest benthic graptolite faunas from the Road River Formation of Northern Yukon. The fauna is poorly preserved and does not show details of their thecal development, making a precise taxonomic identification impossible. The described specimens can only be referred to form-genera, based on the shape of the colonies, and not to biologically meaningful taxonomic entities. The Upper Cambrian record of dendroid graptolites is extremely poor (see Rickards & Durman 2006) and little can be inferred for their early evolution.

Even though the general consensus suggests the origin of the planktic *Rhabdinopora* from a benthic ancestor close to the genus *Dictyonema* (e.g. Erdtmann 1982a; Fortey & Cooper 1986; Mitchell et al. 2013; Maletz 2014a), these upper Cambrian

Figure 8.9 Dendrograptidae. (A) *Dendrograptus hallianus* (Prout, 1851), fragment of bushy colony showing thecal style on distal stipes, based on Hall (1865, Fig. on p. 127). (B) *Dictyonema retiformis* (Hall, 1851), conical colony with numerous dissepiments forming a typical meshwork (after Hall 1865, Fig. 10). (C) *Dictyonema estlandicum* Bulman, 1933, fragment showing thecal development and dissepiments (after Bulman 1933, pl. 7, Fig. 5). Illustrations not to scale.

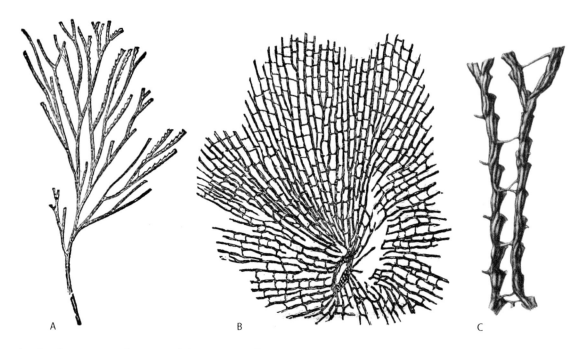

A   B   C

dendroids remain elusive and have never been described in detail. Information on the tubarium construction of benthic graptoloids originates largely from the Middle Ordovician species found in the limestones of the island of Öland (Holm 1890; Wiman 1896b; Bulman 1933, 1936). Bulman (1955, 1970a) discussed and illustrated the development of auto-and bithecae as well as the modifications of thecal apertures in dendroid graptolites. All chemically isolated Middle Ordovician dendroids bear typical triad budding, with small bithecae and larger autothecae. Dissepiments connecting adjacent stipes are common, but the fragmentary preservation often does not allow the colony shape to be understood. Specimens of the genus *Dictyonema* may be fan-shaped or conical, while *Desmograptus* and *Dendrograptus* appear to be bushy and branch in a three-dimensional fashion. Fan-shaped, conical and bushy colonies can also be differentiated in the Ordovician and Silurian dendroids described by Bouček (1957) from Bohemia, and the Carboniferous dendroids of Belgium (Ubaghs, 1941). Thus, the patterns of branching and tubarium development did not change much during the long period of existence of the benthic, dendroid graptolites.

Rickards et al. (1994) described the species *Dictyonema ghodsiae* (Figure 8.10D) from the late Dapingian of Iran, and termed it a "planktonic dendroid", opening the question to the multiple evolution of planktic graptoloids. Did planktic graptoloids evolve only once, or is there a record indicating a repetitive pattern of evolution of the planktic lifestyle in the graptolites? Even though it is the consensus that most known planktic graptoloids evolved from an ancestor of *Rhabdinopora* type, a number of "nematophorous" graptolites have been described from the Ordovician, indicating that several attempts to achieve a planktic lifestyle occurred. *Calyxdendrum graptoloides* Kozłowski, 1960 (Figure 8.10A–C) and *Callograptus* (*Pseudocallograptus*)? sp. cf. C. (*P.*) *salteri* (Hall), as described by Skevington (1963a), have to be noted for their free nemata and may also have to be interpreted as planktic, as do a number of species

Figure 8.10 Planktic dendroids. (A–C) *Calyxdendrum graptoloides* Kozłowski, 1960 (after Kozłowski 1960). (A) Distal fragment showing thecal development. (B) Proximal end with nematophorus sicula. (C) Isolated sicula with initial part of first theca. (D) *Dictyonema ghodisiae* Rickards, Hamedi & Wright, 1994, showing nematophorus sicula and vesicular bodies or nematularium(?), Kerman District, Iran (after Rickards et al. 1994).

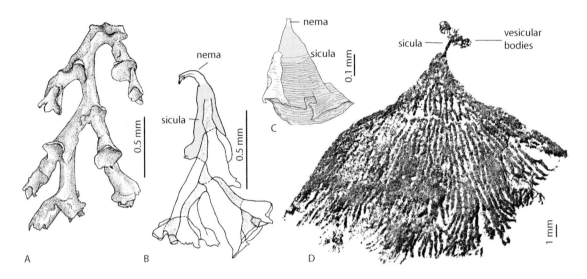

described from the Ordovician succession of the Czech Republic (Kraft & Kraft 2007). The question about a polyphyletic origin of planktic graptolites is not answered, as a detailed analysis of the tubarium construction is not available for most specimens discussed herein Thus, a comparison with the most probably monophyletic Graptoloidea (see Fortey & Cooper 1986) is not possible.

The Acanthograptidae

The Acanthograptidae (Plate 3A–B; Figure 8.11) is one of the strangest groups of the benthic graptolites, and again, a group poorly known from well-preserved material. The family Acanthograptidae includes multiramous, possibly exclusively erect-growing benthic graptolites with a complex construction of stipes including numerous slender thecal tubes (Maletz 2014a). The stipes show a ropy surface pattern with irregularly placed individual thecal tubes or complex twigs originating from the stem. The thecal development is based on the triad budding system, and fusellar construction has been verified in a number of specimens. Maletz and Kozłowska (2013) illustrated fragments of *?Acanthograptus sinensis* Hsü and Ma, 1948 from the Tremadocian of China (Figure 8.11A) with a clear triad budding system and many irregularities in the development of the individual thecal tubes. Wiman (1901) earlier described and illustrated isolated material of *Acanthograptus musciformis* (as *Inocaulis musciformis*) from the Middle Ordovician of Öland, Sweden, and Bulman and Rickards (1966) revised this material, providing detailed illustrations of many specimens. Typical tubular thecae are also present in *Acanthograptus divergens* Skevington, 1963, a slender form in which the isolated thecal tubes are not developed as twigs (Figure 8.11B–C). The development of the Acanthograptidae may easily be misinterpreted in poor material, and Kendrick et al. (1999) documented the true identity of the genus *Boiophyton* Obrehl, 1959, initially described as an early land plant, as a member of the Acanthograptidae.

The Extinction

Very little is known of the disappearance and near extinction of the benthic graptolites. Even though the encrusting rhabdopleurids are present even

Figure 8.11 Acanthograptidae. (A) *Acanthograptus sinensis* Hsü and Ma, 1948, fragment preserved in partial relief, showing thecal tubes, Tremadocian, China. (B–C). *Acanthograptus divergens* Skevington, 1963, fragments in relief, SEM photos, Darriwilian, Middle Ordovician, Öland, Sweden. The scale bar indicates 1 mm in each photo.

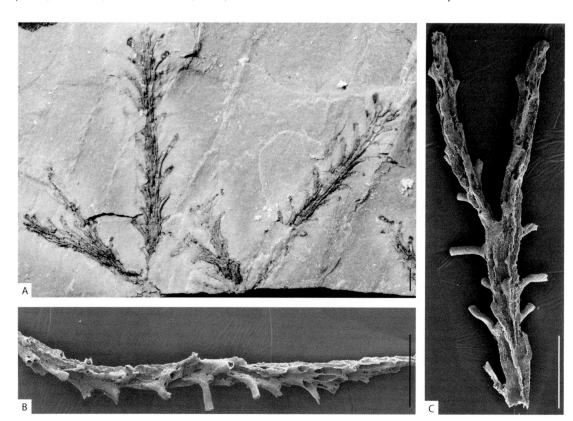

today, the erect-growing, benthic graptolites are completely extinct. One of their latest and best-known fossil occurrences may be in the fauna of the Molignée Formation from the "black marbles" of Denée (Figure 8.12), described by Ubaghs (1941). This apparently still diverse assemblage of benthic taxa includes a number of genera and species. The fauna is of Moliniacian, lower Visean age (Mottequin et al. 2015). Graptolites of Carboniferous age have also been discovered in other regions, from Britain (Hind 1907; Chapman et al. 1993), North America (Ruedemann & Lochmann 1942) and China (Mu et al. 1981), and appear to show a considerable diversity, but are usually based on poorly preserved and fragmentary material. Chapman et al. (1993) discussed the British and Irish faunas of Chadian (lower

Moliniacian) to Arnsbergian (upper Serpulkhovian) age (see Davydov et al. 2012 for Carboniferous chronostratigraphy). This material includes the youngest record of the Dendroidea.

The graptolite fauna from the basal Englewood Formation of the Black Hills of South Dakota (Ruedemann & Lochmann 1942) was initially referred to the Mississippian (Carboniferous), but may belong to the upper Devonian according to the conodont record (Klapper & Furnish 1962), leaving *Dictyonema blairi* Gurley, 1896 as the only Carboniferous graptolite of North America.

The only Chinese record of Carboniferous graptolites (Mu et al. 1981) describes poorly preserved and inconclusive material from the Island of Hainan. The specimens from the Sanglingshan Formation of late early Carboniferous are referred to three species of *Dictyonema*, of which two are new and only found at this locality. The Permian graptolites from the island of Hainan (Deng 1985) may refer to the same material, but have never been described or illustrated.

Algae or Graptolites?

The identification of branched, organically preserved fossils as graptolites is quite difficult, as quite a number of misidentifications and questionable assignments indicate (Plate 1A). Especially the genera *Inocaulis* Hall, 1851 (Figure 8.13A) and *Thallograptus* Ruedemann, 1925 from the Silurian of North America are discussed in various ways in the literature, and their assignment to the benthic Acanthograptidae (see Bulman 1970a) has been questioned (e.g. Hewitt & Birkler 1986; Mierzejewski 1986). Similarities can be seen to the Acanthograptidae in the massiveness of the stipes and the apparent isolated thecal tubes, but fusellar construction is not available for the material, and the organic film of most described specimens is very thin and not indicative of a pterobranch origin.

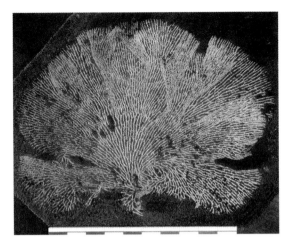

Figure 8.12 The Carboniferous graptolite *Ptiograptus fournieri* Ubaghs, 1941, holotype, Molignée Formation, Visean, Denée, Belgium (photo provided by B. Mottequin, 2015). Scale indicates 10 cm.

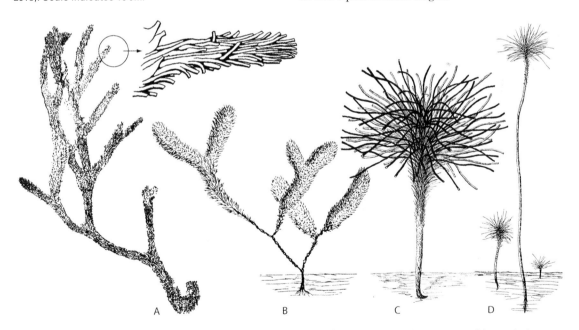

Figure 8.13 Algae and graptolites. (A) *Inocaulis plumulosa* Hall, 1852 (adapted from Bulman 1970a, with permission from the Paleontological Institute). (B) *Diplospirograptus goldringae* Ruedemann, 1925 (after Ruedemann 1947, pl. 41). (C) *Medusaegraptus mirabilis* Ruedemann, 1925 (after Ruedemann 1947, pl. 42). (D) *Medusaegraptus graminiformis* (Pohlmann, 1887) (after Ruedemann 1947, pl. 41). Reconstructions not to scale.

One of the best examples of the non-graptolitic nature of a benthic, organically preserved fossil may be *Medusaegraptus mirabilis* Ruedemann, 1925 (Figure 8.13C). Ruedemann described this species from the Middle Silurian Lockport Group of New York State, USA, as a dendroid graptolite. The species is present in numerous localities in the region and is often associated with dendroid graptolites. Even though Bulman (1938) already questioned the assignment and suggested an algal affinity, it was LoDuca (1990) who identified *Medusaegraptus mirabilis* as a non-calcified dasycladacean alga. Another dasycladacean alga is *Diplospirograptus goldringae* Ruedemann, 1925 (Figure 8.13B) from the same locality (see Mierzejewski 1986 for identification). Non-calcified dasycladacean algae are not uncommon in the Lower Paleozoic and have recently been described as possessing considerable diversity (LoDuca et al. 2011).

The unusual *Leveillites hartnageli* Foerste, 1923 was initially described as a possible alga, but Ruedemann (1947) included it in his monographic work on the North American graptolites. Tinn et al. (2009), however, described a thallophytic algal flora from the Lower Silurian (Aeronian, Llandovery) of Kalana, Estonia, including a species very similar to *Leveillites hartnageli*, suggesting an algal assignment for the taxon. Similarly, Mierzejewski (1991) voiced his concern at the identification of *Estoniocaulis* Obut et Rotsk, 1958 and *Rhadinograptus* Obut, 1960 as graptolites, and preferred to identify

them as possible algal remains of uncertain relationships.

The recognition of many specimens previously referred to the possible alga *Yuknessia simplex* Walcott, 1919 as pterobranch hemichordates or benthic graptolites by Maletz and Steiner (2014, 2015) also adds to the confusion of the recognition of fossil remains in the benthic ecosystem. Here, however, possible algae have been identified as genuine pterobranch fossils, as was discussed earlier.

Outlook

Since the identification of *Rhabdopleura* as a modern graptolite by Mitchell et al. (2013), the race is on to look for graptolite relatives in the fossil record and among living organisms. We finally start to understand the graptolites in a modern context and are able to see connections. We can estimate their evolution on the Paleozoic sea floor and their movement into the planktic realm by looking at the fossil record. Then we find the gaps in the fossil record and have to estimate and interpolate the details, recognizing that there is still a lot to search for and that the answers we are waiting for are not as close as we hope. Where is the origin of the planktic lifestyle? How did the change happen? Were they the first? We know so little, and the fossil record of the Upper Cambrian does not really offer an answer at the moment. So the search has to go on – we need more light.

9

The Planktic Revolution

Jörg Maletz

One of the major developments in the marine ecosystem after the origination and diversification of life during the Cambrian explosion was the exploration of the world's oceans by freely moving organisms. It marks the point in Earth's history when animals and plants conquered most marine environments and when the large-scale diversification of life took place. Before this, organisms were attached to the seafloor, or at least restricted to the water/sediment interface and unable to move around freely. Initially, biomats evolved and provided food for larger organisms. Food was restricted to the bottom of the oceans and the water column was an uninhabitable expanse, not dangerous in terms of predation, but impossible to conquer without a useful source of nourishment. Thus the exploration of the ocean waters was only possible when initially small-scale organisms, most probably forms like acritarchs and other unicellular organisms, abundantly populated the water column and became the food source for the few larger taxa. Now the exploration of the oceans and the population of completely new ecological niches was possible. Planktic larvae evolved in many groups of animals, but are rarely preserved in the fossil record. So we do not know much about the life in the water column of those times.

At some time in this strange world of the late Cambrian, the first graptolite explored the water column and became a planktic organism. The challenge was met and this newly-won freedom changed the direction of graptolite evolution forever. It led to an unexpected bloom that ended only with the demise and final extinction of its planktic clade in the early Devonian. Here, at the base of the Ordovician time interval, the graptolites succeeded in reaching into the water column and became the earliest macro-plankton of the planet of which we have a decent fossil record. Unfortunately, most planktic organisms probably did not leave much of a fossil record as they were soft-bodied organisms. However, it took the graptolites quite a while to reach their goal, from their first appearance probably in the Middle Cambrian to the basal Ordovician more than 20 million years later.

Graptolite Paleobiology, First Edition. Jörg Maletz.
© 2017 John Wiley & Sons Ltd. Published 2017 by John Wiley & Sons Ltd.

Why Move into the Water Column?

The earliest graptolites are poorly known (Maletz 2014b; Maletz & Steiner 2015) and it is uncertain in what exact time interval they originated. What is generally not disputed is the identity of the earliest planktic graptolite. It was *Rhabdinopora flabelliformis* from the basal Ordovician (Moberg 1890, 1900), at the time identified as *Dictyonema flabelliforme*. A number of species have been identified as the first planktic graptoloid, namely *Rhabdinopora proparabola* (Lin, 1986) (Plate 4E), *Rhabdinopora praeparabola* (Erdtmann, 1982a), *Rhabdinopora parabola* (Bulman, 1954) (Plate 4D), and *Rhabdinopora flabelliformis* (Eichwald, 1840), all of these species being from a small clade and very difficult to differentiate. The identification of these taxa as genuine graptoloid species is sometimes questioned, and the taxonomy of these early planktic graptolites is still under discussion (Erdtmann 1982a). Erdtmann (1986b, 1988) dubbed most taxa as "form species" and with this statement indicated the existing problems of species differentiation within the group. *Rhabdinopora*, as the first planktic taxon of the graptolite clade and the ancestor of all planktic graptoloids found in the Lower Paleozoic, took advantage of the still open ecological space provided by the oceanic water column. It has to be expected that benthic graptolites possessed a planktic larval stage, as does the only extant graptolite, *Rhabdopleura* and its pterobranch cousin *Cephalodiscus*. Thus the idea that a planktic pterobranch larva does not settle, but remains in the water column and eventually evolves into a planktic graptolite, is feasible. Details, however, are not available, and the origin of the planktic lifestyle remains a mystery. A number of benthic macro-organism groups invaded the planktic realm during the Paleozoic (Rickards 1990; Rigby & Milsom 1996, 2000), after the micro-plankton was already well established (Vidal & Knoll 1983). These included mainly small organisms, and some of the groups did not possess a preservable part and are difficult to find in the fossil record. Organisms like the enigmatic Burgess Shale *Eldonia* (see Walcott 1911a; Chen et al. 1995; Caron et al. 2010) and the record of cnidarian medusae in the Cambrian (Hagadorn et al. 2002; Hagadorn & Belt 2008; Young &

Hagadorn 2010) indicate a diversity of soft-bodied organisms in the water column, but the composition of the early planktic faunas still leaves much space for speculation. The development of a modern-style pelagic ecosystem in the early Cambrian has been suggested by Hu et al. (2007), but Vannier et al. (2009, p. 2572) regarded it as possible that "the pelagic ecospace remained largely uninhabited by animals during the Cambrian period".

The planktic lifestyle has some advantages for the pterobranchs, especially the gain of new ecospace to explore. The benthic community was already crowded by the early Paleozoic with numerous organisms, as demonstrated by the Cambrian Burgess Shale (Whittington 1985; Briggs et al. 1994) and Chengjiang biota (Chen et al. 1997; Hou et al. 2004), among others. Many organisms were actively moving on the Paleozoic seafloor, but a number of groups also explored a sessile lifestyle and were in direct competition with the dendroid graptolites. Actively moving arthropods, including the trilobites, were a common component of the fauna and evolved into numerous taxa. There were algae, sponges and cnidarians as sessile organisms, and priapulid, phoronid and enteropneust worm-like organisms as well as brachiopods that were less active and usually preferred or were constructionally restricted to a stationary lifestyle. Of these, the sessile groups might have been the most important competitors for the benthic graptolites. The Bryozoa are not found in the earliest Paleozoic, but may have become the most important competitors during Ordovician times. Cambrian records of bryozoans cannot be verified (Taylor et al. 2013) and an early competition with the graptolites is unlikely. Another group, the corals, also evolved and diversified only later during the Ordovician Period, and thus were no serious competition for early benthic graptolites.

Graptolites were slow-growing organisms and the growth of larger colonies might have taken years to complete (Rigby & Dilly 1993). With increasing competition in the Paleozoic seas, even the benthic graptolites might have been outcompeted by faster-growing organisms, but this probably only took place in the Carboniferous, when most of the benthic graptolites suddenly disappeared. Ubaghs (1941) described one of the youngest diverse graptolite faunas from the Carboniferous of Belgium. Similar faunas are known from the British Carboniferous (Chapman et al. 1993) and to a lesser degree in other regions. The only surviving benthic graptolite today is represented by the genus *Rhabdopleura* (see Chapter 8).

Ascent into the water column might have provided the graptolites a shelter from benthic predation and the opportunity for further development and evolution in new directions – a protection from the competition on the sea floor. It is in a way similar to the exploration of the air by the pterosaurs and later on the birds, moving into a largely uninhabited environment that is open for a fresh start. As long as the graptolites were able to keep up in the water column, they were safe from competition and were able to live their protected slow life and enjoy growth to enormous size. The invasion of smaller, quickly growing and reproducing organisms into the water column, however, was the start of the demise of the graptolites. As soon as they became the food source for other organisms, their fragile ecosystem became a trap. The faster-moving nektic organisms started to feed on them and the graptolites were unable to survive. This was the end of the only colonial macro-plankton in Earth history that possessed the challenge of secreting a complex housing construction and used enormous energy and resources for this purpose. In the end, they disappeared, as they were unable to compete within a dramatically changing world. The origin of the planktic lifestyle could, however, be just by chance, without a driving factor or need, but simply filling an ecological niche that was available.

What has Changed in Colony Development?

At first sight, little difference is seen between the benthic late Cambrian *Dictyonema* Hall, 1851 (Figure 9.1A) and the early Ordovician planktic *Rhabdinopora* Eichwald, 1855 (Figure 9.1C), and variations are based on the robustness of the stipes and the density of branchings. For a long time these net-like graptolite colonies have been incor-porated into a single genus, named *Dictyonema*. Erdtmann (1982a) finally sorted out the two groups and resurrected the genus *Rhabdinopora* for the planktic, taxa, essentially based on the proximal end with the nematophorous sicula. Erdtmann (1982b) provided a list of records and a biogeographical interpretation of basal Tremadocian *Rhabdinopora*, citing about 130 ref-erences, but not listing the benthic dictyonemids. The benthic *Dictyonema* and its planktic descend-ant *Rhabdinopora* resemble the bryozoan *Fenestella* Lonsdale in Murchison, 1839, but the bryozoan has a mineral skeleton made from cal-cium carbonate, while the pterobrach tubaria are secreted from an organic compound. *Dictyonema* is quite variable in its colony shape, ranging from fan-shaped to cone-shaped erect colonies, attached to the ground by a short stem. The planktic *Rhabdinopora* is much more limited in its colony shape and invariably is cone-shaped (Figure 9.1C). However, due to the common transport of grapto-lite fragments and deposition outside of the origi-nal environment, graptolite colonies are generally broken and fragmented, and the original shape and other details are often difficult to recognize.

Dictyonema and *Rhabdinopora* initially shared the development of dissepiments to keep stipes at a safe distance (Figure 9.2A), so there is no interference of the zooids from different stipes for feeding (Fortey & Bell 1987). Dissepiments are often quite regularly distributed, probably secreted from the bithecal zoo-ids (Urbanek & Mierzejewski 2009), but this rela-tionship might be more complex. During the early Tremadocian time interval, the dissepiments were reduced in the Anisograptidae and eventually were eliminated. The only graptolites still bearing dissepi-ments by Upper Tremadocian to Lower Floian times was the genus *Sagenograptus* Obut and Sobolevskaya, 1962 (= *Araneograptus* Erdtmann and VandenBerg 1985) (Figure 9.2B; Plate 4D).

Anastomosis (Figure 9.2C–D) is another way of keeping strict distances between the stipes and stabilizing the graptolite tubaria. Anastomosis can be found in the tubarium of a few erect, ben-thic graptolite genera such as *Desmograptus* Hopkinson in Hopkinson and Lapworth, 1875, *Koremagraptus* Bulman, 1927b or *Paleodictyota* Whitfield, 1902. The details of this feature are not well understood. Two types of anastomosis can be

Figure 9.1 (A) *Dictyonema* sp., benthic graptoloid, Burgberg section, Germany, Middle Devonian. (B) *Rhabdinopora parabola* (Bulman), Dayangcha, Jilin, China, details of stipes showing thecae. (C) *Rhabdinopora parabola* (Bulman), planktic graptoloid, Dayangcha, Jilin, China, complete specimen showing colony shape and irregularly placed dissepiments.

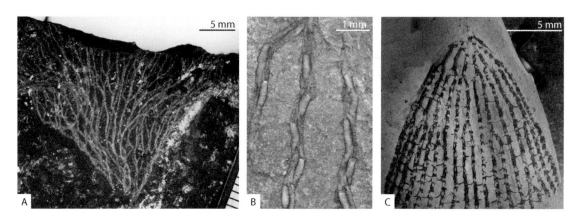

Figure 9.2 (A) *Rhabdinopora parabola* (Bulman), Dayangcha, fragment showing irregular dissepiments. (B) *Sagenograptus macgillivrayi* (Hall, 1897), NMVP 13096, top view of proximal end with regular dissepiments. (C) *Callograptus elegans* (Hall, 1865), GSC 956a, cotype, latex cast, showing anastomosis (arrows). (D) *Callograptus salteri* (Hall 1865, pl. 19, Fig. 7), GSC 955, cotype, showing short bridges or dissepiments (1) and anastomosis (2). Scale in each photo 1 mm, except for (B) in which it is 10 mm.

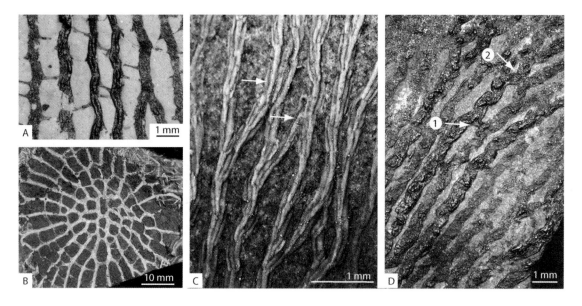

differentiated. Rickards and Lane (1997) discussed anastomosis in the genus *Koremagraptus* and differentiated between pseudanastomosis as anastomosis with transfer of thecae, and anastomosis as temporary connection of stipes without thecal transfer. Bulman (1945) described anastomosis in *Koremagraptus kozlowskii* Bulman, 1945 from the Upper Ordovician of Britain, based on isolated, bleached material. Anastomosis has not been recognized in the planktic Graptoloidea.

Figure 9.3 (A) *Epigraptus* sp., juvenile with dome and initial (sicular) tube (adapted from Kozłowski 1971, Fig. 1, under the terms of the Creative Commons Attribution Licence 4.0 – CC-BY-4.0). (B) *Dendrotubus* sp. with a distal helical line in the prosicula (based on Kozłowski 1971, Fig. 5). (C) *Dendrograptus communis* Kozłowski, 1949, tube-like prosicula with helical line (based on Kozłowski 1949, Fig. 1). (D) Conus and cauda (based on Hutt 1974a). (E–G): *Adelograptus tenellus* (Linnarsson, 1871), based on Hutt (1974a). (E) Conus and cauda in incomplete sicula, showing resorption foramen of first theca in prosicula. (F) Juvenile with complete sicula, part of th1[1] and crossing canal of th1[2], reverse view. (G) Sicula with complex nema. (D–G): adapted from Hutt (1974a) with permission from John Wiley & Sons. Illustrations not to scale.

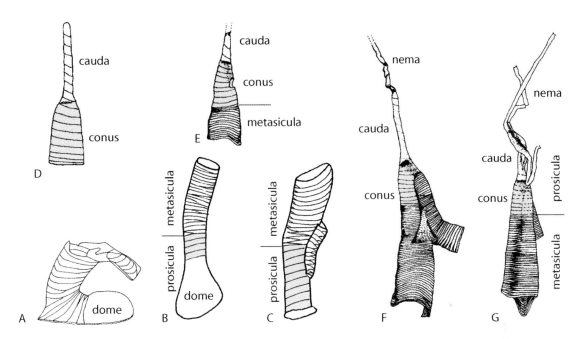

Attachment and the Free Nema

All benthic graptolites were attached to the substrate and fixed to a certain place for life. The larvae attached themselves to the ground by secreting a dome (Figure 9.3A–B) or a prosicula (Figure 9.3C) from organic material before they started to form the remaining parts of the colonies. The dome in the extant *Rhabdopleura* (see Chapter 3) is the only form of initial development known from modern pterobranchs, but it differs considerably from the prosicula of most graptolites (Figure 9.3). The dome of *Rhabdopleura normani* and *R. compacta* was an elongated body formed from a featureless membrane and constructed by glands in the epidermis of the larva (Lester 1988b). The Ordovician genus *Epigraptus* Eisenack, 1941 secreted a dome as in *Rhabdopleura* in place of a prosicula (Kozłowski

1971; Andres 1977) (Figure 9.3A), but otherwise this feature has not been recognized in the Graptolithina. The bottle-shaped, erect prosicula of *Dendrotubus* (Figure 9.3B) had a flat base and was provided with a primary opening, an aperture through which the sicular zooid emerged and started to secrete the metasicular fuselli (Kozłowski 1971). It also had a helical line in the distal part, probably homologous to the helical line in the prosicula of planktic graptolites (Mitchell et al. 2013).

Kozłowski (1971) differentiated the discophorous sicula (Figure 9.3C) of the benthic taxa from the nematophorous sicula of the planktic forms (Figure 9.3D–G), based on the presence or absence of a free nema. Many isolated planktic graptolites are available to show the construction of the sicula with the free, "nematophorous" prosicula, and thus this feature is well known. Kraft

Figure 9.4 (A, B) *Rhabdinopora proparabola* (Lin, 1986), Dayangcha section, Jilin, China, showing nematularium with possible fusellar increments (B). (C) *Staurograptus dichotomous* Emmons, 1855, St Paul's Bridge section, western Newfoundland, coll. Erdtmann. (D, E) *Rhabdinopora parabola* (Bulman, 1954), Dayangcha section, Jilin, China, specimen with multiple branched nemata. (F) *Rhabdinopora parabola* (Bulman, 1954), Shangsonggang section, Jilin, China, specimen with distally branched nema.

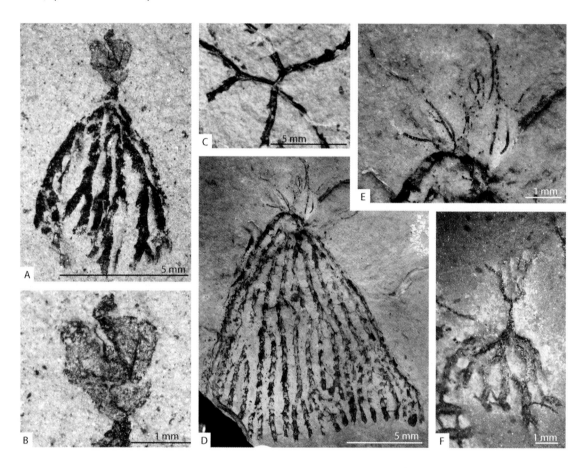

(1926) described the detailed development of the prosicula with its distinct helical line in *Rectograptus gracilis* (Roemer, 1861). Hutt (1974a) differentiated the prosicula into the conus and cauda (Figure 9.3D) based on the strong constriction at the tip of the conus and the presence of a diaphragm.

The nema appears as a thin rod growing from the cauda and its presence is regarded as evidence of a planktic lifestyle (Kozłowski 1971; Rickards 1996). Many biserial graptolites include the nema in the centre of the colony, from where it grows in advance of the thecae in the axonophoran colony. In early planktic graptolites, however, the nema is either a single rod or a branched one. Multiple branchings of the nema (Figure 9.4D–F) are common in the basal Ordovician *Rhabdinopora* from China (Lin 1986, 1988), but have also been discovered in material from Belgium (Bulman 1970b) and Victoria, Australia (Harris & Keble 1928). The reason for this development is unknown. The development of the nema differs considerably from species to species. Very early on, the nema is combined with some probably planar features identified as nematularia. These are rounded or elongated lobes found at the tip of the prosicula or the nema. They develop into very large structures in later graptoloids (see biserial Axonophora, Chapter 11).

Lin (1986) described basal *Rhabdinopora propa-rabola* from the Chinese Cambrian/Ordovician boundary section at Dayangcha with a "float structure" (Figure 9.4A–B). He interpreted this as a three-vaned construction, even though little detail is available from the material. Growth lines as delimited by Lin (1986) are quite difficult to recognize in the material and are visible in few specimens (Figure 9.4B). This feature is unknown from other localities and has not been recorded from chemically isolated material. Thus, the construction of this feature is uncertain.

Jackson (1974) illustrated small nematic vanes in *Kiaerograptus*(?) *peelensis* from the Tremadocian of Yukon, Canada. Vanes are not consistently associated with Tremadocian graptolites and have not been described in more detail. Størmer (1933) discussed a number of features as floating organs and illustrated a possible hollow float in *Rhabdinopora flabelliformis*. As the material is based on a single flattened specimen, it is impossible to judge this feature or to verify its construction as a vesicular body.

Increase in Symmetry

Benthic, dendroid graptolite colonies possess a relatively unordered, irregular colony shape based on environmental conditions. This is most apparent in the encrusting colonies, but these are basically unknown as larger colonies from the fossil record. However, the extant benthic *Rhabdopleura* can provide a good example of the development (Figure 3.3). Even though the individual tubes are symmetrically built and reflect the bilateral symmetry of the involved zooids, the colonies form irregular masses of interconnected tubes, either in dense mats in *Rhabdopleura compacta* (Hincks 1880; Mierzejewski & Kulicki 2003a), or in loosely branching colonies in *Rhabdopleura normani* (Lankester, 1884). The organization remains fairly irregular in erect colonies of dendroid graptolites due to interactions with other organisms and the influence of special environmental conditions, but an increased regularity may be noted.

As soon as the graptolites appear in the water column, an increase in symmetry can be noted. Colonies are initially bell-shaped and fairly regu-larly developed. Distinct intervals of coordinated branching may be noted in early planktic grapto-loids (Bulman 1950). The tubaria evolve into umbrella-shaped colonies with long stipes, but the stipe generations are generally of approximately equal length (Figure 9.5B), indicating a combined effort and interaction of the zooids in the colony to provide a symmetrical construction. This may be necessary to attain a stable orientation in the water column, while increased growth in one direction may have forced the colonies to flip over to one side due to instabilities. Certainly, life as a planktic colony in the water column increased the need for a perfect balance and thus initiated greater interaction of the zooids. It also led to a considerable reduction in the number of stipes and a dramatic change in colony shapes (Figure 9.5C–E). Graptoloids generally grew as bilateral symmetrical colonies with the sicula as the centre of the colony. Thus, stipe length and branching intervals were fairly regular and equal on both sides. Irregularities were rare and may potentially have been fatal, but graptolites were masterful in repairing damaged colonies and regaining the desired colony shapes (see Chapter 4).

The "Dendroid" Bithecae

All derived benthic graptolites, the Dendroidea of Maletz (2014b), possess the typical triad budding system with alternating bithecae on the sides of the stipes associated with all thecae (Figure 9.6D; Plate 4 F). This feature is kept in the Anisograptidae (Bulman 1970a; Maletz 1992a), but is subsequently lost in a number of lineages (Lindholm 1991). The development is difficult to analyse in detail, as especially well-preserved material is needed to see the delicate structure. Legrand (1974) described the proximal development of a quadriradiate *Rhabdinopora* from North Africa based on chemically isolated material and set the standard for description of anisograptid graptolites. Hutt (1974a) described similarly preserved specimens of *Adelograptus tenellus* from the Shineton Shales of Shropshire, Britain. This is the only material of *Adelograptus* described from isolated material in detail after Legrand's (1964) documentation of two *Adelograptus* species from the Algerian Sahara.

Figure 9.5 Increase in symmetry and regularity. (A) *Anisograptus matanensis* Ruedemann, 1937, Matane, Quebec, Canada, adapted from Bulman (1950) from Geological Society London. (B) *Clonograptus rigidus* (Hall, 1858), Levis, Quebec, Canada, after Lindholm and Maletz (1989). (C) *Expansograptus grandis* (Monsen, 1937). (D) *Isograptus lunatus* Harris, 1933. (E) *Archiclimacograptus* sp. All illustrations are reconstructions, not to scale.

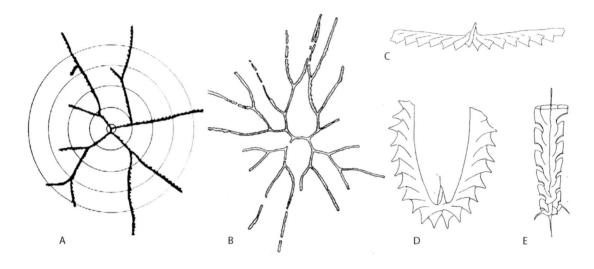

Figure 9.6 (A, B) *Hunnegraptus copiosus* Lindholm, Hunneberg, Sweden, proximal end in obverse (A) and reverse (B) views, showing sicular bitheca (arrow) and stipe development. (C) ?*Paratemnograptus magnificus* (Pritchard, 1892), MG 1967, latex cast, stipe fragment showing plaited thecal overlap without bithecae, Fezouata Biota, Morocco. (D) *Kiaerograptus kiaeri* (Monsen, 1925), PMO 72833, fragment showing long bithecae (arrows). Scale indicated by 1 mm long bar in each photo.

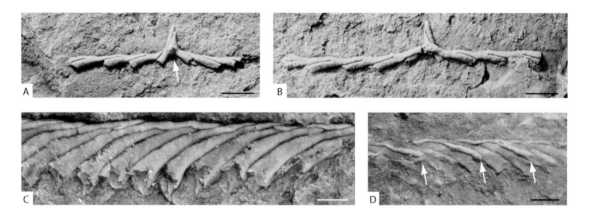

The position of the bithecae is irregular in some later anisograptids like *Kiaerograptus* (Spjeldnaes 1963), and eventually they disappear completely, initially leaving the plaited thecal overlap as a last reminder of their presence (Figure 9.6C). In *Hunnegraptus copiosus* Lindholm, 1991, a single bitheca, the sicular bitheca, is still preserved, while the stipes do not show any indications of bithecae (Figure 9.6A). A sicular bitheca is also found in *Paradelograptus* specimens (Jackson & Lenz 2000) from the Upper Tremadocian of Yukon, Canada, and in specimens of the genus *Ancoragraptus*

(Plate 4B). The reason and processes for the loss of bithecae is unknown. Even though the evidence indicates an independent loss of the bithecae along several lineages, too little material of Upper Tremadocian graptolites is available to trace this change. The suggestion of Hutt (in Palmer & Rickards 1991, p. 50) that bithecae are sexually differentiated would not make much sense, as a considerable reorganization of sexual reproduction would be necessary for the colonies if bithecae had to be eliminated.

The idea of Kirk (1973) might be more reasonable. She compared the graptolite colonies with those of bryozoans, suggesting bithecae to be cleaning individuals. These cleaning individuals might be superfluous in a planktic colony, as it is less likely to be covered by debris. On the other hand, a zooidal differentiation into autothecal and bithecal zooids is not present in modern benthic rhabdopleurids.

Increasing Diversity and Disparity

One of the most obvious changes in graptolite colonies after the origination of a planktic lifestyle was the increase in diversity and disparity of graptoloid colony morphologies (Figures 9.7 & 9.8). Initially the colonies stayed conical and pendent, but quickly subhorizontal to horizontal colonies evolved as we can see in anisograptids of the genera *Staurograptus*, *Anisograptus* and *Adelograptus*. Another step followed in the higher part of the Tremadocian with the reclined psigraptids. However, these had the genus *Triramograptus* Erdtmann (in Cooper et al. 1998) as a predecessor in the *Rhabdinopora flabelliformis parabola* Biozone of western Newfoundland.

The reclination of the stipes appeared in the genus *Psigraptus* (Plate 4C), and due to the difficulty in recognizing and differentiating species of this genus, a high number of genus and species taxa have been erected (Zhao & Zhang 1985). These, however, may have to be regarded as synonyms of a single genus *Psigraptus* (Rickards et al. 1991; Wang & Chen 1996). The closest relatives of *Psigraptus* include the species of the genera *Ancoragraptus* and *Kiaerograptus*, both sharing the partly isolated metathecae in the proximal end. Isolated thecal apertures may be present in

Toyenograptus (Li, 1984), but *Toyenograptus isolatus* (Bulman, 1954) is known from only a few poorly preserved specimens and its development is unknown. Altogether quite a number of new characters appear in the Tremadocian Anisograptidae, but their impact was fully explored only much later during the Floian time interval.

The Graptoloidea

The concept of the order Graptoloidea Lapworth (in Hopkinson & Lapworth 1875) changed considerably during recent decades, as Fortey and Cooper (1986) demonstrated with their cladistic analysis that the planktic Anisograptidae should be included, a view followed by Maletz (2014a). Previously, Bulman (1955, 1970a) included the Anisograptidae in the Dendroidea due to the development of bithecae and the triad budding. Mu and Lin (in Lin 1981) introduced the Graptodendroidina to include the planktic Anisograptidae as the basal suborder of the Graptoloidea. According to Maletz (2014a), it is a paraphyletic group from which all planktic graptoloids can be derived.

Anisograptidae as Inventors

The Anisograptidae are at the starting point of the explosive evolution of planktic graptolites. Therefore, they have a special position in our investigation. Tracing their evolutionary changes through the Tremadocian time interval might help to understand the evolutionary changes in later time periods, and to figure out the reasons for these dramatic changes. In general, it seems that the Anisograptidae were a fairly conservative group of graptolites and changes were implemented slowly. A major problem is that the detailed proximal development of the ancestors, the Dendrograptidae, is not well understood, and even though the origin from a dendroid ancestor similar to the genus *Dictyonema* is reasonable, this ancestry cannot be traced in detail. A number of constructional changes were necessary to make an efficient planktic graptolite out of a clumsy, irregularly developing benthic ancestor.

Figure 9.7 (A) Quadriradiate (*Rhabdinopora*, *Staurograptus*), (B) triradiate (*Anisograptus*, *Triograptus*) and (C) biradiate (*Adelograptus*, *Kiaerograptus*) proximal development in dorsal view (based on Maletz 1992a). Distal dicalycal thecae labelled DD.

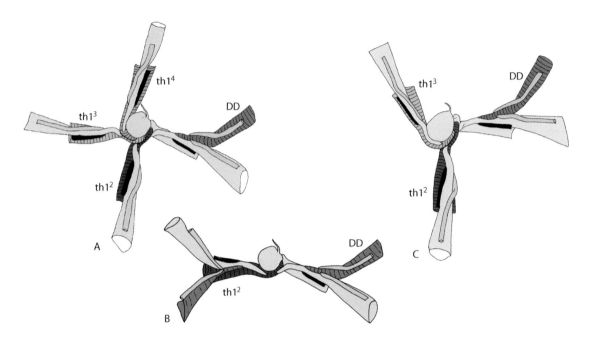

Proximal Development

The proximal development in the Anisograptidae is very stable, and early members all bear a quadriradiate development (Figure 9.7A) in which four stipes are formed from a close succession of dicalycal thecae in the proximal end. Legrand (1974) described this development in some detail and Maletz (1992a) provided a blueprint for the development and the changes necessary to derive the triradiate and biradiate development of younger anisograptids. The analysis shows that the proximal development has a first dicalycal theca at $th1^2$, followed by another two successive dicalycal thecae, forming three successive (proximal) dicalycal thecae (Figure 9.7A). These form the four stipes diverging from the sicula. Loss of these successive dicalycal thecae led to the development of triradiate and biradiate taxa (Figure 9.7B-C). A dicalycal theca separated from the sicula or from another dicalycal theca by a normal autotheca is called a distal dicalycal theca (Figure 9.7). Invariably, when a branching

is formed by a dicalycal theca, the usually associated bitheca is replaced by an autotheca. Thus, bithecae are lacking at the branching points. This development is actually quite simple, but results in a complex structure for the proximal end of the Anisograptidae, which is difficult to see in most specimens due to flattening or distortion.

The development of the ancestral dendroids is not known in comparable detail. The presence and position of dicalycal thecae in the proximal ends of benthic graptolites remains unexplored. It seems clear, however, that the direct ancestor of *Rhabdinopora* was not a benthic taxon with a distinct stem like a bushy *Dendrograptus* species.

Evolutionary Changes and Biostratigraphy

Due to the simple tubarium construction, the evolutionary relationships of the Anisograptidae are easily understood. A single lineage from the

Figure 9.8 Lower Tremadocian graptolite biostratigraphy (based on Cooper et al. 1998). A general succession of quadri-, tri- and biradiate horizontal taxa can be differentiated in addition to the more precise biostratigraphical zonation. The *Rhabdinopora* lineage remains quadriradiate through the whole Lower Tremadocian. Graptolite illustrations from various sources, not to scale.

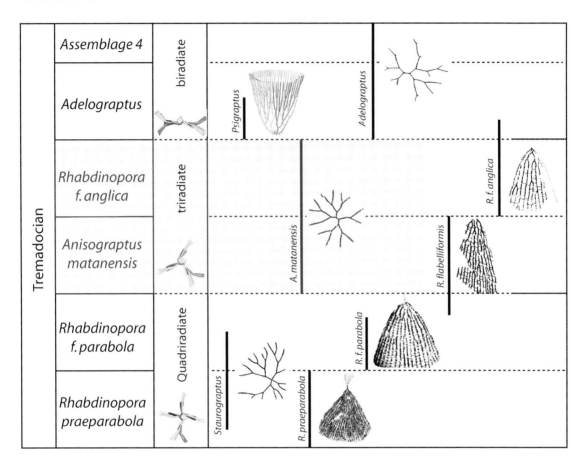

quadriradiate *Rhabdinopora* and *Staurograptus* led to the triradiate *Anisograpus* and finally to the biradiate *Adelograptus* (Cooper et al. 1998) (Plate 4A). After the loss of dissepiments in the colonies, the proximal stipes or first-order stipes became more loosely developed and some taxa even reduced the number to four (*Aletograptus*) or even three (*Triograptus*) stipes. These changes can be used for biostratigraphical purposes, and the genus *Anisograptus* represents a good index for a special graptolite assemblage, followed by the *Adelograptus* assemblage (Cooper et al. 1998) (Figure 9.8). A finer subdivision using species of the genus *Rhabdinopora* has been proposed, but is often difficult to identify due to the fragmentary preservation of the material. Cooper et al. (1998) differentiated the individual species through statistical analysis based on larger populations. A tendency is seen to develop more robust tubaria and larger thecae with a lower thecal count along the stipes, and also the number of dissepiments decreases along this gradient.

Turnover in the Late Tremadocian

In the higher part of the Tremadocian, a dramatic reorganization of the colonies happened in the graptolite world, of which we have only a poor

Figure 9.9 Upper Tremadocian psigraptids. (A) *Chigraptus supinus* Jackson & Lenz, 1999, GSC 117666, Yukon Territory, Canada. (B) *Ancoragraptus bulmani* (Spjeldnaes, 1963), GSC 123191, Yukon Territory, Canada. (C–D) *Psigraptus jacksoni* Rickards & Stait, 1984, TM 01, Yeongwol area, Korea, one juvenile and one mature specimen associated on a slab. (E–F) *Psigraptus arcticus* Jackson, 1967, JM 45, Erdaopu Section, Jilin, China. Scale indicated by 1 mm long bar in each photo.

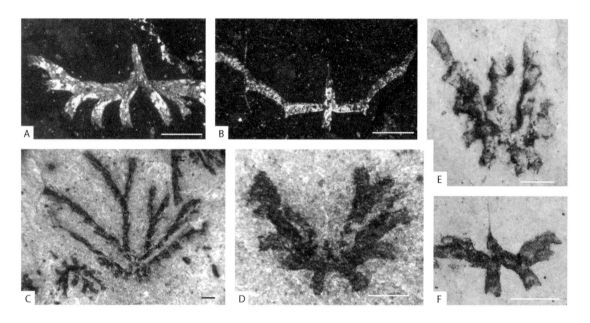

insight from few available faunas. The interval from the *Adelograptus tenellus* to *Hunnegraptus copiosus* Biozone in the late Tremadocian is poorly investigated, and very few faunas of this interval have been described in detail. The best-known fauna is the one from the *Sagenograptus murrayi* (= *Araneograptus murrayi* in older literature) Biozone of Victoria, Australia, last revised by VandenBerg and Cooper (1992). The faunas of the lower Ordovician Fezouata Konservat-Lagerstätte of Morocco, with its exceptional preservation of marine fossils (Martin et al. 2015), has recently been dated due to the presence of the graptolite *Sagenograptus murrayi* (Plate 4D) and a number of associated faunal elements.

The Australasian *Sagenograptus murrayi* Biozone includes a fauna representing a minor part of the whole interval, but does not provide any information for a precise biostratigraphical subdivision, however. The fauna is diverse and includes a number of newly evolved taxa that may

be regarded as transitional to the typical Floian and younger graptoloid faunas.

The slightly older *Psigraptus* Biozone is now found on several continents, from Australasia (VandenBerg & Cooper 1992), North America (Jackson 1967) and China (Zhao & Zhang 1985) to South Korea (Kim et al. 2006), but equivalent faunas have not been recovered in Baltoscandia or South America. *Psigraptus* (Figure 9.9C–F) is a spectacular new faunal element with a multiramous to two-stiped, reclined tubarium. The thecae are tubular and at least in the proximal end show apertural isolation. *Psigraptus* appears to be phylogenetically related to the genera *Kiaerograptus*, *Chigraptus* (Figure 9.9A) and *Ancoragraptus* (Figure 9.9B). These also show the aperturally free sicula, but do not have a strongly reclined tubarium. All species are multiramous, but the number of the stipes might be reduced and it is often difficult to relate juvenile specimens (Figure 9.9 F) to the mature ones (Figure 9.9C).

Figure 9.10 Transitional graptoloids. (A) *Sagenograptus macgillivrayi* (Hall, 1897), Victoria, Australia. (B) *Hunnegraptus copiosus* Lindholm, 1991, Hunneberg, Sweden. (C) *Hunnegraptus novus* (Berry, 1960a), juvenile, Texas, USA. (D) *Paratemnograptus magnificus* (Pritchard, 1892), Victoria, Australia. (E) *Paradelograptus mosseboensis* Erdtmann, Maletz and Gutierrez-Marco, 1987, Hunneberg, Sweden. (F) *Paradelograptus* sp., Cow Head Group, western Newfoundland, Canada. (G) *Aorograptus victoriae* (Hall, 1899), Cow Head Group, western Newfoundland, Canada. Illustrations not to scale.

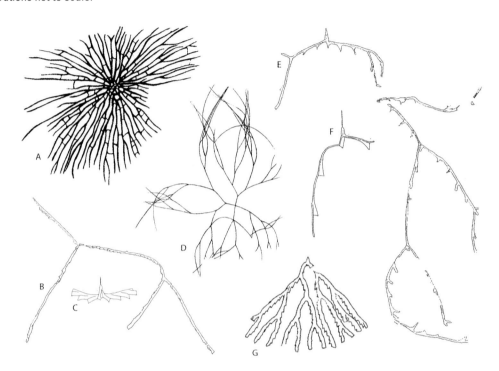

The transition of the Anisograptidae to the Dichograptina and Sinograpta is entirely conjectural, and faunas from the transitional interval are virtually unknown, except for the Australasia La2 or *Sagenograptus murrayi* Biozone fauna. Only a few taxa provide a glimpse into the actual evolutionary changes present in this interval. Lindholm (1991) suggested a transition to the Dichograptina through several lineages, supported by the analysis of Fortey and Cooper (1986).

Sagenograptus macgillivrayi (Hall, 1897) (Figure 9.10A) is one of the largest multiramous colonies from the late Tremadocian time interval and recalls the genus *Rhabdinopora* as it possesses dissepiments. It differs from other species of the genus *Sagenograptus* through a subhorizontal colony shape. Erdtmann and Vandenberg (1985) suggested that these large meshed graptoloids originated from the genus *Clonograptus* and are not related to *Rhabdinopora*.

The faunas of the *Aorograptus victoriae* Biozone of western Newfoundland may shed some light on the late Tremadocian transition by providing information on colony shapes and thecal style. *Paratemnograptus magnificus* (Hall, 1892) (=*P. isolatus* Williams and Stevens, 1991) (Figure 9.10D) shows a sicular bitheca, but appears to have no bithecae along the stipes. Illustrations show that the species has at least the plaited thecal overlap with lateral thecal origins as seen in the Scandinavian *Clonograptus* (*Clonograptus*) sp. aff. *C.* (*C.*) *multiplex* (Nicholson) (Lindholm & Maletz 1989) and in ?*Paratemnograptus magnificus* (Pritchard, 1892) from the Fezouata Biota of Morocco (Figure 9.6C).

The most spectacular species from the transition interval is *Hunnegraptus copiosus* Lindholm,

1991, a large, multiramous graptoloid (Figure 9.10B) with a small, obliquely placed sicula, similar to those of the Anisograptidae. Some specimens have numerous stipes and reach a colony diameter of more than 50 cm. The biradiate proximal end bears a prominent sicular bitheca (Figure 9.6A). Bithecae are lacking on the stipes and the thecal origins are dorsal, not producing a plaited overlap. Specimens that belong to this species are typical of this interval in Scandinavia (Lindholm 1991), Bolivia (Maletz & Egenhoff 2001), and the Taurus Mountains of Turkey (Sachanski et al. 2006), but the species is also found in fragments in Canada (Jackson & Lenz 2003). A few small specimens of *Hunnegraptus novus* (Berry 1960a) from Texas show the sicular bitheca (Figure 9.10C), but the final size and development of this species is unknown (Maletz 2006b).

A clear differentiation of the thecal style is seen in *Paradelograptus*, a genus including numerous two-stiped to multiramous species of Upper Tremadocian age (Erdtmann et al. 1987). The stipes do not show bithecae or plaited thecal over-lap, but a sicular bitheca (Figure 10.3A) has been found in a number of specimens (Jackson & Lenz 2000). They provide the first glimpse into the thecal variation seen in younger Ordovician graptoloid faunas.

Outlook

The exploration of the oceanic environment by the planktic graptolites is one of the most interesting and important events in the evolution of marine life. It sets the stage for the exploration of the nektic lifestyle by many other organisms. Still, very little is known about this evolutionary step, and even though the invasion of the graptolites into the oceanic environment is well known by paleontologists, few take the next step and ask for the basis of this behaviour and its evolutionary implications. We are content to see the planktic graptolites as key for Paleozoic biostratigraphy, but they have much more to offer.

10

Early Ordovician Diversity Burst

Jörg Maletz and Yuandong Zhang

The Ordovician represents a special time in evolution as many areas on the planet Earth were still uninhabited, and exploration of new environments was of high priority. This was the time of additional evolutionary novelties, after the Cambrian explosion fostered the design of the main organismic *Bauplans* and was responsible for the resulting early diversity of marine life. The vast expanses of the world's oceans were free of life, or at least of complex or advanced life. Unicellular organisms like acritarchs and larval stages of marine animals may have already inhabited this realm prior to the Ordovician. Graptolites were among the first larger organisms to explore the water column at the dawn of the Ordovician time interval, and quickly made this vast and nearly empty environment their home by establishing a new, planktic lifestyle.

At some time during the late Tremadocian a considerable change took place in the planktic marine ecosystem, not just through an increase in biodiversity, but also through utilization of new constructional developments that shaped and changed the planktic ecosystem. Graptolites were forced to evolve in different directions and explore new colony designs and lifestyles. While a planktic lifestyle became the norm in the life of graptolites, this innovation did not coincide exactly with changes in colony structures. Most of the Anisograptidae (see Chapter 9) still adhered to the old concept of the benthic graptolites, developing autothecae and bithecae in a regular fashion, branching in fairly regular intervals and producing multiramous to pauciramous bell-shaped to umbrella-shaped colonies. Well into the Ordovician, the Great Ordovician Biodiversification Event (GOBE) successfully led to dramatic changes in the colony development of the graptolites. The graptolite architects began to explore the possibilities of new colony shapes and different use of

Graptolite Paleobiology, First Edition. Jörg Maletz.
© 2017 John Wiley & Sons Ltd. Published 2017 by John Wiley & Sons Ltd.

their building material. This did not just influence the colony shapes, but also the individual thecae. Spines, complex elaboration of thecal apertures and even new thecal shapes began to emerge and make it easier for us as paleontologists to identify and determine graptolite species for geological purposes.

While the graptolite diversity was low during the Tremadocian, with few morphologically similar species dominating the marine expanses of the Ordovician seas, the graptolites entered a new era at the base of the Floian time interval. Numerous inventions demonstrated the newly developed abilities and previously unknown architectural features of the planktic graptolites, and eventually led to the first and largest increase in graptolite diversity that the world ever witnessed, in the early Ordovician. Never again – with the possible exception of the Lower Silurian evolution of the monograptids in the Aeronian and Telychian – was the diversity of graptolites higher.

The Great Ordovician Biodiversification Event

The Great Ordovician Biodiversification Event, or GOBE (Webby et al. 2004a), is seen as one of the major biotic events in Paleozoic history. "During this time interval of about 45 million years, an extraordinarily varied range of evolutionary radiations of Cambrian-, Paleozoic- and Modern-type biotas appeared" (Webby 2004, p. 1). Life evolved rapidly and spread to occupy all marine environments, from shallow-water platforms to deep-water oceanic sites. There were crawling organisms on the bottom of the seas, and planktic and nektic, free-swimming organisms in the wide expanses of the world's oceans. The first plants and arthropods conquered the land areas and started the speedy exploration of terrestrial environments, preparing the soil for the vertebrates to come.

The influence of the Great Ordovician Biodiversification Event on the evolution of the graptolites can be recognized and interpreted only indirectly from the fossil record. We clearly see the diversification of the dichograptids in the early Ordovician (Figure 10.1), but we cannot be sure about the factors leading to this increase in their diversity. Any interpretation would be speculative, but definitively has to take into account the availability of ecological space.

Sadler et al. (2011) recently provided diversity curves for the Ordovician and Silurian graptolites, showing the extent of the Ordovician diversification event, which is highly spectacular. The spindle diagrams for the Ordovician (Figure 10.1) also show an apparent near-extinction of the Anisograptidae and a subsequent recovery at the base of the Upper Tremadocian. The biostratigraphical level of this recovery is the same as the origin of the Dichograptidae. The reason for the extinction event in the history of the Anisograptidae is unknown as yet, but it appears likely that much of it is based on a biased fossil record and may not represent a true extinction at all. The Upper Tremadocian time interval is poorly represented by rock successions, and only a few graptolite faunas have been described documenting the interval.

The Sinograpta (= Sigmagraptidae + Sinograptidae + Abrograptidae) and Glossograptina (= Isograptidae + Glossograptidae) originated during the Floian–Dapingian time interval, possibly as descendants from the Dichograptina. Sadler et al. (2011) did not separate the Sinograptidae from the Sigmagraptidae in their analysis (Figure 10.1), which is represented by the second diversity maximum of the Sigmagraptidae in the Darriwilian. The early sigmagraptids, taxa related to the genus *Paradelograptus*, appear actually in the late Tremadocian, and are represented by the younger Anisograptidae after the assumed extinction in the

Figure 10.1 Ordovician graptolite diversity diagrams, modified from Sadler et al. (2011, Fig. 14). Three main events are shown by horizontal lines. (1) Origin of Dichograptina in the late Tremadocian and diversity burst in late Floian. (2) Origin of Isograptidae in basal Dapingian with subsequent increase in diversity. (3) Origin of Axonophora (Normalograptidae, Diplograptidae, Lasiograptidae) at the base of the Darriwilian and subsequent diversification.

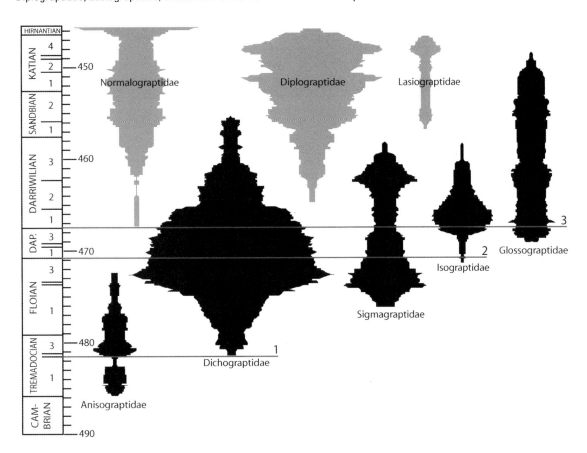

mid Tremadocian. From there, the Sigmagraptidae ranged upwards into the Sandbian, where the last members are represented by robust *Acrograptus* specimens (*"Didymograptus" serratulus* J. Hall, 1847 and related taxa). While the Isograptidae died out in the late Darriwilian, the closely related Glossograptidae survived into the Katian.

Change in the Colony Design

Early planktic graptolites were multiramous, with pendent to declined colonies, strongly mimicking their benthic ancestors (Chapter 8). Now, the number of stipes in the graptolite colonies decreased

and pauciramous to two-stiped taxa (Figure 10.2C, D) – the didymograptids – appeared, but they were still associated with quite a number of multiramously branching taxa (Plates 5–6). A few taxa, generally identified as *Azygograptus* (see Beckly & Maletz 1991), even lost all stipes except for one, and in this respect are more similar to the Silurian monograptids. Thus, a direction of graptolite architectural evolution is already visible and demonstrated in these forms (Figure 7.6). At the same time, the colony size increased dramatically in certain taxa, and graptolite colonies with a diameter of up to 1 m swarmed the oceans (Figure 10.2A).

One of the most important changes from a constructional point of view may be the loss of

Figure 10.2 Loss of stipes in the Dichograptina. (A) *Paratemnograptus magnificus* (Pritchard, 1892), large multiramous graptoloid, Victoria, Australia. (B) *Tetragraptus amii* (Elles & Wood, 1902), side view showing three out of four stipes, latex cast, Hunneberg, Sweden. (C) *Expansograptus validus* (Törnquist, 1901), two-stiped, flattened, Hunneberg, Sweden. (D) *Expansograptus* sp., latex cast, Tøyen, Norway. Magnification indicated by 1 mm long bar in (B–D) and 10 cm long bar in (A).

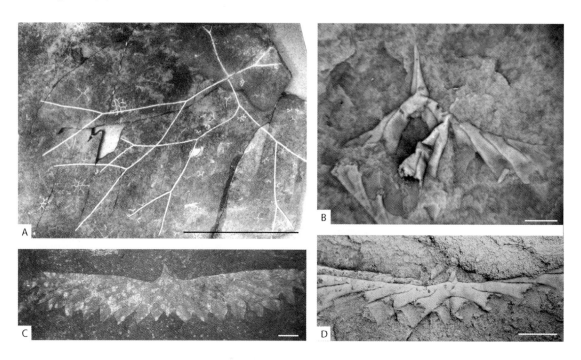

bithecae (Figure 10.3), which are associated with autothecae in the typical dendroid graptolites. Bithecae are present in most derived benthic graptolites and are regularly developed in the early planktic taxa of the Anisograptidae. In the Upper Tremadocian, these bithecae appear to be reduced and are subsequently lost (see Chapter 9).

Fortey and Cooper (1986) for the first time suggested that the evolution of the planktic taxa led to an early differentiation of several groups from the Anisograptidae, and that the concept of the Graptoloidea used at the time (e.g. Bulman 1955, 1970a) was polyphyletic. The loss of the dendroid bithecae was recognized to have occurred several times independently in the early planktic graptoloids (Lindholm 1991). Some taxa kept the sicular bitheca, the earliest developed bitheca of the colony, longer than the bithecae along the stipes, as is seen in the sigmagraptine genera *Hunnegraptus* and *Paradelograptus* (Figure 10.3A).

A dramatic change in colony shape, typified by the derivation of reclined to scandent stipes, happened in the Middle Ordovician. Earlier attempts at reclination of stipes can be seen in the rare Tremadocian genera *Psigraptus* and *Triramograptus* (see Chapter 9). These lineages, however, did not survive long, but died out rapidly without descendants. The reclination of stipes became one of the substantial major innovations in graptolite colony shape only during the Middle Ordovician, preceded by the appearance of *Tetragraptus phyllograptoides* in the basal Floian. The evolutionary origins of *Tetragraptus phyllograptoides* are uncertain as the species suddenly emerged and older reclined tetragraptids are unknown, even though four-stiped and three-stiped horizontal *Tetragraptus* species have been described from the latest Tremadocian *Hunnegraptus copiosus* Biozone (Lindholm 1991).

Figure 10.3 The loss of bithecae. (A) *Paradelograptus antiquus* (T.S. Hall, 1899), Yukon, Canada, showing sicular bitheca. (B) *Kiaerograptus supremus* Lindholm 1991, Sweden, specimen in reverse view with regular bithecae along stipes. (C) *Baltograptus vacillans* (Tullberg, 1880), latex cast in reverse view, Hunneberg, Sweden, bithecae are absent throughout the stipes, and the origin of thecae is from the dorsal side of stipes. Bithecae indicated by white arrows in (A–B).

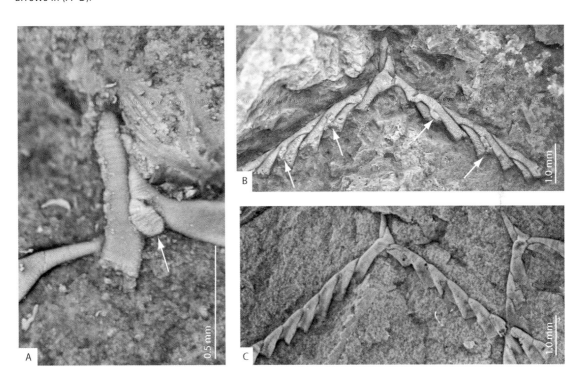

Suborder Sinograpta

The differentiation of the early planktic grapto-lites is hampered by the lack of well-preserved material. Maletz et al. (2009) attempted a cladistic analysis of many early planktic graptoloids in order to separate them into major groups and to trace their evolutionary origins. The analysis shows the Dichograptina and Sinograpta as sister groups (Figure 10.4). These two main groups can be differentiated through the position and devel-opment of the crossing canals and the position and orientation of the sicula. The Sinograpta rep-resent a group of multiramous to pauciramous graptoloids that at first sight have few differences from its sister group, the Dichograptina. The first noticeable difference can be found in the proximal development of the colonies (Figure 10.5). The proximal development of the Sigmagaptidae, the earlier forms of Sinograpta, was distinguished by prominent asymmetry of the two first-order stipes, which is marked by an obliquely placed sicula, a high origination of th1^1 and the first stipe from the sicula, and the oblique extension of the second stipe, starting from th1^2, close to the sicu-lar aperture. This asymmetry in the proximal development of Sigmagraptidae is retained from the Tremadocian Anisograptidae (Figure 10.4A), but the development is symmetrical in the Dichograptina, in which the sicula is now placed vertically between the stipes (Figure 10.4B).

Cooper and Fortey (1982) initially erected the Sigmagraptinae as a subfamily of the Dichograptidae, but did not consider the possible relationships of the Sigmagraptidae to the Sinograptidae, a family defined by Mu (1957) by

Figure 10.4 Revised interpretation of the Maletz et al. (2009) analysis, using preferred taxon names. Insets show proximal development of the (A) Anisograptidae (*Adelograptus*). (B) Sigmagraptidae (*Paradelograptus*). (C) Didymograptidae (*Didymograptellus nitidus*). (D) Isograptidae (*Isograptus*). Graptolite illustrations not to scale.

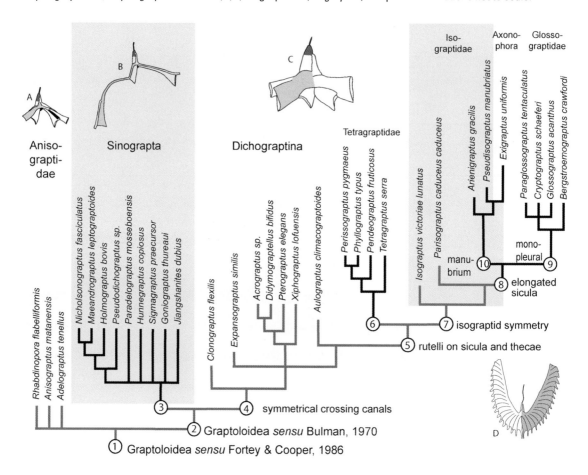

the characteristic prothecal folds of its members. Fortey and Cooper (1986), in their phylogenetic analysis of the Graptoloidea, included the families Sinograptidae, Sigmagraptidae and Dichograptidae in the suborder Dichograptacea of the order Graptoloidea. Maletz et al. (2009) provided a clear differentiation of the Dichograptina and Sinograpta in the form of the proximal end symmetry, the orientation and shape of the sicula, and the crossing canals. A differentiation of the Sinograptidae and the Sigmagraptidae, however, was not achieved in the investigation due to the low resolution of the analysis and the few included taxa.

Taxonomic and biostratigraphical data (e.g. Jackson & Lenz 2000, 2003) indicate the origin of

the Sigmagraptidae in the upper Tremadocian from an anisograptid ancestor close to the genus *Paradelograptus* (Figure 10.3A) through the change of the sicula from a widening cone to a parallel-sided shape with a proportionally large prosicula. The slender crossing canals retain their asymmetry as a symplesiomorphic characteristic (Figure 10.5B).

Family Sigmagraptidae

The origin of the Sigmagraptidae (Plate 5) is still not traced in detail, especially as there is little knowledge of the late Tremadocian graptolite faunas or the origin of the major clades. Early taxa

Figure 10.5 The proximal development of the Anisograptidae (A), Sinograpta (B–C) and Dichograptina (D–F). (A) *Anisograptus matanensis* Ruedemann, 1937, NGPA 216/07, oblique sicula and asymmetrical development. (B) Sigmagraptine indet., SPI 63, vertical sicula and asymmetrical development. (C). *Sigmagraptus* sp. with elongated, slender sicula, CHN 11.4E, nearly symmetrical development. (D) *Xiphograptus* sp., GSC 133392. (E–F) *Expansograptus hirundo* (Salter, 1863) in obverse (E) and reverse (F) views, Tøyen Shale, Slemmestad, Norway. (A–D) flattened specimens, Cow Head Group, western Newfoundland. Magnification for all specimens provided by 1 mm long bar in each photo.

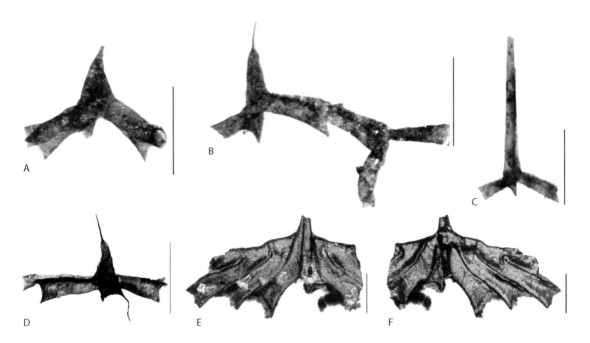

of the Sigmagraptidae appear in the *Hunnegraptus copiosus* Biozone of latest Tremadocian in Scandinavia (Lindholm 1991), and in the *Paradelograptus antiquus* to *Paradelograptus kinnegraptoides* biozones of late Tremadocian age in Arctic Canada (Jackson & Lenz 2000, 2003).

The Sigmagraptidae (Figure 10.6) embraces initially multiramous taxa with a highly irregular development of the stipes. However, a considerable regularity is achieved in several taxa, especially in the Dapingian to Lower Darriwilian species of the genus *Sigmagraptus* (Figure 10.6; Plate 5B), in which the monoprogressive branching produces strikingly regular geometries. A considerable variety of colony shapes is present in the Sigmagraptidae, and an extreme example is that two- and one-stiped taxa appear even in the early forms of the family. Many of the sigmagraptine genera (see complete list in Maletz 2014a) are easily misidentified as dichograptids, as they follow the same patterns of branching in multiramous taxa. Thus, *Paradelograptus* colonies are very similar to the dichograptid *Clonograptus* or even to the anisograptid *Adelograptus* (see Chapter 9). These taxa can only be differentiated by the details of their proximal end development and thecal style.

Family Abrograptidae

The Abrograptidae is a small family of Middle Ordovician graptolites with a strongly reduced fusellum. Their colonies are preserved often as a few thin rods only, which may have been considerably deformed during preservation. However, the sicula is completely preserved in all genera and provides the information that the specimens

Figure 10.6 Colony design of the Sigmagraptidae. (A) *Paradelograptus smithi* (Harris & Thomas, 1938a). (B) *Paradelograptus mosseboensis* Erdtmann, Maletz and Gutierrez-Marco, 1987, proximal end. (C) *Yushanograptus* sp. (D) *Goniograptus* sp., GSC 125786, proximal end. (E) *Sigmagraptus praecursor* Ruedemann, 1904, GSC 79889, western Newfoundland. (F) *Etagraptus tenuissimus* (Harris & Thomas, 1942), holotype. (G) ?*Goniograptus* sp., GSC 125768, proximal end in reverse view. (H) Sigmagraptine indet with single stipe, GSC 125815. (I) *Trichograptus dilaceratus* Herrmann, 1885. (J) *Goniograptus* sp., GSC 125788, juvenile. (K) Sigmagraptine indet, GSC 125806. Various magnifications.

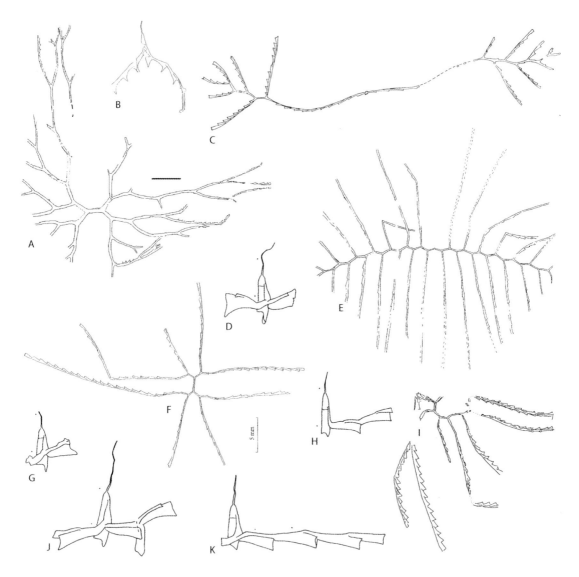

can actually be referred to the Sinograpta. Maletz (2014a) included the genera *Abrograptus*, *Dinemagraptus*, *Jiangshanites*, *Metabrograptus* and *Parabrograptus* into the Abrograptidae and questioned the inclusion of *Reteograptus*, as its sicula clearly shows a ventral virgellar spine (see Finney 1980). Most of the abrograptids are found as flattened specimens in shale, but a few are in

Figure 10.7 Specialized tubaria in Abrograptidae. (A) *Jiangshanites*(?) *dubius* Maletz, 1993, reconstruction (based on Maletz 1993). (B) *Jiangshanites*(?) *dubius*, GSC 102774. (C) *Jiangshanites*(?) *dubius* Maletz, 1993, GSC 102779, holotype. (D) *Abrograptus formosus* Mu, 1958. (E) *Dinemagraptus warkae* Kozłowski, 1951. (F) *Parabrograptus tribrachiatus* Mu & Qiao, 1962. (B, C) flattened, isolated specimens, digitally cleaned, see also Maletz (1993); (D–F) modified from Finney (1980, Fig. 11). Arrows indicate preservation of part of first theca in *Jiangshanites*(?) *dubius*. Various magnifications.

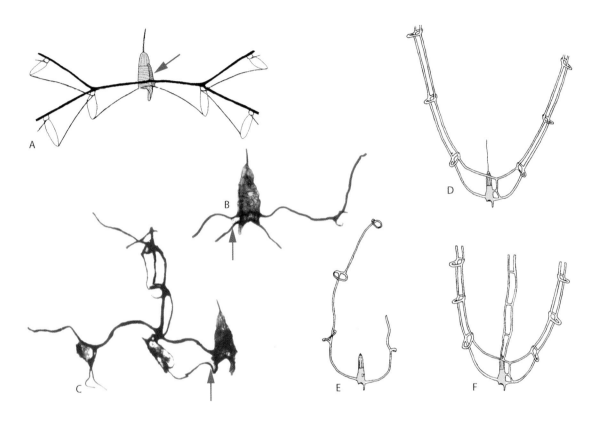

three dimensions and have successfully been chemically isolated. Kozłowski (1951) described *Dinemagraptus warkae* (Figure 10.7E) on a single specimen of Middle Ordovician age, which was derived from a glacial boulder. The only other abrograptids described on isolated material from the Darriwilian of western Newfoundland belong to *Jiangshanites*(?) *dubius* Maletz, 1993 (Figure 10.7A–C). The species has a small parallel-sided sicula with a strong rutellum and shows a prosicular origin of the first theca (Figure 10.7A). It is a multiramous species with possibly horizontally extended stipes.

Other abrograptids possess two or possibly three reclined stipes. *Parabrograptus* Strachan, 1990 appears to have a biserial proximal end and two uniserial distal stipes, and thus appears to be similar to the dicranograptid colony shape. The stipes of the abrograptids are preserved as one or two thin rods with a series of apertural rings (Figure 10.7D–F). Details of the thecal development have not been recognized.

Thecal Complexity

Species of the Sinograptidae (Plate 6) appear in the uppermost Dapingian or basal Darriwilian, but the precise biostratigraphical constraint on the origin of this group is uncertain. Earliest taxa

Figure 10.8 Sinograptidae. (A) *Anomalograptus reliquus* Clark, 1924, NIGP 8847, small specimen with comparably few stipes, thecal details not available. (B) *Allograptus mirus* Mu, NIGP 8868, paratype. (C) *Anomalograptus reliquus* Clark, 1924, NIGP 8852, large specimen with typical dichotomous branching and long funicle. (D) *Anomalograptus reliquus* Clark, 1924, wb2.34-42b. (E) *Anomalograptus reliquus* Clark, 1924, wb2.34.29a. (D, E) are flattened isolated specimens from shales of the Cow Head Group, western Newfoundland, Canada. Scale indicates 5 mm (A–C) and 1 mm (D–E).

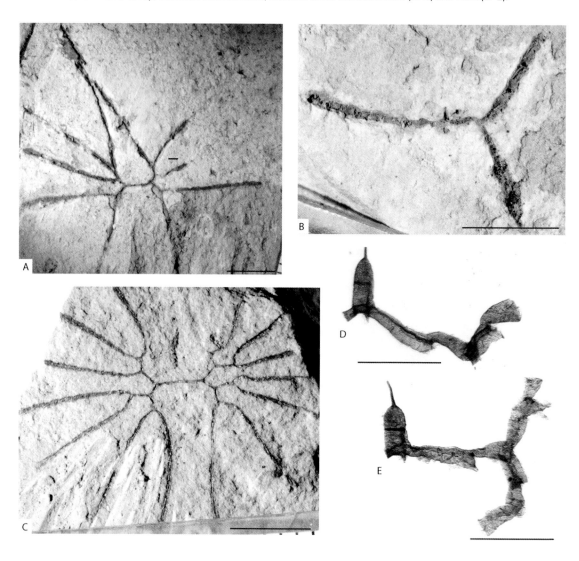

already possess the typical parallel-sided sinograptid sicula with a distinct ventral rutellum. In younger taxa the rutellum on the ventral side is complemented with a dorsal rutellum (Figure 10.8D–E). The family includes multiramous to single-stiped taxa, which were identified under a number of generic names indicating a differentiation from their homologous dichograptid counterparts (e.g. *Pseudologanograptus, Pseudodichograptus, Pseudotetragraptus*).

The main characteristic initially defining the group is the prothecal fold of the thecae, but prothecal folds are not developed in all taxa. Mu (1957) recognized the prothecal folds in specimens of the

Figure 10.9 Sinograptidae. (A–B) *Holmograptus callotheca* (Bulman, 1932), holotype, isolated material, Öland, Sweden. (C) *Holmograptus lentus* (Törnquist, 1892), LO 3260 t, holotype, Scania, Sweden. (D) *Holmograptus* sp., apertural view of theca (from Kozłowski 1954). (E) *Holmograptus bovis* Williams & Stevens, 1988, reverse view (from Bulman 1936). (F–G) *Nicholsonograptus fasciculatus* (Nicholson, 1869), Table Head Group, western Newfoundland. Scale indicates 1 mm in each photo.

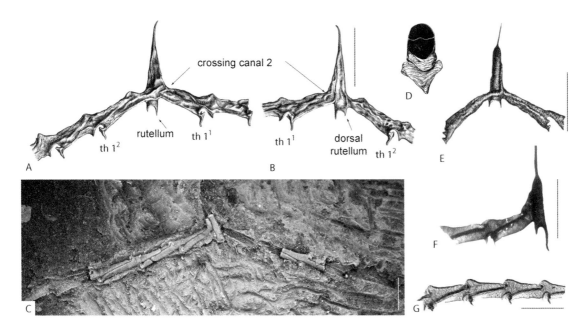

genera *Tylograptus* and *Sinograptus*, the latter being the name-giver for the family Sinograptidae. At the time, the proximal development was not considered relevant for taxonomy and the thecal style was regarded as the dominating characteristic. Recent investigations showed that the Sinograptidae include a number of multiramous to single-stiped taxa united by their proximal development (Maletz 2014a). The number of stipes may be quite variable even within a single species, and in the usual dorsoventral preservation the specimens can easily be misidentified as dichograptids as, for example, in *Anomalograptus*. Some Darriwilian specimens that belonged to this genus were often inappropriately identified as *Loganograptus* or *Dichograptus* in the past (see Figure 10.8).

As most material of the Sinograptidae is flattened, thecal details are rarely available for investigation, and many details are still uncertain or open to speculation. Kozłowski (1954) based the genus *Holmograptus* (Figure 10.9) on a single chemically isolated stipe fragment with incomplete thecae, and referred the isolated material described by Bulman (1932b, 1936) to *Holmograptus callotheca*. Mu (1957) erected the genus *Tylograptus* based on pyritic specimens from the Middle Ordovician of China, but subsequently Skevington (1965) synonymized the genus with *Holmograptus*. Jaanusson (1960) suggested that *H. callotheca* (Figure 10.9A–D) is a synonym of *Holmograptus lentus* (Törnquist, 1892). The latter species, together with *Nicholsonograptus fasciculatus* (Figure 10.9 F–G; Plate 6D–E), were interpreted by Skevington (1967) as indicating a case of genetic polymorphism, adding to the confusion. Zhang and Fortey (2001), however, did not accept the synonymy and resurrected the genus *Tylograptus* for the Chinese material. Williams and Stevens (1988) recognized that Bulman's (1936) specimens of *Holmograptus callotheca* belonged to two different species, *H. callotheca* and *Holmograptus bovis* Williams and Stevens,

Figure 10.10 Complex thecal structure of the Sinograptidae. (A) *Sinograptus typicalis* Mu, 1957, holotype, weathered pyritic internal cast (right stipe) and high relief imprint in light coloured shale (left stipe). (B) *Sinograptus*, thecal reconstruction with prothecal and metathecal folds. (C) *Holmograptus*, thecal reconstruction with prothecal folds. Reconstructions not to scale.

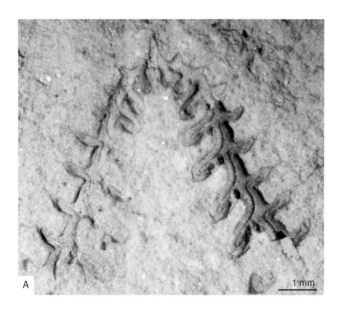

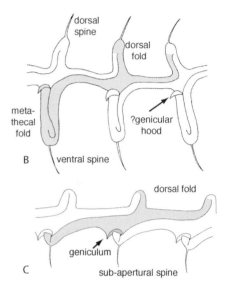

1988 (Figure 10.9E). Thus, the complex taxonomy of the sinograptids is not yet fully resolved, and a detailed investigation of the available and additional material is needed.

The genus *Sinograptus*

The species of the genus *Sinograptus* (Plate 6B–C) appear to be some of the most complex and also most beautiful graptoloids of the Ordovician. They possess spectacularly elongated and complexly folded thecae (Figure 10.10), of which the development could not have been recognized in flattened material. Fortunately, a number of specimens preserved in full relief as pyritic internal casts have been discovered (Figure 10.10A). These specimens show the thecal development in detail and provide us with the information necessary for a structural analysis and phylogenetic interpretation. The thecal development has been described a number of times based on Chinese material (Mu 1957; Zhang & Fortey 2001) and the genus is originally named after its occurrence in China (Sino-, also spelt as Cina, is an ancient name for China used for 1800 years). It was initially suggested to

be an endemic faunal element restricted to the South China plate, but *Sinograptus* species have also been discovered in the Darriwilian of the Canadian Cordillera of Yukon, western Canada (Jackson 1966; Lenz 1977; Lenz & Jackson 1986) and in Washington State, USA (Carter 1989), indicating a wider distribution.

Suborder Dichograptina

The Dichograptina (Plates 7–8) include the most characteristic "dichograptids" of Lapworth (1873a), the multiramous taxa ancestral to all the derived biserial and uniserial graptoloids that dominate Late Ordovician to Early Devonian times, and a number of pauciramous to single-stiped taxa as well. It is still difficult to establish their phylogenetic relationships, as little detail is available for reconstructing their proximal development and thecal architecture. Multiramous taxa are usually preserved in the sediment with their thecae pointing into the mud. The thecal style and the development of the important proximal

Figure 10.11 Various branching patterns in Dichograptina (A–D, F) and Sinograpta (E). (A) *Clonograptus*, progressive, dichotomous branching. (B) *Schizograptus*, dichotomous branching with lateral origin of stipes. (C) *Holograptus*, dichotomous, lateral branching with fairly irregular branching. (D) *Triaenograptus*, branching in triads. (E) *Goniograptus*, monoprogressive branching. (F) *Dichograptus*, dichotomous branching proximally, specimen with large proximal web. Illustrations not to scale.

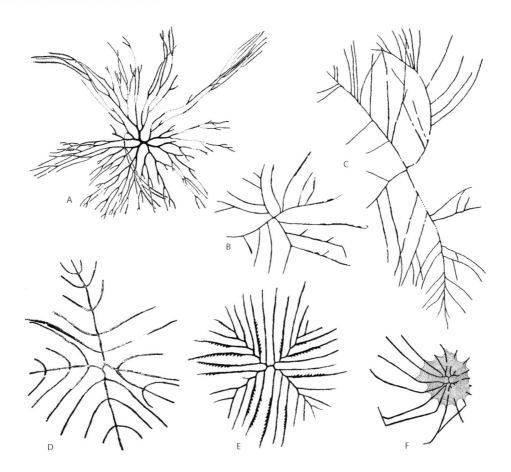

end of the colony are therefore difficult to see. Thus, the inclusion of many taxa in Dichograptina is still tentative. This is certainly the reason why multiramous taxa have rarely been treated taxonomically in detail, and most of the genera have been differentiated based on the branching pattern of the stipes only.

The latest major revision of dichograptid taxa was by Rickards and Chapman (1991), who based their revisions on the branching pattern, and differentiated dichotomous and lateral branching in the Dichograptidae, but did not focus on the proximal developments as the revision was mainly based on the redescription of the flattened material from the Victoria, Australia, succession. Maletz (2014a) differentiated four families of the Dichograptina on proximal developments and the architecture of the tubaria, but confessed at the same time that at least the Dichograptidae may represent a paraphyletic taxon, from which some of the other groups originated.

The Dichograptidae (Figure 10.11) include most of the multiramous, horizontal to subhorizontal taxa with their proximal ends showing biradiate, tetragraptid construction and distal dichotomous branching. Lindholm and Maletz (1989) described and illustrated the proximal development of *Clonograptus multiplex* (Nicholson) from relief material, and the species is one of the few from which the development is known in some detail. It differs little from the development in *Tetragraptus* (Figure 10.2B), except that its sicula might be shorter and wider.

Family Tetragraptidae

The Tetragraptidae are represented in the fossil record by a small number of genera, but abundant species and subspecies. They range from the basal Floian to the Middle Ordovician, and the genus *Pseudophyllograptus* from the higher Darriwilian was probably one of the last surviving members of the family. Initially, the Tetragraptidae were represented by horizontal to slightly declined or reclined four-stiped species (Figure 10.12A). These species evolved quickly, during the Lower Floian time interval, into reclined to scandent members, and into pendent forms like *Pendeograptus fruticosus* (Hall, 1865), a common and biostratigraphically important member of the family. Geh (1964) described the variability of *Tetragraptus* species from the Ningkuo Shale of Zhejiang, China, in some detail. Cooper and Fortey (1982) revised the type species and redefined the genus *Tetragraptus* based mainly on chemically isolated material from Spitsbergen, Norway. Maletz et al. (2009) and Maletz (2014a) discussed the Tetragraptidae

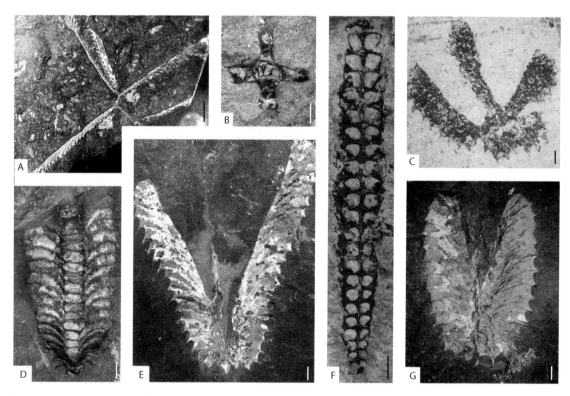

Figure 10.12 Tetragraptidae. (A) *Tetragraptus amii* Elles and Wood, 1902, horizontal. (B) *Pseudophyllograptus* sp., cross-section. (C) *Tetragraptus serra* (Brongniart, 1828), reclined. (D) *Pseudophyllograptus densus* (Hall, 1865). (E) *Tetragraptus phyllograptoides* Strandmark, 1902, showing three out of four stipes. (F) *Phyllograptus* sp., showing central columella. (G) *Tetragraptus phyllograptoides* Strandmark, 1902, showing proximally united stipes. Scale indicates 1 mm in all photos, except (A), where it indicates 5 mm.

and suggested a list of included genera, in which the inclusion of *Pendeograptus* and its derivative *Corymbograptus* are most noticeable. All these taxa are united by their proximal end construction with a robust sicula, compact crossing canals, an isograptid-type proximal development (Figure 10.2B) and a simple thecal style.

Scandency was one of the very new developments, found in the Floian for the first time. It makes a fundamental difference to the appearance of a graptolite colony and is achieved quite easily in the Tetragraptidae (Figures 10.12, 10.13). The stipes of these taxa, the genera *Phyllograptus* (Plate 8D), *Pseudophyllograptus* and *Pseudotrigonograptus* became scandent in a back-to-back manner and enclosed the sicula, forming a quadriserial colony with a characteristic cross-section (Figure 10.12B). Commonly the median septum between these four thecal series is modified into the columella with characteristic openings (Figure 10.12 F). An earlier and intermediate form, probably ancestral to phyllograptids, is represented by *Tetragraptus phyllograptoides* (Strandmark 1902) (Figures 10.12E, 10.13C, D) or *Pseudophyllograptus archaios* (Braithwaite 1976) (Figure 10.13G) of earliest Floian age, in which two or four stipes are initially connected with their dorsal sides, but distally separate (Strandmark

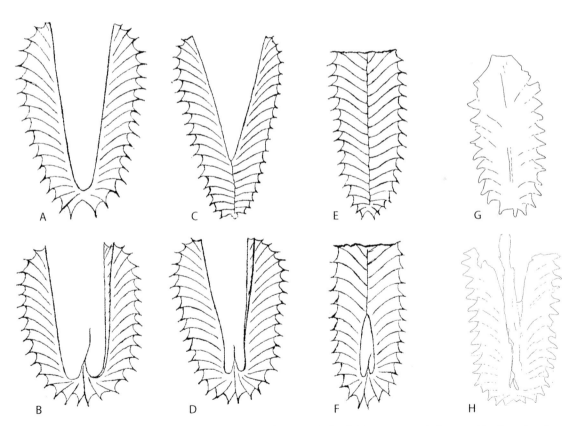

Figure 10.13 Tetragraptidae. (A–B) *Tetragraptus bigsbyi* (Hall, 1865). (C–D) *Tetragraptus phyllograptoides* Strandmark, 1902. (E–F) *Tetragraptus cor* (Strandmark, 1902). (A–F from Strandmark, 1902). (G–H) *Tetragraptus archaios* (Braithwaite, 1976), Trail Creek, Idaho (see Maletz et al., 2005), showing distally united stipes. (A, C, E, G) in a-b orientation; (B, D, F, H) in 1-2 orientation (see Chapter 3). Illustrations not to scale.

1902; Cooper & Lindholm 1985; Maletz et al. 2005a). In the younger, early Darriwilian taxon *Tetragraptus cor* (often identified as *Pseudophyllograptus cor*), the development is advanced and the four stipes initially leave an open space for the sicula (Figure 10.13 F), before they are distally united into a quadriserial colony. In the 1–2 preservation the stipes of the second order are separate initially (Figure 10.13D, F), while in a-b view, the two stipes are united from the start (Figure 10.13C, E). The quadriserial development forms one of the most robust and massive graptolite colonies known from the fossil record, using enormous amounts of material to secrete their tubaria.

Through a loss of two of the four stipes, the tetragraptids evolved into the biostratigraphically important isograptids and pseudisograptids, roughly at the beginning of the Dapingian age (see Glossograptina). A second, even more important change was achieved in the symmetry of the colony. In the earlier Dichograptina the sicula is situated in the centre and the stipes are developed symmetrically around it. In the descendent Isograptidae the sicula becomes less "special", and forms a symmetrical pair with th1[1]. With this change in the symmetry of the proximal end, a dramatic increase in diversity can be noted and the isograptids quickly evolved into numerous species.

Family Didymograptidae

When Mu (1950) introduced the Didymograptidae (Plate 8A–C, E), he probably did not consider a phylogenetic concept, but still employed it as a kind of form taxon for two-stiped graptoloids (see also Mu et al. 2002). Maletz (2014a) discussed a likely monophyletic origin for the group, based on the development of the proximal end. Didymograptids in a general, non-phylogenetic sense as two-stiped dichograptids are characteristic for the Floian time interval, and numerous species have been described (Figure 10.14) and are employed for biostratigraphical purposes. Generally, the tubarium shape was used to differentiate a number of genera. Pendent species were often included in *Didymograptus*, horizontal ones

in *Expansograptus*, and deflexed species in *Corymbograptus*, but this differentiation became more blurred with the understanding of proximal development types. Maletz (1994a) discussed the pendent didymograptids and concluded that they may be referred to a number of genera based on their proximal end construction. The misidentification of pendent didymograptids led to one of the great controversies in Ordovician biostratigraphical correlation (e.g. Berry 1960b, 1967, 1968; Skevington 1963b, 1968; Jackson 1964; Bergström & Cooper 1973). It is known that there has been more than one diversification event in the evolution of pendent didymograptids (see Cooper & Fortey 1982), through detailed investigation of the proximal developments of pendent didymograptids and based on the revision of the mistaken correlation of the Floian *Didymograptellus bifidus* Biozone interval of North America with the Darriwilian *Didymograptus artus* Biozone of Britain and Scandinavia (Fortey et al. 1990; Maletz 1994a). Another point complicating the story from a taxonomic point of view is the recognition of *Didymograptellus bifidus* and the closely related *Yutagraptus mantuanus* as members of the Pterograptidae (Maletz 2010a).

Didymograptid species are differentiated mainly through details in their proximal development types (Figure 10.14). Thus, well-preserved relief specimens or even chemically isolated specimens are necessary for a precise identification. The proximal development is generally of isograptid type with th1[2] as the dicalycal theca (see Chapter 3). This development is combined with a prosicular origin of the first theca (th1[1]), high on the sicula (Figure 10.14A), but in derived taxa the origin may become much lower on the sicula as in *Baltograptus* (Figure 10.14B) and is very low in *Didymograptus* (Figure 10.14 F, G). The progressive lowering of the origination position of the first theca (th1[1]) from the sicula seems to be a critical evolutionary direction for the didymograptids in the early and middle Ordovician. The proximal development evolved into an *artus*-type development, in which th1[1] is the dicalycal theca, as examplified in *Didymograptus murchisoni* from the middle Darriwilian (Figure 10.14E). However, the general tubarium outline of the didymograptid species did not change significantly, and the dif-

Figure 10.14 Didymograptidae. (A) *Expansograptus holmi* (Törnquist, 1901), latex cast, showing the high (prosicular) origin of the first theca (th1[1]) from the sicula, Diabasbrottet, Hunneberg, Sweden. (B) *Baltograptus vacillans* (Tullberg, 1880), latex cast, showing the origin of the th1[1] from middle part of the sicula, Diabasbrottet, Hunneberg, Sweden. (C) *Expansograptus praenuntius* (Törnquist, 1901), LO 1611 t, latex cast, Flagabro, Scania, Sweden. (D) *Expansograptus grandis* (Monsen, 1937), latex cast, Slemmestad, Norway. (E) *Didymograptus murchisoni* (Beck in Murchison, 1839), *artus* type proximal development typified by th1[1] as the dicalycal theca, Ebbe anticline, Germany. (F–G) *Jenkinsograptus spinulosus* (Perner, 1895), isograptid-type proximal development typified by th1[2] as the dicalycal theca, Krapperup, Scania, Sweden. All specimens in reverse view, except for (C–D) in obverse view. Bar indicates 1 mm in each photo.

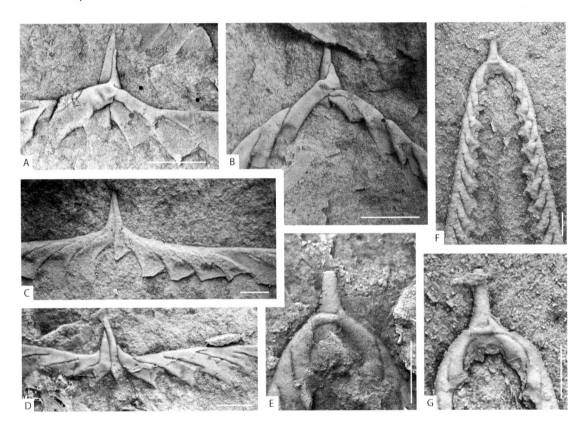

ferentiations of the species are usually difficult, especially when the development type cannot be clearly identified (see *Jenkinsograptus spinulosus*: Figure 10.14G).

Family Pterograptidae

When Fortey and Cooper (1986) proposed the Suborder Virgellina for all virgellinate graptoloids, they were probably not aware that the virgellar spine is not a homologous feature in all taxa, but originated independently several times. Maletz (2010a) investigated in some detail the origin and evolution of the virgellar spine, and differentiated the dorsal and the ventral virgellar spines. Conventionally, the sicular side with virgella is oriented as the ventral side of the sicula, and the other side as the dorsal side (Bulman 1970a). In some cases, spines are developed from the dorsal side of the sicula at the sicular aperture, and are generally termed dorsal spines. Accordingly, the origin of the first theca (th1[1]) from the dorsal side of the sicula, for example in *Xiphograptus*

Figure 10.15 Growth stages and adults of some species in the Pterograptidae. *Pterograptus elegans* Holm, 1881: (A) reconstruction of colony. Adapted from Bulman (1970a) with permission from the Paleontological Institute, and also from Bulman (1970b). (C) Fragmented proximal end showing cladial origin of theca (arrow). (D) Proximal end showing artus type development and metasicular origin of th1[1]. *Xiphograptus formosus* (Bulman, 1936): (B) Proximal end. (E–F) Juveniles showing construction and growth of virgellar spine (adapted from Skevington 1965). (G) *Didymograptellus bifidus* (Hall, 1865), juvenile showing strong virgellar spine. *Yutagraptus mantuanus* Riva, 1994: (H) Holotype (Riva 1994). Reproduced with permission from John F. Riva. (I) Isolated juvenile showing virgellar spine attached to ventral wall of theca 1[2] (arrow). Illustrations not to scale.

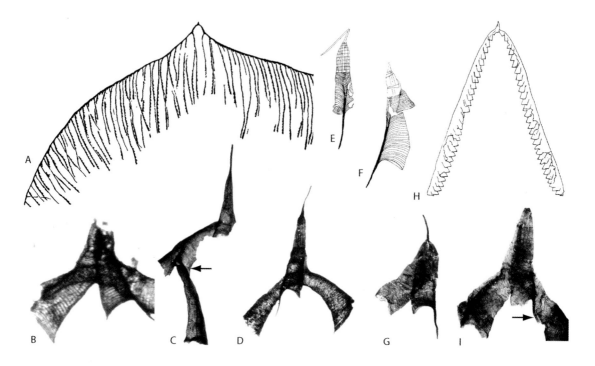

(Figure 10.15B) and *Yutagraptus*, is described as the antivirgellar origin (Cooper & Fortey 1982). Recently, Maletz (2010a) defined the sicular side that derives the first theca (th1[1]) – where the initial resorption foramen is – as the ventral side. He identified a virgella positioned on the same side of the sicula as the origin of th1[1] as a ventral virgella, and a virgella positioned opposite to the origin of th1[1] as a dorsal virgella.

One of the most characteristic groups bearing a dorsal virgellar spine is the two-stiped genus *Xiphograptus* and its relatives. Bulman (1936) illustrated the dorsal virgellar spine of *Xiphograptus formosus* (Figure 10.15B) based on material etched by Gerhard Holm from limestone samples collected from the Middle Ordovician succession of

Öland, Sweden, but he did not seem to consider the precise position of the virgella on the sicula as an important feature. Maletz (2014a) suggested use of the name Pterograptidae Mu, 1950 for this group, as the genus *Pterograptus* clearly could be interpreted as a relative of *Xiphograptus* based on the presence of a prominent dorsal virgellar spine (Figure 10.15D) and some additional constructional features.

Early members of the genus *Didymograptellus* differ considerably from *Xiphograptus*, as they possess a fairly large, parallel-sided sicula with a huge, parallel-sided prosicula (Figure 10.15G). A representative of the early forms, *Didymograptellus primus*, occurred early in the Floian (Maletz 2010a, Fig. 7) and is significantly older than any later virgellate graptoloids. The species is succeeded by *Didymograptellus bifidus* (Hall, 1865), the most common and well-known pendent didymograptid in the upper Floian. There has been a consensus that the species of *Didymograptellus* are not phylogenetically related to the pendent didymograptids of the genus *Didymograptus* occurring in the mid-Darriwilian (Fortey et al. 1990). Only quite recently has the genus *Yutagraptus* (Figure 10.15H, I) from North America been separated from the diverse pendent didymograptids (Riva 1994) and was recognized as a fairly long-ranging taxon of the late Floian to mid Darriwilian time interval (Maletz 2010a).

The genus *Pterograptus* is the most unusual taxon of the Pterograptidae, as it is a multiramous graptoloid in which the secondary stipes originate laterally and alternately from the two main stipes (Figure 10.15A). Isolated material of *Pterograptus* from Australia and Canada (Skwarko 1974; Maletz 1994b) indicated that these secondary stipes are actually cladia growing from their mother thecal apertures (Figure 10.15C) and are not comparable in development to the stipes in other dichograptids. They represent the earliest development of cladial stipes in the evolutionary history of graptolites, and are morphologically comparable to a few Upper Ordovician dicranograptids (e.g. *Syndyograptus* and *Tangyagraptus*) and some Silurian to Lower Devonian monograptids (e.g. *Cyrtograptus* and *Sinodiversograptus*). The cladia in the Silurian monograptids differ in their construction by the presence of a secondary nema along the dorsal side of the cladial stipes (see Chapter 13).

Symmetry and the Glossograptina

The Glossograptina are one of the biostratigraphically most important Ordovician graptolite groups, including the characteristic Isograptidae and their descendants. The Isograptidae (Plate 9A, D) are usually easily recognized by the perfect bilateral symmetry of the two-stiped, reclined tubaria (Figure 10.16B). Their proximal development, however, can be quite complex, as is seen in some relief specimens, especially those of *Parisograptus* and the pseudisograptids. The typical isograptid symmetry is characterized by a line of symmetry that lies between the sicula and the first theca

(th1¹) (Figure 10.16B), while in most graptoloids the line of symmetry passes directly through the sicula and indicates a maeandrograptid symmetry of th1¹ and th1² (Cooper & Fortey 1982; Figure 10.16A).

Family Isograptidae

Harris (1933) initially recognized the use of the isograptids and their rapid evolutionary changes as key to the biostratigraphy of the Dapingian to Lower Darriwilian time interval, especially the regional Castlemainian Stage of the Australasian succession (cf. VandenBerg & Cooper 1992). Cooper (1973) used the stipe width and number of pendent thecae in the proximal parts of isograptids to establish a detailed biozonation for this time interval that is still used today (Figure 10.17). Any fossil collector may easily recognize the distinctive reclined to nearly scandent arrangement of the stipes in isograptids and their close relatives, no matter how poor the preservation. The elegantly reclined, symmetrical stipes are reminiscent of the wings of bats or angels. One species, based on specimens from Wales, was even named *Arienigraptus angel* (Jenkins, 1982) to depict the impressive shape of the colony.

What can be expected from the isograptids with their reclined stipes? They quickly evolved into scandent, biserial colonies. Initially, the two stipes became reclined to scandent, but still left an opening around the sicula (*Proncograptus, Procardiograptus*: Xiao et al. 1985). Distally, the tubaria kept their two separate stipes. These intermediate forms, especially at their juvenile stages, may be easily misidentified as species of the genus *Isograptus*. The opening quickly closed and a Y-shaped colony formed, which received the taxonomic name *Oncograptus*. The closely related *Cardiograptus* bears an entirely biserial, dipleural colony, and its juvenile specimens resemble *Oncograptus*.

The Manubrium

Flattened specimens of the genera *Pseudisograptus* and *Arienigraptus* do not differ much in shape from the usual isograptids, even though the proximal development of the former two is a bit more

Figure 10.16 Different symmetries in Tetragraptidae and Isograptidae. The maeandrograptid (A) and isograptid (B–D) symmetry. (A) *Tetragraptus reclinatus*. (B) *Isograptus victoriae*. (C) *Parisograptus caduceus*. (D) *Arienigraptus zhejiangensis*. Sicula, dicalycal theca (th1^2) and downward growing part of manubriate thecae (in D) is shaded. Reconstructions based on Maletz & Zhang (2003); Maletz (2011d).

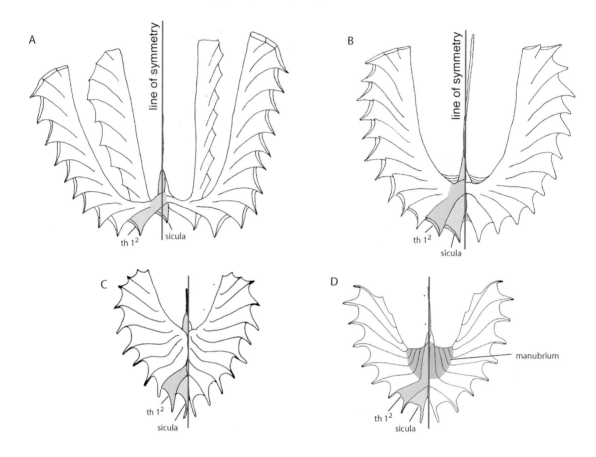

complex. A big surprise comes only when specimens of these taxa are preserved in three dimensions and their development is shown (Figure 10.18). We suddenly see a number of thecae growing in close succession downwards, parallel to the sicula and bending outwards close to their apertures. This initial downward growth of the proximal thecae forms a massive construction, the manubrium (Figure 10.18B, E). Then we start to ask ourselves: what is the reason for the evolution of this complex structure? Obviously, the manubrium would include a lot of building material to make the tubarium stronger and more resistant to any potential attacks and, on the other hand, would change the centre of gravity of the colony. Is this important for the graptolite? We do not know yet why this was achieved, but can only speculate.

The genus *Arienigraptus* first appears in the upper part of the Dapingian, represented by the species *Arienigraptus hastatus*, a robust isograptid with a somewhat elongated sicula (Figure 10.17). A number of species-level taxa evolved quickly during the Dapingian and led to many biostratigraphically useful taxa. Cooper and Ni (1986) were the first to recognize and show the complex construction of the proximal end in *Arienigraptus* and *Pseudisograptus* through the findings of a number of exquisite Chinese specimens.

Figure 10.17 Middle Ordovician biostratigraphy based on isograptids and pseudisograptids (based on Cooper 1973; Cooper & Ni 1986). Graptolite specimens from various sources, not to scale.

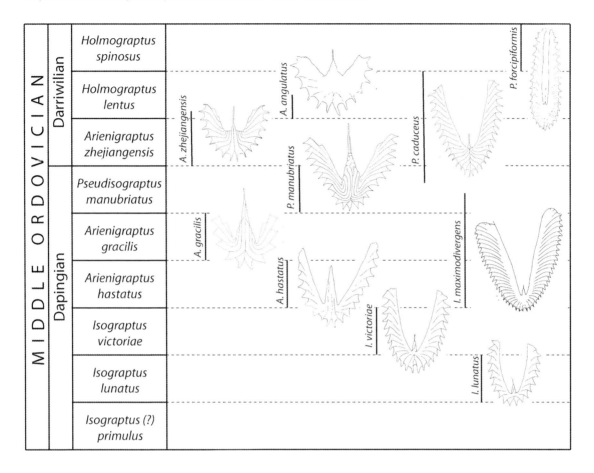

The pseudisograptid manubrium (Figure 10.18) consists of a number of tightly-folded and compacted proximal thecae, which grow parallel downwards along the sicula for a variable length, before they bend outwards. Thus, all these thecae are strongly elongated and can reach lengths of up to 10 mm in some *Pseudisograptus* species. The colony shape of the pseudisograptids can be quite variable, even within a single species or subspecies as shown in *Pseudisograptus manubriatus koi*, making species differentiation somewhat difficult. Cooper (1973) and Cooper and Ni (1986) described the high intraspecific variation of the pseudisograptids and tried to differentiate species and subspecies by using statistical methods. The variation includes not only dimensional factors, but also the construction style of the manubrium and the orientation of the stipes. The colony shapes range from widely open to strongly reclined and nearly scandent, while the stipe width varies considerably and the maximum width is more than double the width in slender specimens (Cooper & Ni 1986).

Scandency and Biserial Tubaria

Scandent graptolite colonies are most commonly identified as biserial graptolites, grouped as Axonophora. However, a number of smaller groups have achieved scandency independently (Figure 10.19). Scandency invariably involves the

Figure 10.18 The manubriate isograptids. (A) *Pseudisograptus manubriatus janus* Cooper and Ni, 1986, GSC 82977, flattened specimen, Melville Island, Arctic Canada. (B) *Arienigraptus* sp., LO 12244t, relief specimen in reverse view, coated with ammonium chlorite, Krapperup drillcore, Scania, Sweden. (C) *Arienigraptus* sp., GSC 82979, flattened, Melville Island, Arctic Canada. (D) *Arienigraptus dumosus* Harris, 1933, GSC 82978, flattened, Melville Island, Arctic Canada. (E) *Pseudisograptus manubriatus* ssp., KR-5b-2, latex cast in reverse view, Krapperup drillcore, Scania, Sweden. Magnification indicated by 1 mm long bar in each photo.

supradorsal connection of the stipes, which leads to enclosure of the sicula in the colony. We have discussed this feature already in the quadriserial phyllograptids, derived from the genus *Tetragraptus*. The axonophorans, however, were the most successful scandent graptolites. Instead of a four-stiped colony, they developed with two stipes united back to back, the biserial condition. The origin of the biserial, axonophoran graptolites has been a matter of debate for a long time and no substantive consensus has yet emerged (Jenkins 1980; Fortey & Cooper 1986; Mitchell 1990; Fortey et al. 2005). This situation may have changed more recently as you will see in Chapter 11. A glimpse into the evolution of biserial graptolites based on latest studies is provided in Figure 10.19. It shows, in the form of a cladistic diagram, the evolutionary relationships of the Glossograptina and provides insight into the multiple origination of biserial graptolites within the group. Biserial, dipleural graptolites originated independently from the genus *Isograptus* at least twice. The *Isograptus victoriae–Isograptus mobergi* lineage led to the genera *Oncograptus* and *Cardiograptus*, but ended quickly with the extinction of the latter two genera in the early Darriwilian (Figure 10.19C–E). A more successful lineage led through *Arienigraptus* and *Pseudisograptus* to the first axonophoran, *Exigraptus* (Figure 10.19 F–G). From the beginning of the Darriwilian on, this group formed the leading branch of the Graptolithina until their extinction in the Lower Devonian.

The detailed proximal end construction of the genus *Parisograptus* (Figure 10.20), known only in a few relief specimens from the Hengtang section of Zhejiang Province, China (Chen & Zhang 1996; Maletz & Zhang 2003), is proximally biserial and

Figure 10.19 Evolution of scandency in the Glossograptina. (A) *Tetragraptus reclinatus*. (B) *Isograptus lunatus*. (C) *Isograptus mobergi*. (D) *Oncograptus*. (E) *Cardiograptus*. (F) *Pseudisograptus*. (G) *Exigraptus*. (H) *Bergstroemograptus*. (I) *Cryptograptus*. (J) *Glossograptus*. (K) *Kalpinograptus*. Scandency seen in (E) (Isograptidae, through *Isograptus*), (G) (Isograptidae, through *Pseudisograptus*, leading to the Axonophora), (H–J) (Glossograptidae, through *Isograptus*; (K) lost scandency secondarily). Illustrations not to scale, based on various sources.

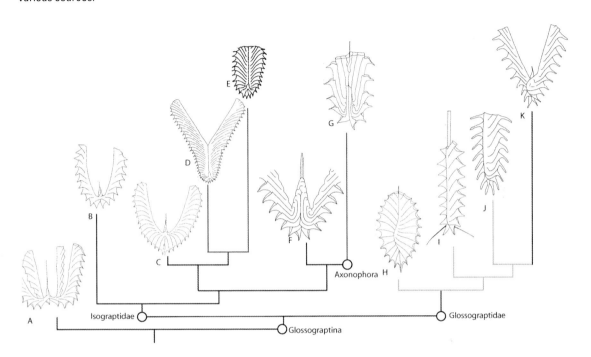

dipleural. The specimens show the origins of the proximal thecae like a string of pearls in a vertical sequence in which the two stipes are completely separated by a median suture (Figure 10.16C, 10.20). The construction superficially resembles that of the genus *Arienigraptus*, but differs in the vertically successive thecal origins (vs. horizontally successive in the latter) (Figure 10.16C–D). The example nicely explains how difficult it may be to reconstruct three-dimensionally the astogeny based only on flattened material, and how important it is to get the best available information from the specimens for a proper taxonomic treatment.

Family Glossograptidae

The Glossograptidae (Plate 9B, C, E) include a small group of compact graptolites with a biserial, monopleural development of the colony (Figure 10.19I–K).

The thecae are simple, tubular in shape with distinct ventral rutellae in most species. The characteristic paired, lateral apertural spines growing from the sicular aperture are typically elongated (Figure 10.21D). In derived forms, lateral apertural spines on distal thecae are also common (e.g. *Glossograptus*, *Paraglossograptus*). A meshwork of lists on the two sides of the thecal apertures is secreted in species of the genus *Paraglossograptus* (Figure 10.21E). This feature, called a lacinia, is formed from round bars creating a meshwork of lists, which resembles the lacinia of the Lasiograptidae in the Upper Ordovician (see Chapter 11). The development of these two types of lacinia is phylogenetically unrelated and independently achieved in discrete time intervals, probably as a result of evolutionary convergence.

Due to the complex overlap of the two stipes of the colony, the proximal development is difficult, if at all possible, to see and understand, as during

Figure 10.20 *Parisograptus caduceus* (Salter in Bigsby, 1853), (A) NIGP 126527, obverse view; (B, D) NIGP 12523, reverse view. (C) NIGP 126522, obverse view. Scale indicates 1 mm in each illustration.

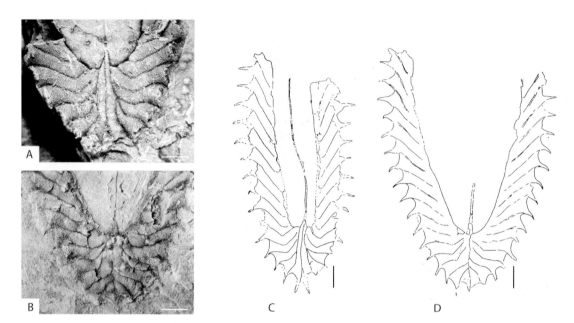

the astogeny the later thecal series cover and enclose the sicula and proximal thecae completely. Therefore, the discussion on the proximal development has been difficult (Bulman 1947; Strachan 1985; Maletz & Mitchell 1996). A series of well-preserved, isolated juveniles of *Cryptograptus* and *Glossograptus* allowed Maletz and Mitchell (1996) to reconstruct the simple, isograptid-type proximal development in these two genera, and infer the subsequent development and growth of thecae.

The origination of the biserial, monopleural colony apparently took place at approximately the same time as that of the biserial, dipleural colony. Thus, they can be regarded as coincidentally similar, representing an analogous development, or showing an inherent trend in the evolution of the graptoloids.

Glossograptids proved successful in the marine planktic ecosystem of the Middle Ordovician, and a few taxa even survived into the late Katian (Upper Ordovician). However, they lost their footing, became rare, and eventually went extinct in the late Katian. The youngest member of this group is *Sinoretiograptus mirabilis* from the Upper Ordovician *Dicellograptus complexus* Biozone of South China (Mu et al. 1974) and Australia (VandenBerg & Cooper 1992). The biserial, monopleural Glossograptidae, thus, were by far outlived by the biserial, dipleural Axonophora, which survived into the Upper Silurian and were ancestral to the monograptids that dominated the Silurian to Lower Devonian.

Little is known of the evolutionary origins of the Glossograptidae, even though it is clear from details of the colony construction that they originated from an isograptid ancestor through the development of a monopleural colony. The best candidate to show part of the early transition is the genus *Parisograptus* with its initially biserial, dipleural development and the vertically successive origin of the proximal thecae (Figure 10.16C). In *Parisograptus* the two stipes form an initially biserial, dipleural colony, but there is no indication of how this could develop into a monopleural construction. The change might include a considerable and dramatic change in the growth direction of one of the two stipes. The second stipe has to be transferred dextrally around the sicula to the obverse side of the sicula to achieve a monopleural

Figure 10.21 The Glossograptidae. (A) *Cryptograptus schaeferi* Lapworth, 1880, Table Head Group, western Newfoundland. (B–C) *Nanograptus lapworthi* Hadding, 1915. (B) Specimen on GSM 5495 with lectotype of *Rogercooperia phylloides* (Elles & Wood, 1908). (C) LO 2743t (paratype). (D) *Cryptograptus* sp., SPS 28, western Newfoundland. (E) *Paraglossograptus tentaculatus* (Hall, 1865), showing partially preserved lacinia. (F) *Glossograptus hincksii* (Hopkinson, 1872), LO 2370t, Scania, Sweden (Hadding 1913, pl. 2, Fig. 6). Scale indicated by 1 mm long bar in each photo.

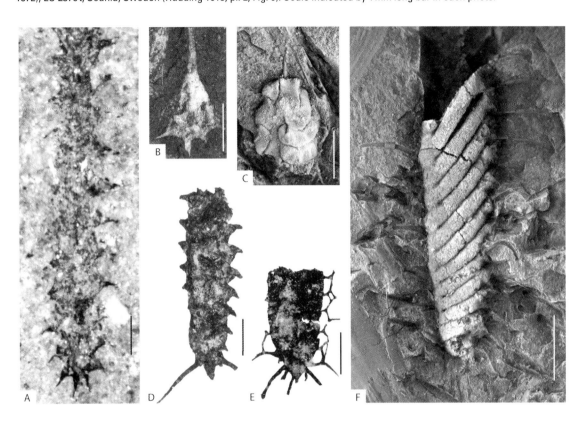

development (Ni & Cooper 1994; Maletz & Mitchell 1996). A partial monopleural development is first seen in *Bergstroemograptus crawfordi* (Figure 10.19H). The species is morphologically similar to *Skiagraptus*, but their relationships are unknown as the latter is recorded only in flattened specimens. Whittington and Rickards (1969) identified a few isolated proximal ends of *Bergstroemograptus* as *Skiagraptus* sp., which demonstrate the partial monopleural development of the colony and the complete covering of the sicula on both sides by the two stipes. The specimens clearly show a dextral mode of displacement of the two stipes that led to the monopleural colonies in derived taxa of the Glossograptidae (Figures 10.21, 10.22).

Both *Glossograptus* and *Cryptograptus* have been recognized from the basal Darriwilian throughout to the Lower Katian. They suddenly appear as fully developed new taxa in the earliest Darriwilian, suggesting that their potential ancestor may be found in the Dapingian. Unfortunately, as faunas representing the upper Dapingian (the Yapeenian in the Australasian chronostratigraphy) are rarely found and documented, the origins of the Glossograptidae cannot be traced yet.

One of the more common design features of the Glossograptina was the attenuation of the fusellum. It is first seen in the genera *Isograptus* and *Parisograptus*. Many species of these two genera possessed a fairly thin fusellum, and the specimens appear nearly transparent in flattened shale

Figure 10.22 Proximal development of the Glossograptidae. (A) *Cryptograptus insectiformis* Ruedemann, 1908 (based on Maletz & Mitchell 1996). (B) Reconstruction of sicula and first theca in *Cryptograptus schaeferi* (based on Maletz & Mitchell 1996). (C) *Cryptograptus schaeferi* Lapworth, 1880, Table Head Group, western Newfoundland, Canada, juvenile showing complete sicula with lateral apertural spines and part of first theca; note the thickened rim around the sicular aperture that will become part of the list structure of younger cryptograptids. (D) *Glossograptus acanthus* Elles & Wood, 1908, proximal end in relief. Adapted from Ni and Cooper (1994, Fig. 1), with permission from Taylor & Francis. (E) Thecal diagram of *Glossograptus*, showing monopleural arrangement (after Maletz & Mitchell 1996). Illustrations not to scale.

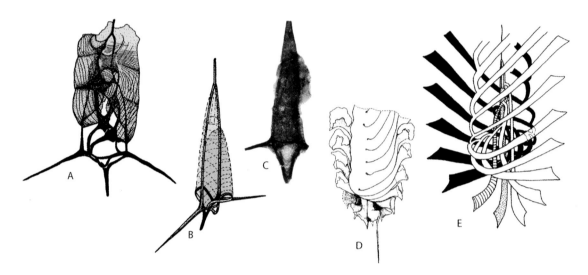

specimens. This may also explain the reason why they have been so rare in isolated material (Maletz 2011d), despite the fact that more robust taxa can be found to abound in chemical residues. Thin fusellum is also an important characteristic of the genus *Bergstroemograptus*, and the specimens described by Whittington and Rickards (1969) represent only the most robust parts of the colonies, the sicula and the surrounding thecae. The attenuation of the fusellum (Figure 10.21a; 10.22) may also account for the near non-preservation of the sicula and the first few thecae in some derived species of *Cryptograptus* from the Upper Ordovician (Maletz & Mitchell 1996). The sicula and the first two thecae of *Cryptograptus insectiformis* (Figure 10.22A) are preserved as a number of lists and looped lists outlining the thecae, but the fusellum is not found. In the slightly older Middle Ordovician species *Cryptograptus schaeferi* (Figure 10.22B, C), the sicula and first thecal pair still bear a complete but thin fusellum. Based on this material, Maletz and Mitchell (1996)

were able to interpret reasonably the proximal development of the genera *Cryptograptus* and *Glossograptus*.

It is noticeable that many species of the Glossograptidae developed thecal spines in various positions. Initially, two lateral apertural spines are attached to the sicular aperture, as is seen in *Cryptograptus* and *Glossograptus* (Figure 10.23B, C), but soon lateral apertural spines appear in the thecae of *Glossograptus*, and regularly developed lateral colony spines were also formed (Figure 10.23C). Whittington and Rickards (1969) for the first time described the position and development of these spines from isolated material, as the spines are often broken or displaced in flattened shale material and their precise positions cannot be ascertained. The lateral apertural spines of the thecae are connected through some strips of cortical tissue and eventually form a somewhat irregular meshwork in the shape of four ladder-like constructions, the lacinia, in the genus *Paraglossograptus* (Figure 10.23 D).

Figure 10.23 Tubarium reconstruction of the Glossograptidae. (A) *Bergstroemograptus*. (B) *Cryptograptus*. (C) *Glossograptus*. (D) *Paraglossograptus*. (E) *Corynoides*. (F) *Corynites*. (G) *Kalpinograptus*. Reconstructions (JM) not to scale.

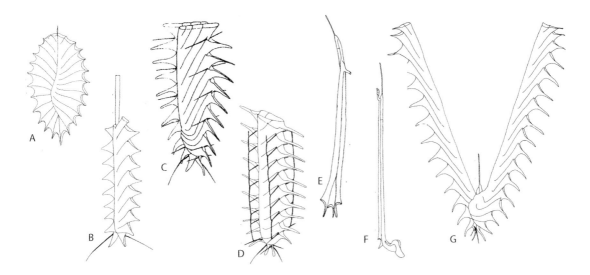

The lacinia development varies considerably among the species referred to *Paraglossograptus* and is largely lacking in some species.

During the later evolutionary history of the Glossograptidae, additional changes occurred. Secondarily, the biserial, monopleural colony shape was modified into a monopleural-reclined colony in the genus *Kalpinograptus* (Figures 10.19 K, 10.23 G), mimicking the (at the time already extinct) genus *Isograptus* and its relatives. The differentiation of a Middle Ordovician *Kalpinograptus* from an *Isograptus* specimen is quite difficult in shale material. The finding of relief specimens (Jiao 1977; Finney 1978; Maletz & Mitchell 1996) allowed an astogenetic reconstruction of the taxon and a differentiation from *Isograptus*. Except for the distal, two-stiped colony, *Kalpinograptus* is structurally identical to *Glossograptus*, but there is no information on the presence of lateral apertural thecal spines, except for those at the sicula of the likely synonymous *Apoglossograptus* (Finney 1978).

A major problem arises from the inclusion of the Upper Ordovician genera *Corynoides* (Figure 9.23 E) and *Corynites* (Figure 10.23 F) in the Glossograptidae, a suggestion first advocated by Maletz and Mitchell (1996). Bulman (1944)

established the family Corynoididae on the genus *Corynoides*, acknowledging the uncertain relationships of this strange graptolite. Both genera possess only a small number of thecae, and thus show considerably reduced colonies. This reduction in size leads to a reduction also in available colony features, and makes an interpretation and assignment of them to any certain group of graptoloids difficult. The long and slender siculae with the high prosicular origin of the first theca indicates it is likely to be derived from a glossograptid ancestor, while an origin from a dichograptid cannot be excluded. However, all dichograptids had been extinct for quite a while by that time, making a dichograptid origin unlikely. The only available group with a development remotely comparable to the corynoidids are the glossograptids and especially the genera *Glossograptus* and *Kalpinograptus*. The species of the genus *Cryptograptus* possess a reduced proximal fusellum and – more importantly – a metasicular origin of the first theca, which is shared with some *Glossograptus* taxa, but differs from the high, prosicular origin of the first theca in *Corynoides*. A considerable elongation of thecae is present in some glossograptids, but not as extreme as in *Corynoides*. The thecal elongation in *Corynoides* is only comparable with the elongation of the siculae in species of the genus *Pseudisograptus* and especially in the Llandovery (Silurian) monograptid genus *Coronograptus* and its relatives (see Chapter 13).

Outlook

The Ordovician Dichograptina is one of the most diverse and most disparate groups of the graptolites that ever evolved. The estimation of its diversity and the variation in tubarium construction has only been explored in part, and astonishing new constructions emerge from time to time. The reclined isograptids and pseudisograptids, the pendent didymograptids, and other groups have long been known to be extremely useful in biostratigraphy and paleobiogeography, when their morphologies and phylogenetic relationships are well understood. Thus, the biostratigraphical significance of these taxa is quite well known, and relevant successions have been established. Further research on the Dichograptina may concentrate not only on the improvement of their taxonomy and biostratigraphical resolution, but also on the biogeographical implications of their distribution.

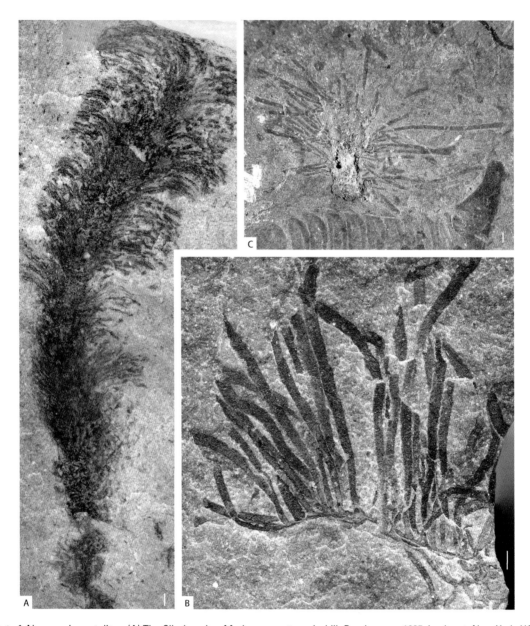

Plate 1 Algae and graptolites. (A) The Silurian alga *Medusaegraptus mirabilis* Ruedemann, 1925, Lockport, New York, USA. (B) *Sphenoecium wheelerensis* Maletz & Steiner, 2015, an early rhabdopleurid graptolite, Marjum Fm., Utah, USA. (C) *Yuknessia simplex* Walcott, 1919, holotype, an early pterobranch, Burgess Shale, British Columbia, Canada. Bar indicates 1 mm in each photo.

Graptolite Paleobiology, First Edition. Jörg Maletz.
© 2017 John Wiley & Sons Ltd. Published 2017 by John Wiley & Sons Ltd.

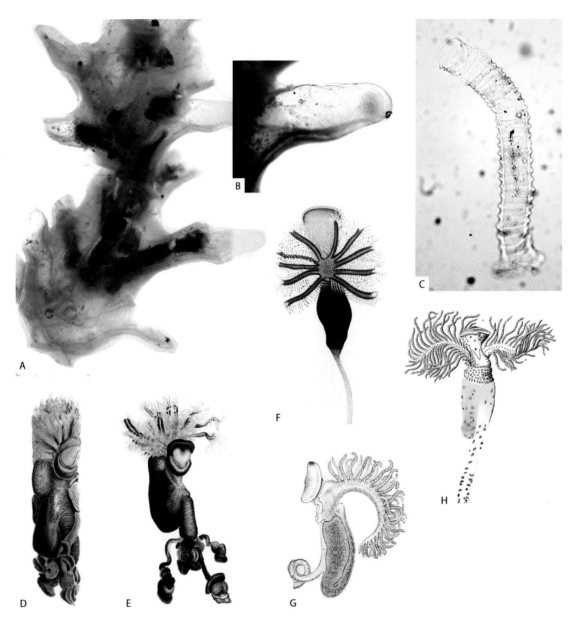

Plate 2 (A, B) *Cephalodiscus (Idiothecia) levinseni* (Harmer, 1905), part of colony (A) and enlargement of aperture showing fuselli (B), origin uncertain, recent. (C) *Rhabdopleura normani* Allman, 1869, single empty tube, Bergen, Norway. (D, E) *Cephalodiscus* Ridewood, 1907, zooid with immature juveniles (from Ridewood 1907, pl. 3, Figs. 7–8). (F) *Cephalodiscus inaequatus* Andersson, 1907, female zooid (from Andersson 1907, pl. 1). (G) *Rhabdopleura normani*, zooid in lateral view (from Schepotieff 1906, pl. 23). (H) *Rhabdopleura normani*, zooid in dorsal view (from Lankester 1884, pl. 38).

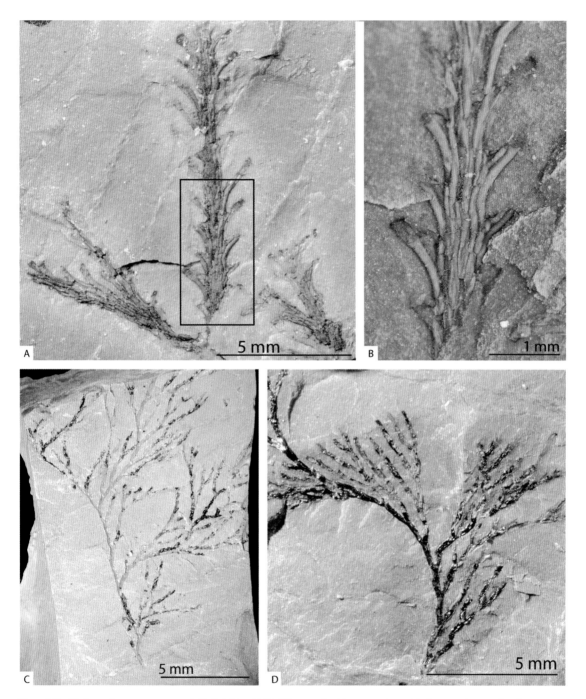

Plate 3 (A–B) *Acanthograptus sinensis* Hsü & Ma, 1948, (B) coated with ammonium chlorite to enhance details. (C) *Dendrograptus*(?) *mui* Yu et al., 1985. (D) *Aspidograptus*(?) *uniflexilis* Yu et al., 1985. All specimens from the Tremadocian of China (photos by A. Kozłowska and J. Maletz).

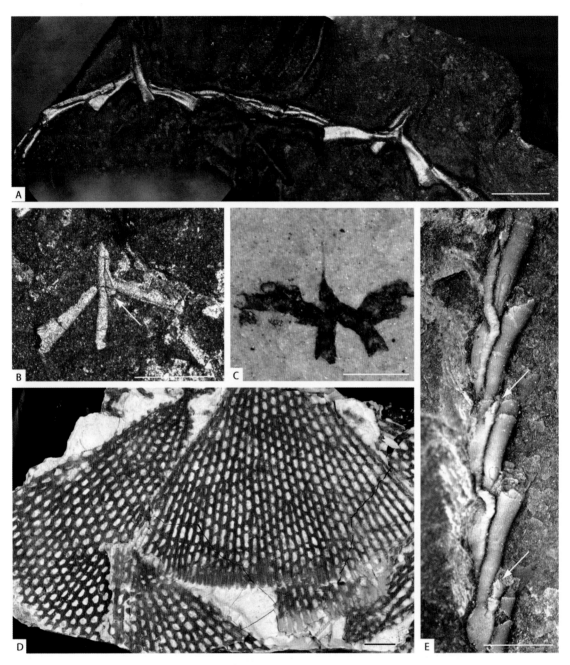

Plate 4 (A) *Adelograptus tenellus* (Linnarsson, 1871), Scania, Sweden, LO 2257 t-2258 t. (B) *Ancoragraptus bulmani* (Spjeldnaes, 1963), PMO 214.030, Slemmestad, Norway, showing sicular bitheca (arrow). (C) *Psigraptus lenzi* (Jackson, 1967), juvenile, Jilin, China. (D) *Sagenograptus murrayi* (Hall, 1865), Fezouata Biota, Zagora area, Morocco (photo provided by J.C. Gutiérrez-Marco). (E) Anisograptid fragment in relief, showing bithecae (arrows), Green Point, western Newfoundland. Scale indicated by 1 mm long bar in A–C, E, 10 mm in D.

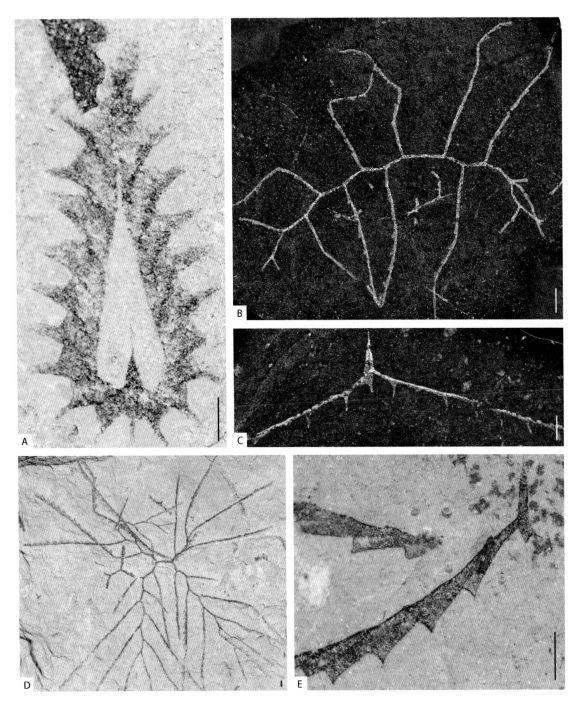

Plate 5 (A) *'Didymograptus' eocaduceus* (Harris, 1933), NMVP 319254, Victoria, Australia. (B) *Sigmagraptus praecursor* Ruedemann, 1904, NMVP 320445B. (C) *Kinnegraptus* sp., NMVP 318595, Victoria, Australia. (D) *Goniograptus thureaui* M'Coy, 1876, NMVP 315040, Victoria, Australia. (E) *Azygograptus lapworthi* Nicholson, 1875, NIGP 20973, South China. Scale indicated by 1 mm long bar in each photo. (A–D by A.H.M. VandenBerg.)

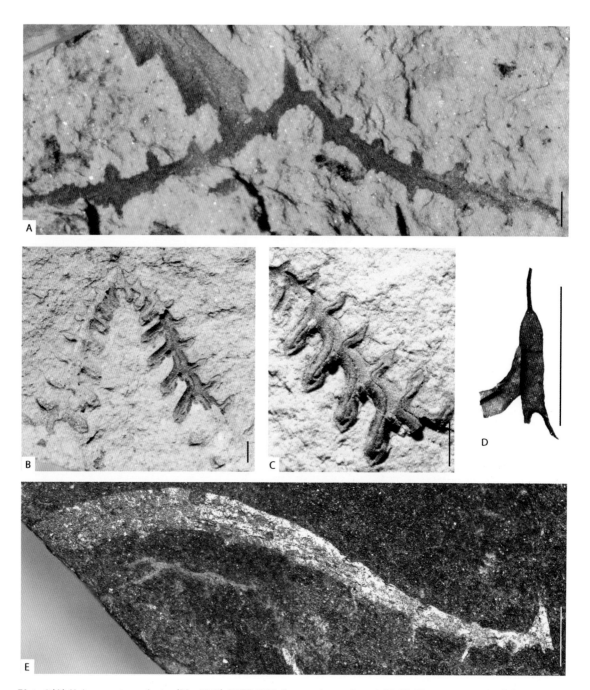

Plate 6 (A) *Holmograptus spinatus* (Mu, 1957), NIGP 8904, flattened, weathered. (B–C) *Sinograptus typicalis* Mu, 1957, holotype, NIGP 8909, partial relief (B) and detail (C) of thecal construction. (D) *Nicholsonograptus fasciculatus* (Nicholson, 1869), isolated proximal end, showing prosicula and metasicula, Table Head Group, western Newfoundland. (E) *Nicholsonograptus fasciculatus* (Nicholson, 1869), LO 3315 T (holotype of *Azygograptus falciformis* Ekström, 1937), flattened. Scale indicated by 1 mm long bar in each photo.

Plate 7 *Orthodichograptus robbinsi* Thomas, 1973, NMVP 73827 (left specimen, holotype) and NMVP 83089 (right specimen, paratype), Victoria, Australia. (Photo by A.H.M. VandenBerg.)

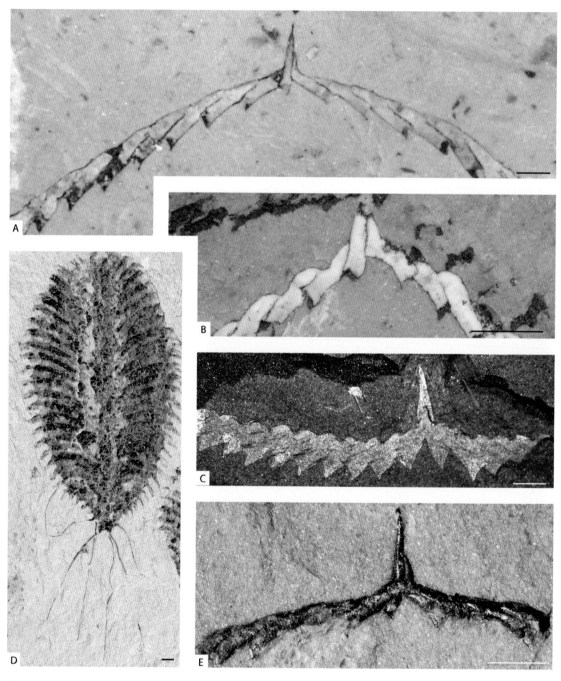

Plate 8 (A) *Cymatograptus bidextro* Toro and Maletz, 2008, holotype in obverse view, Eastern Cordillera, Argentina. (B) *Cymatograptus bidextro* Toro and Maletz, 2008, specimen showing strong prothecal folds, Eastern Cordillera, Argentina. (C) *Cymatograptus undulatus* (Törnquist, 1901), Hunneberg, Sweden. (D) ?*Phyllograptus typus* Hall, 1865, NMVP 318956, specimen with long proximal spines, Victoria, Australia (photo by A.H.M. VandenBerg). (E) *Baltograptus geometricus* (Törnquist, 1901), NIGP 32160, China. Scale indicated by 1 mm long bar in each photo.

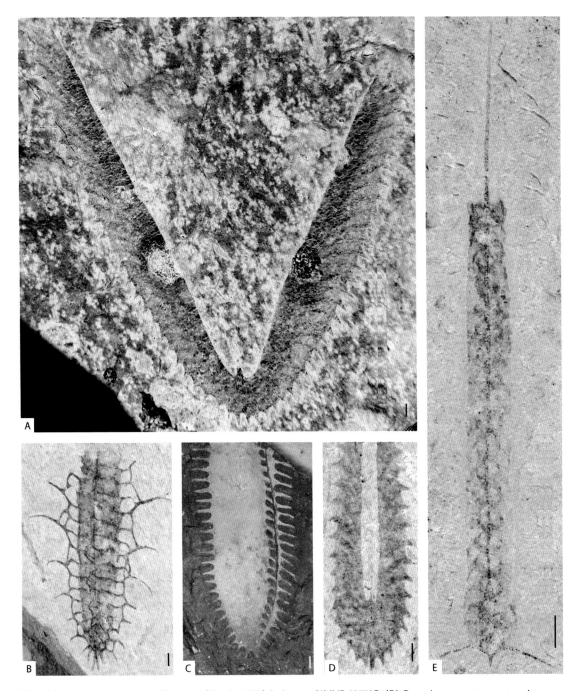

Plate 9 (A) *Isograptus maximodivergens* (Harris, 1933), holotype, NMVP 28770B. (B) *Paraglossograptus tentaculatus* (Hall, 1865), NMVP 319302. (C) *Glossograptus* sp., holotype of *Phyllograptus typus parallelus* Bulman, 1931, specimen was painted white on the slab to enhance visibility, background painted black. (D) *Parisograptus forcipiformis* (Ruedemann, 1904), NMVP 34855. (E) *Cryptograptus schaeferi* Lapworth, 1880, NMVP 56066. Scale indicated by 1 mm long bar in each photo. Specimens from Victoria, Australia, except (C) from Korpa, Bolivia. Photos by A.H.M. VandenBerg, (C) by E.D. Brussa.

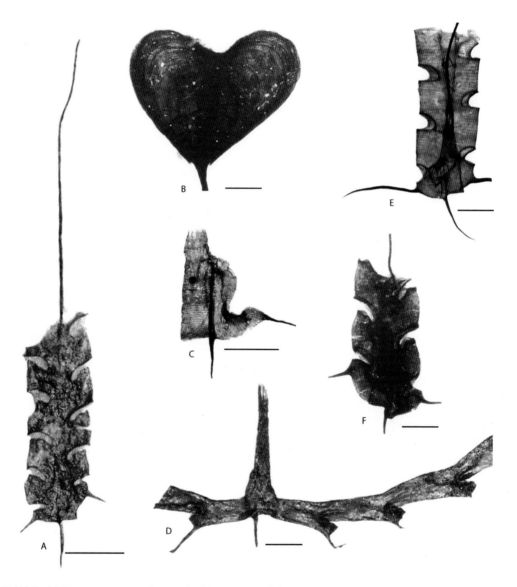

Plate 10 (A) *Archiclimacograptus* sp., flattened, with long nema. (B) *Archiclimacograptus decoratus* (Harris & Thomas, 1935), nematularium. (C) *Archiclimacograptus* sp., juvenile, showing fusellar construction. (D) *Dicellograptus flexuosus* Lapworth, 1876, proximal end. (E) *Climacograptus cruciformis* VandenBerg, 1990, bleached specimen. (F) *Archiclimacograptus* sp., bleached, showing thickened thecal rims. Scale is 1 mm in (A) and 0.5 mm in (B–F). (A–C, F) from western Newfoundland, (D–E) Viola Limestone (photos by D. Goldman).

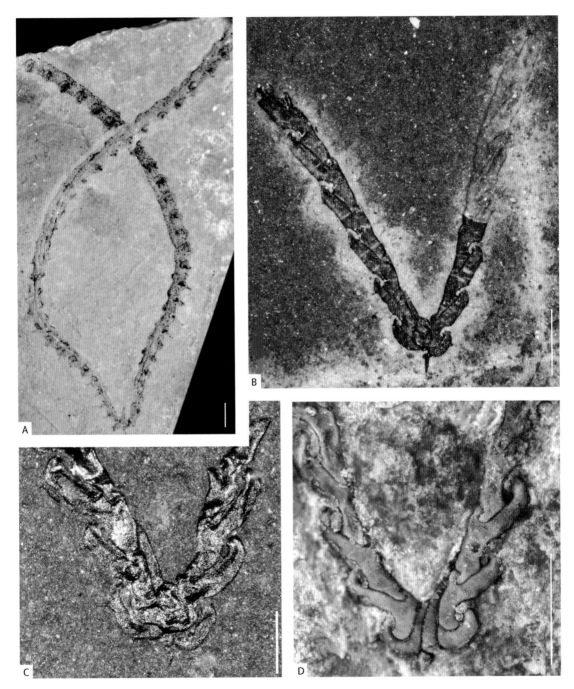

Plate 11 (A) *Dicellograptus caduceus* Lapworth, 1876, NMVP 68810A, Victoria, Australia. (B) *Jiangxigraptus divaricatus* (Hall, 1859), reverse view, Saergan Formation, Kalpin, Xinjiang. (C) *Jiangxigraptus sextans* (Hall, 1847), reverse view, Saergan Formation, Kalpin, Xinjiang. (D) *Jiangxigraptus vagus* (Hadding, 1913), SMF 75781, obverse view, relief, glacial boulder, Laerheide, Germany, coll. Schöning. Scale indicated by 1 mm long bar in each photo. Photos provided by A.H.M. VandenBerg (A) and Yuandong Zhang (B–C).

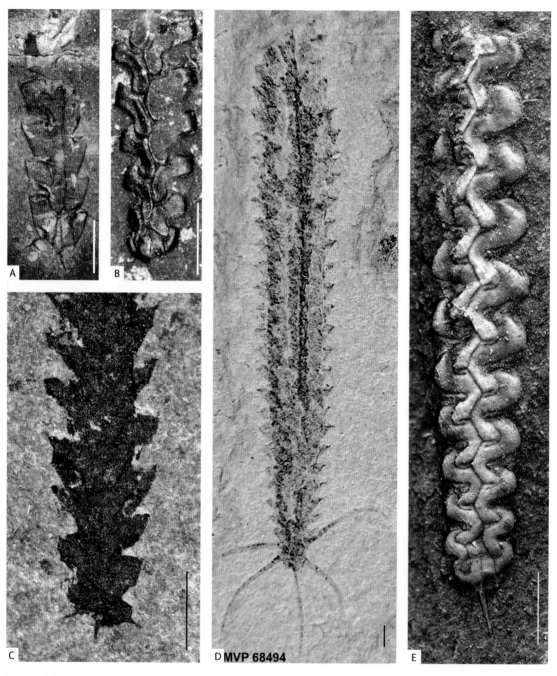

Plate 12 (A) *Hustedograptus teretiusculus* (Hisinger, 1840), Saergan Formation, Subashigou, Xinjiang.
(B) *Haddingograptus eurystoma* (Jaanusson, 1960), Dawangou, Kalpin, Xinjiang. (C) *Rectograptus* sp., L'Egaré
Motel, Neuville, Québec, Canada. (D) *Orthograptus pageanus maximus* Goldman, 1995, NMVP 68494, flattened,
weathered, Victoria, Australia. (E) *Haddingograptus oliveri* (Bouček, 1973), Slemmestad, Norway. Scale indicated
by 1 mm long bar in each photo. Photos provided by A.H.M. VandenBerg (D) and Yuandong Zhang (A–B).

Plate 13 (A) *Metaclimacograptus internexus* Törnquist, 1893, LO 1110t, polished pyritic section in black shale. (B–C) *Skanegraptus janus* Maletz, 2011c, holotype. (D) *Cephalograptus cometa* Geinitz, 1852, LO 1121t, latex cast. (E) *Metaclimacograptus internexus* Törnquist, 1893, LO 1111t, polished section. (F) *Petalolithus palmeus* Barrande, 1850, LO 1119t, polished section. (G) *Petalolithus minor* Elles, 1897, LO 1113t. All specimens from Scania, Sweden. Scale indicated by 1 mm long bar in each photo.

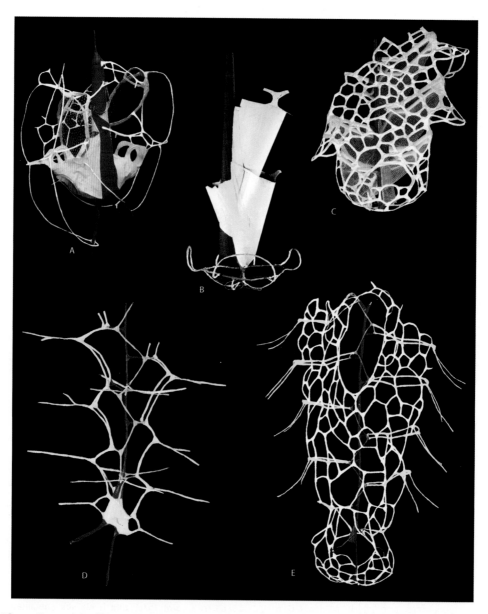

Plate 14 Graptolite models of Nancy H. Kirk and Denis E.B. Bates. (A) *Pipiograptus hesperus*, Lasiograptidae. (B) *Pseudorthograptus insectiformis*, Petalolithinae. (C) *Retiolites geinitzianus*, Retiolitinae. (D) *Orthoretiolites hami*, Lasiograptidae. (E) *Pseudoplegmatograptus obesus*, Retiolitinae. Illustrations not to scale. Photos provided by D.E.B. Bates.

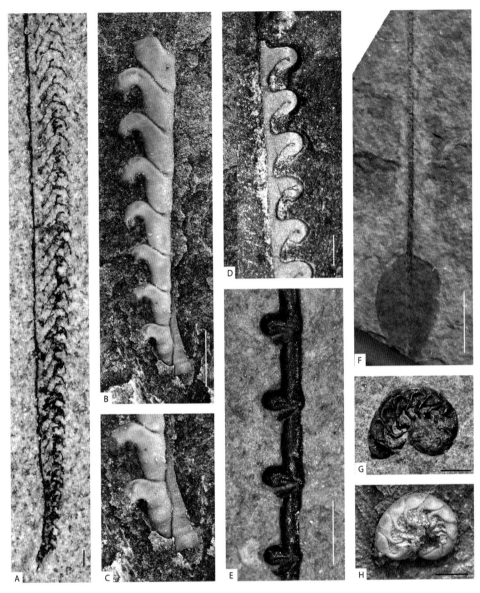

Plate 15 (A) *Monograptus priodon* (Bronn, 1835), Bornholm, JM 60. (B–C) SMF 75780, coll. R. Klafack, specimen in relief (B) and sicula (C) showing growth lines, coated, glacial boulder, Mecklenburg-Vorpommern, Germany. (D) *Campograptus lobiferus* (M'Coy, 1850), LO 1028 T, polished section, Scania, Sweden. (E) *Streptograptus nodifer* (Törnquist, 1881), Dalarna, Sweden. (F) *Monograptus pala* Moberg, 1893, LO 1090 T, counterpart of holotype, Sweden. (G) *Cochlograptus veles* (Richter, 1871), LO 1527 t, Scania, Sweden. (H) *Cochlograptus veles* (Richter, 1871), LO 1071 t, relief specimen, coated, Dalarna, Sweden. Scale indicated by 1 mm long bar in each photo, except for (F), where it is 10 mm.

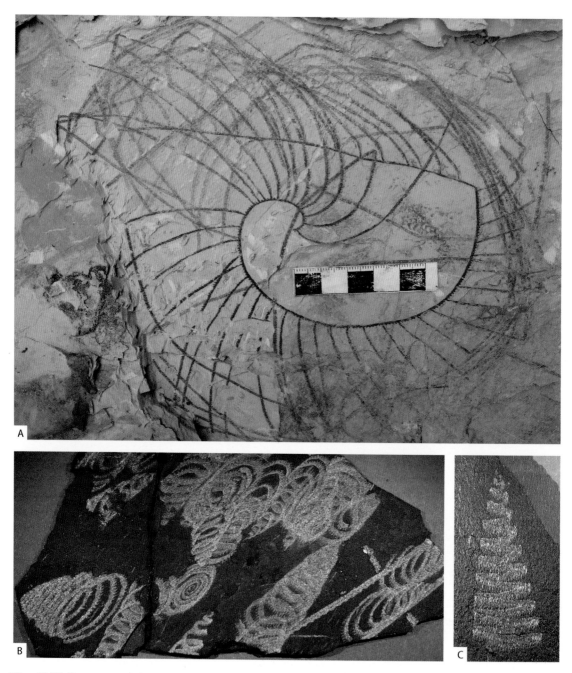

Plate 16 (A) *Cyrtograptus robustus* Fu, 1986, Shaanxi Province, China. (B–C) *Spirograptus turriculatus* (Barrande, 1850), from Thuringia, Germany, coll. R. Hundt. (B) NMG 9900, several tectonically distorted specimens. (C) NMG 10148, tectonically elongated tubarium. Photos provided by Wang Jian (A) and Frank Hrouda (B–C).

The Biserial Graptolites

Jörg Maletz

A dramatic rearrangement of the thecae in the graptolite tubarium was initiated in the early Middle Ordovician. Biserial graptolites with two series of thecae growing back to back along the nema appeared in the latest Dapingian and fairly suddenly evolved into a multitude of genus-level taxa in the early Darriwilian. This explosive evolution of the group probably led to the slow demise of the older dichograptid and sinograptid graptoloids and their eventual extinction. The change may be seen as one of the most dramatic turnovers in the course of graptoloid evolution, and is unprecedented in the history of the graptolites. Never again, except possibly for the emergence of the monograptid graptoloids in the Lower Silurian, can we see a replacement of one fauna by another one at such a magnitude. Almost all previously established graptolite groups disappeared within a relatively short time interval.

So what was the advantage of the biserial axonophorans? Why did they evolve at all? Why were they so successful? Many questions and not a single answer, not even an idea of what may have been the reason for this dramatic takeover. Let's just accept the inevitable and look into the bright future of the axonophoran graptolites.

The Axonophora Concept

Axonophora – quite an unusual name for a group of graptolites, as most graptolite names include an ending like *-graptus* at the genus level or a similar one in higher-level taxonomic names. The term Axonophora describes graptolites with an axis and is based on a translation from the Greek language. Fritz Frech (1897) introduced the name for a group of uniserial and biserial graptolites in which the stipe(s) follow the nema as the main axis (Plate 10A). At least, this is how we understand the structure nowadays. Obviously, the term is taken from the idea that the nema is supporting the colony, which appears to be correct in the derived Monograptidae, but not necessarily in the biserial taxa discussed in this chapter. Actually, this chapter covers only a part of the Axonophora, namely the biserial graptolites and not the derived axonophorans, the monograptids (Chapter 13) and excluding also the unusual retiolitids, a group of biserial axonophorans that are covered in Chapter 12.

The nema has been present in earlier graptoloids, but never attained the central position or importance in the construction of the tubarium as in the Axonophora. It was a short rod at the tip of the sicula, sometimes adorned with a nematularium of some sort, but rarely preserved in the fossil record as a longer and more prominent feature (Figure 11.1A). The axonophoran clade includes biserial, dipleural (Figure 11.1B) and derived, uniserial (Figure 11.1C) graptoloids, all maintaining the elongated nema in their colonies. Frech (1897) introduced the name Axonophora for uniserial and biserial, dipleural graptolites with a nema leading the growth of the stipe, but largely misinterpreted the colony development by accepting Ruedemann's (1895) interpretation of the synrhabdosomes of biserial graptolites with the development of a pneumatophore and gonangia (see Maletz 2015). However, he correctly recognized the central position of the nema in the

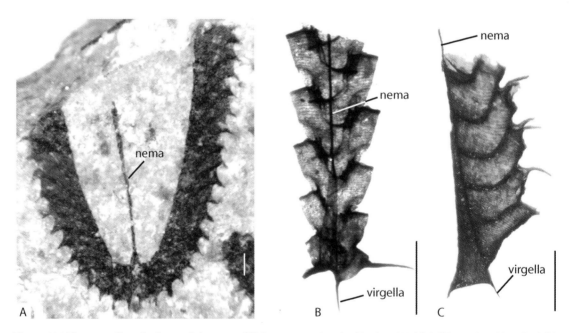

Figure 11.1 The graptolite tubarium and the nema. (A) *Isograptus victoriae* Harris, 1933, Vinini Formation, Nevada, USA, completely free nema between two reclined stipes. (B) *Rectograptus intermedius* (Elles & Wood, 1907), nema incorporated in tubarium, visible through the colony in this infrared photo. (C) *Saetograptus leintwardinensis* (Hopkinson in Lapworth, 1880), glacial boulder, Nienhagen northern Germany, nema visible on dorsal side of stipe and distally of thecae. Scale indicated by 1 mm long bar in each photo.

growth of the graptolite colony as a fundamental feature for tubarium construction. Fortey and Cooper (1986) and Fortey et al. (2005), in discussing the early evolution of the axonophorans, preferred the name Virgellina for this clade, based on another characteristic feature, the virgellar spine on the sicula. Maletz et al. (2009, p. 14) reintroduced the concept of the Axonophora for modern graptolite taxonomy based on a cladistic analysis.

Many problems still remain with the origin and early evolution of the axonophorans, as little information on proximal development and colony structure exists for most early taxa. Mitchell et al. (1995), Fortey et al. (2005) and Maletz (2010a) discussed the general transition of the Arienigraptidae to the Axonophora and provided ample evidence for the understanding of the stepwise change from the reclined isograptids (Figure 11.1A) to the first biserial, dipleural colonies through cladistic analyses. The base of the axonophoran clade was taken at various nodes of the attained cladograms, and a consensus does not exist. Maletz et al. (2009, Fig. 6) separated the early biserials of the *Undulograptus austrodentatus* group (now *Levisograptus*: see Maletz 2011a) as stem-axonophorans and even included the Arienigraptidae sensu Maletz and Mitchell (1996) in the Pan-Axonophora. Using this concept, the "stem-group Pan-Axonophora" included reclined to biserial taxa, bracketing the transition, but not defining the two groups clearly.

Maletz (2014a) preferred to include all biserial, dipleural taxa in the Axonophora and included even the biserial, pseudisograptid-type genera *Exigraptus* and *Apiograptus*. In this way, the Axonophora are easily identified, even in poorly preserved material. The defining synapomorphy of the Axonophora is the biserial, dipleural colony shape engulfing a central nema between the dorsal sides of the two stipes. In the derived Monograptidae, a second stipe is not developed and the nema is leading the growth of the single stipe (Figure 11.1C).

Early Biserial Axonophorans

Our understanding of the evolutionary relationships of the early biserial taxa is still in its infancy. Many species are poorly known from flattened material, and constructional details are unknown. This may seem surprising, especially as the basic evolutionary steps of the transformation can be reconstructed from the fossil record, but an early diversification appears to mislead the available reconstructions. Maletz (2011a) introduced the basal Darriwilian genus *Levisograptus* as probably closest to the origin of the axonophorans. The species *Levisograptus austrodentatus* has often been identified as one of the earliest biserial graptolites (Fortey et al. 1990, 2005; Maletz 1992b; Mitchell & Maletz 1995), and was used to define the base of the Darriwilian Stage of the Ordovician System (Chen & Mitchell 1995; Mitchell & Maletz 1995; Mitchell et al. 1997). The few even older biserial graptolites that have been found include species of the genera *Exigraptus* and *Apiograptus* (see Ni & Xiao 1994) and the robust *Levisograptus sinodentatus* (see Mitchell 1994). Maletz (2011a) provided an overview of the biostratigraphical ranges of many of the early axonophorans (Figure 11.2), but did not discuss the detailed evolutionary relationships of these. There are indications that the evolutionary diversification of the axonophorans was a late Dapingian to early Darriwilian event, as many typical features of later axonophorans appear already in the earlier taxa. The biserial, dipleural tubarium shape was established at the base of the Darriwilian and a diversification had already started. Colony design and thecal apertural developments are quite variable in *Levisograptus* and its closest relative *Exigraptus* (see Mitchell 1994; Maletz 1998b, 2011a). The thecal apertures may be straight with slight rutelli (*Levisograptus sinodentatus*), spined (*Levisograptus sinicus*), with spines on the first thecal pair and distally with introverted apertures and lateral lappets (*Levisograptus austrodentatus*), or rutellate with lateral lappets (*Exigraptus uniformis*). Even a species with geniculate thecae existed in *Levisograptus primus* (see Figure 11.2), but all share a nearly identical proximal development pattern U astogeny. This heterogeneous development may indicate an early diversification of the group that has barely been noted and needs detailed investigation before the evolutionary patterns of early biserial graptolites can be understood.

Figure 11.2 Early biserial axonophorans and their biostratigraphy (based on Maletz 2011a, Fig. 4).

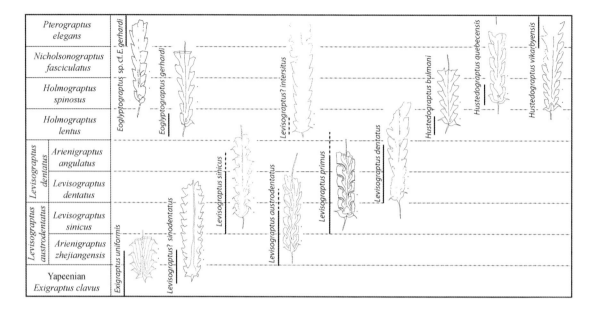

An Axis for Support

The distal development of the nema in graptolite colonies has often been identified as the virgula, especially in the Monograptidae (Wiman 1896a). This differentiation of the nema and virgula (Rickards 1996) cannot be upheld, and the virgula in older literature has to be regarded as homologous to the nema (Bates 1987; Maletz et al. 2014). The construction of the distally elongated nema has been hard to explain, and various interpretations as a hollow rod or a spine exist (Rickards 1996). However, the general model for pterobranch secretion now explains the secretion of the graptolite thecae and of spines on the tubaria. Bates (1987) demonstrated the secretion of the nema from fusellar increments as an extension of the cauda of the prosicula forming a distinct spine. The strong elongation seen in the nema and the added nematularium of many biserials indicate that the zooids must have had a chance to move onto the distalmost part of the colony, the tip of the growing nema, to add fusellar material here. This is still difficult to explain with the rhabdopleurid zooid model, unless the zooids were able to detach from the colony for free movement, or the connecting stolon was able to stretch much more than we can imagine from the rhabdopleurid model, or they possessed considerably longer stolons.

There are very few examples of biserials in which an extended nema as a colony support is not found. One of the rare species is *Climacograptus*(?) *uncinatus* Keble & Harris, 1934 from the upper Katian of Australia and North America (Keble & Harris 1934; Carter 1972). The species shows a short nema, about 1–2 mm long, extending above the tip of the sicula in juveniles before it branches into two diverging rods (Figure 11.3A). The colony proceeds to grow over this point into a normally developing biserial colony, but an internal nema support cannot be noted in the specimens (Figure 11.3B–C) and is not present distal of the thecal part of the colony. Mu (1963, Fig. 12) also illustrates the species *Climacograptus ensiformis* from the Upper Ordovician of Tianzhu, China, with a similar development of two lateral spines originating from the colony (Figure 11.3D). In this case, however, a normally developing nema can be seen through the colony and extends distally of the growing end, and two large membranes develop on the base of these two lateral tubarium spines.

Figure 11.3 Unusual development of the nema. (A–C) *Climacograptus*(?) *uncinatus* Keble & Harris, 1934, all specimens from Trail Creek section, Idaho (see Goldman et al. 2007), drawings by Kristen Paris (UB Buffalo, 2005). (D) *Climacograptus ensiformis* Mu and Zhang in Mu, 1963 with paired spines and lateral membranes (after Mu 1963, Fig. 12d). Scale indicates 1 mm in each illustration.

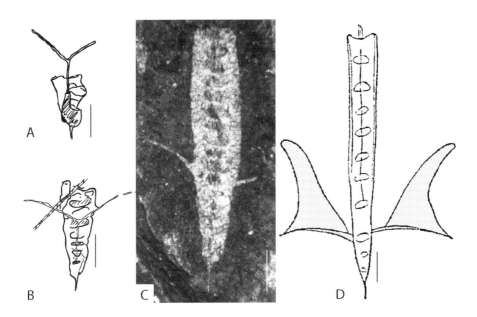

The Axonophoran Sicula

The axonophoran sicula is very similar in its construction to the sicula of the earlier graptoloids. It can be differentiated into the prosicula with its typical spiral line, and the metasicula formed from fuselli (Figure 11.4). Kraft (1926) was able to separate the "nema prosiculae" (cauda) from the "prosicula" (conus) in chemically isolated material of *Diplograptus gracilis* Roemer, 1861 (now identified as *Rectograptus gracilis*) and did the same in an indeterminate Silurian monograptid, possibly *Heisograptus micropoma* (Jaekel, 1889). The conus and cauda may invariably be present in the axonophorans, but have been differentiated in the prosicula of only a few taxa. Kraft (1926) remarked that the nema prosiculae (cauda) is very short in monograptids, but does not differ otherwise from the development in *Rectograptus*. The presence of a cauda has not been recognized in later descriptions of isolated axonophoran graptolites, even though many well-preserved biserial and uniserial graptoloids have been chemically isolated from limestones (e.g. Holm 1895; Urbanek 1958, 1997a; Jaeger 1991; and many more).

There is very little information on the number and development of the typical longitudinal rods (Figure 11.4A) on the prosicula available for most axonophoran taxa. Mitchell (1987) illustrated the presence of longitudinal rods in a number of axonophorans, but did not mention these in his descriptions. The exact number and development of the longitudinal rods in these species is unknown, therefore. Longitudinal rods are common in Silurian and Lower Devonian monograptids (e.g. Urbanek 1997a). Many climacograptids reduce the prosicula to just one or two rods united distally to form a normal nema (Mitchell 1987). These prosicular rods may have been formed originally as longitudinal rods on the surface of the prosicula, as the illustrations of Williams and Clarke (1999, pl. 3) indicate, and would suggest a reduction in the number of longitudinal rods in the climacograptids.

Sicular annuli form darker rings on the inside wall of the sicula and sometimes also in the first theca. They are a typical feature of many

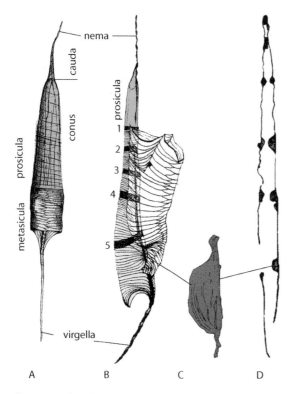

Figure 11.4 Development of the axonophoran sicula. (A) Immature sicula with prosicula showing longitudinal rods and spiral line, several metasicular fusellar rings and development of virgellar spine. (B) Monograptid with complete sicula and five sicular annuli (labelled 1–5). (C) Section through sicular annulus showing construction from cortical layers. (D) Section through sicula showing development of sicular annuli on the inside. (A–B) Monograptid indet., ?*Heisograptus micropoma* (after Kraft 1926). (C–D) *Pseudomonoclimacis dalejensis* (Bouček, 1936) (after Urbanek 1958). Illustrations not to scale.

monograptids (Figure 11.4B–D). Sectioned material clearly shows the secretion of the annuli as cortical additions on the inside of the sicula (Kozłowski 1949; Urbanek 1958). They are typically present in late Homerian (Late Wenlock, Silurian) and younger monograptids (Lenz & Kozłowska-Dawidziuk 1998), but very little is known about sicular annuli in older taxa. According to Lenz and Kozłowska-Dawidziuk (1998), the presence and number of sicular annuli appears to increase considerably through the Silurian, but there are no records from the

Llandovery. Sicular annuli are rare in the Lower Wenlock, but the number of species with annuli increases in the Homerian. In the upper Homerian, not all species bear annuli, but the long-ranging *Colonograptus praedeubeli* shows an increasing number of sicular annuli up the stratigraphic column. From the Ludlowian onwards, all monograptids appear to possess sicular annuli. Sicular annuli appear to be formed in fixed numbers and positions in Upper Silurian to Lower Devonian monograptids (Urbanek 1997a) and can be used to identify juveniles.

Rhythmic dark and light banding in Lower Ordovician (Floian to Dapingian) graptoloids have initially been called annuli (Williams & Stevens 1988), but indicate thickening intervals of the fusellar tissue in the thecal walls. These variations may be interpreted as diurnal cycles in wall secretion by the zooids (Williams et al. 1997), but cannot be homologized with the sicular annuli in the monograptids.

The Virgella

The virgellar spine, or in short the virgella, is one of the most prominent features of all axonophorans, but is not restricted to this clade. Maletz (2010a) discussed the construction of the virgellar spine (Figure 11.5). He compared the construction of the virgella in the axonophorans, the xiphograptids and the phyllograptids, and came to the conclusion that in each of the three groups the virgella evolved separately and independently. One main argument is the position of the virgella on the dorsal (phyllograptids, xiphograptids) or the ventral side (axonophorans) of the sicular aperture. The origin of the ventral virgella of the Axonophora from the extended rutellum of the derived isograptids can be documented in all steps leading to the final apertural virgellar spine (Maletz 2010a). The rutellum evolves into a lamelliform rutellum, a lanceolate virgella, and finally into the true virgella (Figure 11.5A–D). Early axonophorans of the genus *Exigraptus* apparently possess a lamelliform rutellum or an extended rutellum similar to the situation of the isograptids and pseudisograptids (Figure 11.5E–F), but the details are unclear, as no chemically isolated material exists. The closely

Figure 11.5 The virgellar spine development. (A) Rutellum. (B) Lamelliform rutellum. (C) Lanceolate virgella. (D) Virgella (after Maletz 2010a, Fig. 2). Specimens: (E) Isograptid indet., sicula with lamelliform rutellum, wb1 34 22. (F) *Isograptus* sp., chs 13 1 73, flattened proximal end with extended rutelli. (G) *Levisograptus sinicus* (Mu & Lee, 1958), GSC 133381, complete sicula with lanceolate virgella and part of first theca. (H) *Levisograptus sinicus* (Mu & Lee, 1958), GSC 133378, specimen with two thecal pairs, some parts broken. (I) *Levisograptus sinicus* (Mu & Lee, 1958), GSC 133386, small specimen with five thecae and lanceolate virgella. (J) *Archiclimacograptus* sp., bas 123, West Bay Centre Quarry, western Newfoundland. All specimens are flattened, chemically isolated from shales. Scale indicated by 1 mm long bar in each photo.

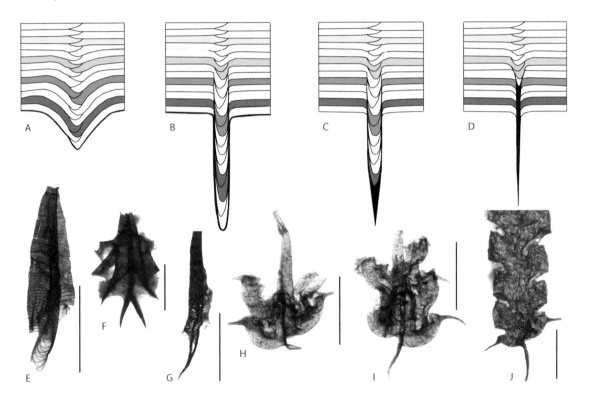

related genus *Levisograptus* has a lanceolate virgella (Figure 11.5G–I), but the first taxon with a true virgellar spine still needs to be determined. The oldest chemically isolated *Archiclimacograptus* specimens showing a true virgella are from the *Holmograptus spinosus* Biozone of early, but not earliest, Darriwilian age (Figure 11.5J).

The virgellar spine is short in most axonophorans, but can be considerably elongated as in *Climacograptus cruciformis* VandenBerg or *Diplacanthograptus lanceolatus* VandenBerg (VandenBerg 1990). In these species, and in a few others, the virgellar spine may be longer than the thecate tubarium of the graptolite and may even be joined by a parasicula. The reason for this elongation of the virgellar spine is uncertain, and it is noticeable that closely related species do not show any elongation of the spine.

Proximal Development Types

Elles (1922) developed the idea of the importance of the proximal development types (Figure 11.6) and proximal end structure, an idea that quickly gained acceptance and is used especially for the understanding of the proximal development in biserial axonophorans today (Mitchell 1987;

Figure 11.6 Examples of the proximal development types of biserial axonophorans. Patterns A to I are shown together with specimens as examples demonstrating the development (based on Mitchell 1987).

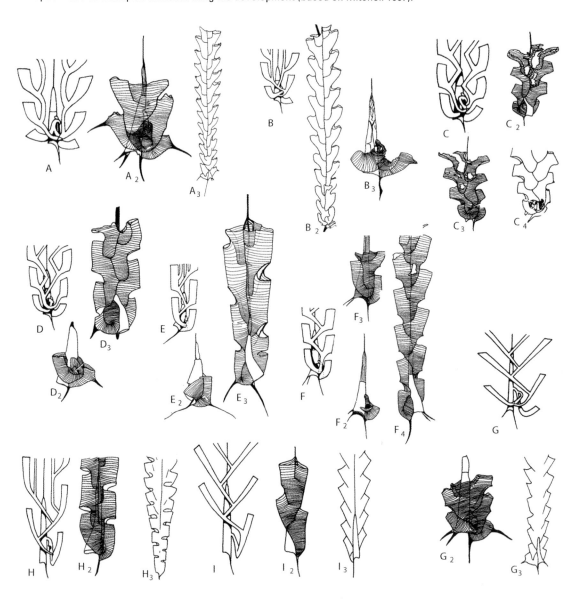

Melchin et al. 2011; Štorch et al. 2011). The proximal development types are regarded as the key to the taxonomic and evolutionary interpretation of the planktic graptolites. The proximal development of the axonophorans is quite complex in early taxa, in which a pseudisograptid manubrium has left its traces, and becomes successively simpler as features are eliminated from the development. The proximal development includes features of the sicular development, thecal growth, branching patterns and apertural modifications of the colonies. All the involved features may change independently and form a complex array of constructional patterns.

Figure 11.7 The transition from the reclined *Pseudisograptus* to the biserial, axonophoran *Archiclimacograptus*. The manubrium, sicula and th1^1 are highlighted to show the changes more clearly. (A) *Pseudisograptus*. (B–C) *Exigraptus*, showing prosicular origin of th1^1 (arrow in B). (D) *Archiclimacograptus*, juvenile, showing metasicula rorigin of th1^1 (arrow). (E) *Levisograptus*. (F) *Archiclimacograptus*. Reconstruction of specimens from various sources, not to scale.

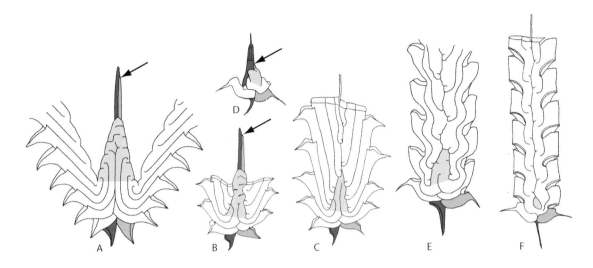

Mitchell (1987) developed the system of proximal development types used for the taxonomic interpretation of the Ordovician biserial graptolites. He differentiated the specialized, primordial thecae from the repetitively developing, unmodified distal thecae, and regarded these primordial thecae as a reliable guide for the interpretation of the evolutionary relationships of the graptolites. Altogether, Mitchell (1987) differentiated nine proximal development types (A–I) and a few derived patterns. Through a cladistic analysis of the Ordovician and Lower Silurian biserial axonophorans, he was able to provide a coherent taxonomic concept for these graptolites and improved our understanding of graptolite taxonomy and evolution. A few development types have been established for Upper Ordovican to Lower Silurian biserial graptolites subsequently (Melchin & Mitchell 1991; Melchin 1998; Melchin et al. 2011), adding useful details to the axonophoran taxonomy.

The proximal development types (Figure 11.6) are based on the origins and growth directions of proximal thecae (Mitchell 1987, 1990; Melchin & Mitchell 1991; Melchin 1998; Mitchell et al. 2007a; Melchin et al. 2011). The proximal development of the earliest taxa of the Axonophora is known from very few relief specimens, but not from chemically isolated material, and thus some of the details are difficult to estimate. Relief specimens of the oldest axonophorans of the genera *Exigraptus* and *Levisograptus* can easily be related to the manubriate isograptids. The typical manubrium of the pseudisograptines as expressed in the genus *Pseudisograptus* (Figure 11.7A) is still present in early axonophorans, as we can see in the genus *Exigraptus* (Figure 11.7B–C). It is clear from the cladistic analyses that the latter genus together with the more poorly known *Apiograptus* are intermediates in the lineage leading to the derived biserial graptolites (Mitchell et al. 1995, 2007a; Fortey et al. 2005). The manubrium is strongly reduced in *Levisograptus* (Figure 11.7E) and lost in *Archiclimacograptus* (Plate 10A, F) except for an exposed patch of the crossing canals (Figure 11.7D, F), and the looped growth of th1^2 can be regarded as the last remnant of this feature. The development of thecal apertural spines is not easy to follow in early axonophorans, but it seems that, early on, the restriction to apertural spines on the first thecal pair can be found in most members of the genus *Levisograptus*. The genus *Exigraptus* still bears lateral apertural lappets and rutelli on all thecae, but the size of these is reduced in distal thecae.

The low prosicular origin of th1[1] as a symplesiomorphic character of *Exigraptus* (Figure 11.7B) connects the genus with the reclined, two-stiped pseudisograptids (Figure 11.7A). The prosicular origin of th1[1] in this genus differs considerably from the typical metasicular origin of th1[1] found in all later axonophorans (Figure 11.7D). A number of further characteristics connect the reclined pseudisograptids with their biserial, dipleural descendants. These include the remains of the pseudisograptid manubrium, the elongated, sinuously bending thecae with intrathecal folding, and the thecal rutelli.

The Thecal Styles

In the literature, thecal shapes are often described by referring to typical genera. The terminology for thecal styles ranges from dichograptid, orthograptid to climacograptid, lasiograptid, and so on. This terminology is somewhat misleading as it may suggest evolutionary relationships between genera sharing a similar thecal style, which is not the case. It is preferred here to understand the thecal features without connection to any named taxa. The thecal descriptions are based on constructional features, according to the presence/absence or style of the geniculum, genicular additions and apertural modifications. The biserial axonophorans develop numerous different thecal styles and thecal modifications, and it is hard to provide a simple overview (e.g. Figure 11.6). It is also not possible from the thecal shape alone to infer the taxonomic relationship of a certain specimen, as the shape may vary considerably even within a single genus. For example, the family Diplograptidae includes numerous species with strongly geniculate thecae, as does the family Climacograptidae, but non-geniculate taxa are also present. Geniculate thecae feature prominently in the climacograptids *Pseudoclimacograptus* (Figure 11.6D) and *Diplacanthograptus* (Figure 11.6E), but also independently in the orthograptine *Geniculograptus* (Figure 11.6F) or the neograptid *Normalograptus* (Figure 11.6H). Taxa with non-geniculate, straight, inclined ventral thecal sides are *Hustedograptus* (Figure 11.6A) and *Eoglyptograptus* (Figure 11.6B),

but also the Silurian *Glyptograptus* (Figure 11.6I), all showing considerable differences in the development of their thecal apertures, the median septum and other features.

Other independently changing thecal characteristics include the thecal length and the length of the interthecal septae. The thecal length and shape varies from long and undulating (Figure 11.8A, F), as in *Undulograptus* or early *Archiclimacograptus* species (Maletz 2011b), to short and straight as in *Normalograptus* (Figure 11.8D) and in distal thecae of *Orthograptus* (Figure 11.8E). Apertures and geniculae may be adorned with ventral or lateral lappets or with spines in many taxa (e.g. Figure 11.8E, I), thus complicating the identification.

The Median Septum

The two stipes of the biserial axonophoran colony fold over the tip of the sicula and grow back-to-back, parallel to each other, with the sicula and the nema embedded between them and completely covered at least on the reverse side of the colony (Figure 11.9). They supposedly form a double-layered wall, the median septum, separating the two stipes and covering most parts of the proximal thecae. The construction is similar to that of the monopleural Glossograptina, in which the nema is embedded between the lateral tubarium walls of the two stipes. Thus, in the axonophorans, the two stipes are connected back-to-back, while they are positioned side-by-side in the Glossograptidae (see Chapter 10). The presence of a double-layered median septum can be demonstrated through the presence of abnormal biserials in which one of the stipes is abandoned. Maletz (2003) described *Normalograptus scalaris* with a poorly preserved dorsal wall of the single developed stipe. In this example, the wall is not covered and thickened with additional cortex, but preserved as a ragged edge where the material is broken off.

Due to the presence of intrathecal folding, the median septum may be strongly undulating or even zigzag shaped in early axonophorans, with short intrathecal septae originating from the median septum (Figure 11.9A, B). Intrathecal folds

Figure 11.8 Comparison of the Neograptina (Normalograptidae) (A–D) and the Diplograptina (Diplograptidae and Lasiograptidae) (E–L) in the Ordovician. (A) *Undulograptus formosus* (Mu & Lee, 1958). (B) *Undulograptus novaki* (Perner, 1895). (C) *Skanegraptus janus* Maletz, 2011c. (D) *Normalograptus antiquus* (Ge, in Ge et al. 1990). (E) *Orthograptus quadrimucronatus* (Hall, 1865). (F) *Levisograptus primus* (Legg, 1976). (G) *Diplograptus pristis* (Hisinger, 1837). (H) *Exigraptus clavus* Mu in Mu et al. 1979. (I) *Paraorthograptus pacificus* (Ruedemann, 1947). (J) *Pipiograptus* sp. (K) *Brevigraptus quadrithecatus* Mitchell, 1988. (L) *Amplexograptus* sp. Graptolite specimens and reconstructions from various sources, not to scale.

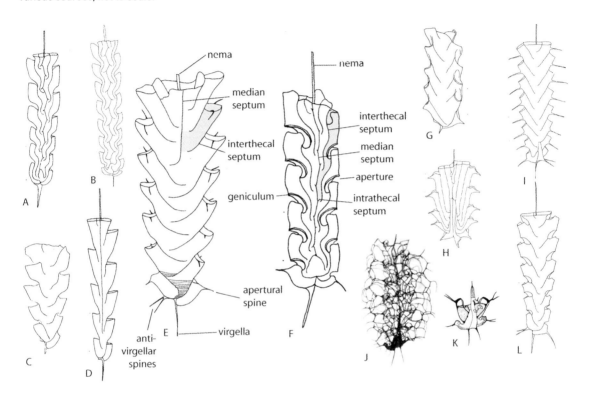

are common, but easily overlooked as they are only recognizable in well preserved relief material. They are already found in the pseudisograptids (see Chapter 10). In taxa with shorter thecae, the intrathecal septae are lacking and the median septum may be straight (Figure 11.9C). A number of derived biserial graptolites develop colonies in which the thecae grow alternately to both sides and a median septum is not formed (Melchin et al. 2003). These colonies are termed unistipular (Figure 11.9D). The nema is entirely free or attached by a short bar to the interthecal septae in the centre of these colonies.

The species *Haddingograptus eurystoma* is one of the few species in which the development is known from isolated material. Bulman (1932b, pl. 1) illustrated specimens showing the connection of the thickened rims of the thecal origins to the nema in the centre of the colony. Bulman (1932b) also described the development of the unistipular colony of *Geniculograptus typicalis* in some detail, and illustrated the connection of the interthecal septae with a centrally positioned bar to the nema.

The Diplograptina

The (Infraorder) Diplograptina Obut, 1957 (Plates 10–12) represents one of the two main clades of the Axonophora (Mitchell et al. 2007a;

Figure 11.9 Median septum development. (A) *Levisograptus* sp., showing intrathecal folds with undulating median septum and long, double sigmoid thecae, Krapperup drill core, Scania, Sweden. (B) *Haddingograptus oliveri* (Bouček, 1973), showing a strongly zigzag median septum, Table Head Group, western Newfoundland. (C) (?)*Archiclimacograptus* sp., showing short thecae with nearly straight median septum, Table Head Group, western Newfoundland. (D) *Petalolithus minor* Elles, 1897, LO 1115t, showing alternating thecae, median septum lacking, Scania, Sweden. (1) indicates the interthecal septum, (2) the intrathecal septum in (A–B). All specimens coated with ammonium chlorite. Scale indicated by 1 mm long bar in each photo.

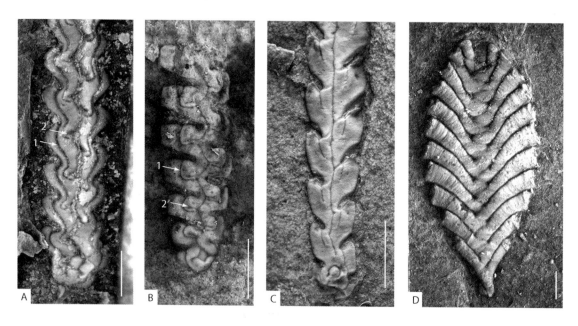

Maletz et al. 2009; Melchin et al. 2011). Štorch et al. (2011) identified the two groups as the Diplograptina and the Neograptina, and Sadler et al. (2011) provided information on the diversity and biostratigraphical ranges of these taxa and some of the included family group units (Figure 11.10). The detailed phylogenetic relationships of both taxa are still uncertain (Maletz 2011a). Štorch et al. (2011, p. 368) recognized the earliest taxon of the Neograptina as *Undulograptus formosus* (Figure 11.8A), a typical axonophoran with a pattern C astogeny (Mitchell et al. 2007a: Fig. 1), most probably derived from a diplograptine ancestor. This leads to the interpretation of the Diplograptina as a paraphyletic taxon from which the Neograptina originated in the Lower Darriwilian (Maletz 2014a). Sadler et al. (2011), however, indicated the family Normalograptidae of the Neograptina to represent the earliest biserial axonophorans.

Luckily, it is quite easy to differentiate the Diplograptina and the Neograptina based on their tubarium outline (Figure 11.10), as they show a few characteristics that are easy to recognize even in poorly preserved material. The proximal end in the early Neograptina (Figure 11.10A) is rounded and possesses only the virgella as a proximal spine. The apertural parts of the first thecal pair grow upwards fairly symmetrically, or the proximal end is pointed and highly asymmetrical in derived taxa. The thecae are simple with a straight, outwards inclined to horizontal aperture and short thecal overlap. The median septum may be undulating, but in most cases is nearly straight or missing in many species. The Diplograptina (Figure 11.10B) are more variable in colony shape and thecal styles. The proximal end is often symmetrical and wide, with the virgellar spine and apertural spines at least on the first thecal pair.

Figure 11.10 The differentiation and evolution of the Axonophora. Graptolite examples: (A) *Normalograptus kukersianus* (Neograptina). (B) *Archiclimacograptus* sp. (Diplograptina). Diversities and ranges based on Sadler et al. (2011).

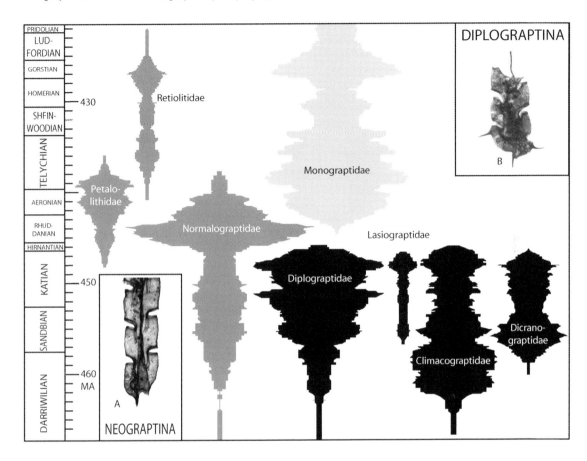

The axonophorans are the main players in the evolutionary history of the Upper Ordovician, where the dramatic patterns of change from the DDO-fauna (Dicranograptidae–Diplograptidae–Orthograptidae) to the N-fauna (Normalograptidae) has been recognized as a stepwise extinction of one group (Melchin & Mitchell 1991; Chen et al. 2003), just to be replaced by another group that was apparently waiting for a chance to step in. While the Diplograptidae and Lasiograptidae gained their highest diversity during the Sandbian and Katian (Figure 11.10), the recovery of the previously low-diversity clade of the Normalograptidae produced a burst in diversity in the post-extinction Hirnantian. It eventually led to the diversification of the monograptids in the lower Silurian and the slow demise of the biserial axonophorans. The Lasiograptidae

(Figure 11.11E–F, J, M), however, never played a larger role in the history of the graptolites, even though they already employed the concept of the fusellum reduction, later used so successfully by the Retiolitidae. They died out during the Hirnantian extinction event (Figure 11.10).

Štorch et al. (2011) differentiated three clades in the Diplograptina, the superfamilies Dicranograptoidea, Climacograptoidea and Diplograptoidea, based on their cladistic analysis of Upper Ordovician axonophorans, but the three groups form an unresolved trichotomy. The differentiation supports largely the earlier analysis of Mitchell et al. (2007a). Maletz (2014a), however, decided to use fewer taxonomic levels and recognized four family-level taxa in the Diplograptina, the Diplograptidae, Lasiograptidae, Climacograptidae and Dicranograptidae, as

Figure 11.11 Examples of the Diplograptidae (A–D, F–I, K–L) and Lasiograptidae (E–F, J, M). (A) *Pseudamplexograptus* Mitchell, 1987. (B) *Urbanekograptus retioloides* (Wiman, 1895). (C) *Amplexograptus* sp. (D) *Hustedograptus uplandicus* (Wiman, 1895). (E) *Nymphograptus velatus* Elles & Wood, 1908. (F) *Orthoretiolites hami* Whittington, 1954. (G) *Diplograptus pristis* (Hisinger, 1837). (H) *Rectograptus gracilis* (Roemer, 1861), neotype. (I) *Rectograptus intermedius* (Elles & Wood, 1907). (J) *Lasiograptus harknessi* (Nicholson, 1867a) (adapted from Bulman 1947, with permission from The Paleontological Society). (K) *Geniculograptus typicalis* (Hall, 1865). (L) *Peiragraptus fallax* Strachan, 1954. (M) *Yinograptus disjunctus* (Yin & Mu, 1945). Illustrations not to scale, based on various sources.

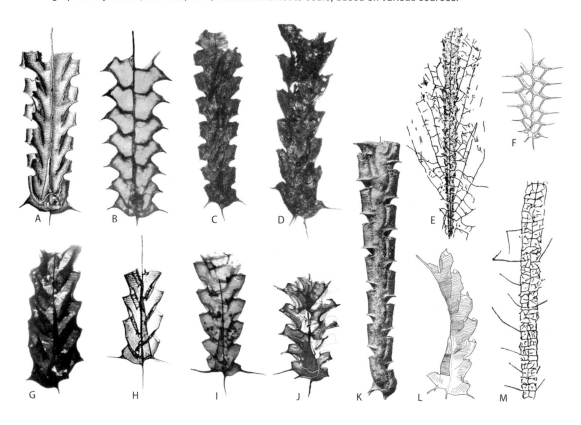

they were also differentiated by Sadler et al. (2011). The precise evolutionary origins of the four families are still uncertain, as many early biserial taxa are known only from poor, flattened material.

Family Diplograptidae

Mitchell et al. (2007a) and Štorch et al. (2011) defined the Diplograptoidea as a monophyletic clade and separated a paraphyletic stem group including the early biserial taxa of the "*Undulograptus austrodentatus* group" with the aim of defining only monophyletic groups or clades. Similarly, Maletz et al. (2009) regarded the manubriate isograptids (*Arienigraptus*, *Pseudisograptus*) and the early biserial taxa (*Exigraptus*, *Levisograptus*, *Undulograptus*) as stem group axonophorans. Their infraorder Axonophora then was differentiated into two monophyletic clades, the "monograptids" and the "diplograptids", a concept similar to the ideas of Štorch et al. (2011). Maletz (2014a), however, decided to base the Axonophora on the biserial colony shape and included the earliest biserial, dipleural, but manubriate graptolites of the genera *Exigraptus* and *Apiograptus* in the family Diplograptidae.

The family Diplograptidae Lapworth, 1873b includes two subfamilies, the Diplograptinae and the Orthograptinae (Maletz 2014a). Both share a general pattern with a parallel-sided to strongly widening tubarium with mostly outwards inclined ventral thecal walls and open, everted apertures, often with lateral apertural lappets or even spines. The proximal end is provided with a ventral virgella on the sicula and apertural to subapertural spines on the first thecal pair. In the Orthograptinae, paired antivirgellar spines are also present on the sicula (cf. Figure 11.8E). Obviously, some variation is possible, and a number of taxa develop additional spines in the colony or modify the shape of the thecae considerably.

Orthograptinae and the Antivirgellar Spine

The Orthograptinae are generally united by the presence of paired antivirgellar spines (cf. Figure 11.8E). These may be difficult to recognize in flattened shale material, but are prominent in all chemically isolated specimens. The antivirgellar spines are lacking in a few early species like *Hustedograptus teretiusculus* (Jaanusson 1960; Mitchell 1987) (Plate 12A), where a slight notch indicates the place where they will be developed in derived taxa. The proximal ends of the Orthograptinae are often quite strongly asymmetrical, and the genera *Amplexograptus* and *Rectograptus* lack the apertural spine on th1^2 (Figure 11.11C, H, I). Mitchell (1987) introduced the family Orthograptidae with three subfamilies, Orthograptinae Mitchell, 1987, Peiragraptinae Jaanusson, 1960 and Lasiograptinae Lapworth, 1879. Goldman (1995) discussed the taxonomy, evolution and biostratigraphy of the *Orthograptus quadrimucronatus* species group (Plate 12D) and provided taxonomic information for this group, but a cladistic analysis of the Orthograptinae does not exist. The Orthograptinae are an important group of Upper Ordovician (Sandbian to Katian) graptolites as they form an integral part of the DDO fauna of Melchin and Mitchell (1991).

The presence of multiple antivirgellar spines in *Prolasiograptus hystrix* (Bulman, 1932b) may represent an independent origin of antivirgellar spines as the proximal development and complex thecal style of this taxon is more comparable with the Dicranograptinae (Mitchell, 1988; Mitchell et al., 2007a). Interestingly, the lasiograptid genus *Brevigraptus* (Mitchell, 1988) bears a single instead of the typically paired antivirgellar spines of the Orthograptinae and Lasiograptidae.

Among the Diplograptidae, *Peiragraptus fallax* Strachan, 1954 (Figure 11.11 L) from the Upper Ordovician of Anticosti Island, Canada, is an unusual species, as it appears to have grown a single stipe only. Closer examination of the species, known from a single population of chemically isolated material, shows that the proximal end follows the development typical of *Amplexograptus* (Figure 11.11C) with a pattern G astogeny (Mitchell 1987). The second stipe is abandoned after the formation of a single theca, and the single stipe is slightly curved. The reason for this strange development is unknown, but cannot be attributed to damage of a single specimen, as the whole population associated with the type specimen shows this feature. The development is otherwise identical to the development of the biserial, unistipular *Rectograptus gracilis* (Figure 11.11H).

Family Lasiograptidae

The Lasiograptidae Lapworth, 1880e include some of the strangest Ordovician graptolites. A reduction of the fusellum is accompanied by the secretion of a meshwork of lists outside the original tubarium. This feature is called a lacinia, and is typical of all derived Lasiograptidae. The lacinia and the reduction of the fusellum produces graptolites that bring to mind the Silurian Retiolitidae, and actually some of the taxa now included in the Lasiograptidae were initially identified as archiretiolitids (see Bulman 1955, 1970a). The Archiretiolitinae of Bulman (1955) were originally thought to be ancestral to the Retiolitidae due to the similarities in their development, a notion that has been proven wrong due to constructional investigations (Bates & Kirk 1987, 1991) and cladistic analyses (Mitchell et al. 2007a; Melchin et al. 2011). In some members, the fusellum is attenuated and the tubarium outline is strengthened by lists and with genical spines added as in *Lasiograptus* (Bulman 1944; Rickards & Bulman 1965) (Figure 11.11J). These spines can be connected with additional rods, forming a complex lacinia around the original tubarium as in *Pipiograptus* (Figure 11.8J; Plate 14A) and *Phormograptus* (Bates & Kirk, 1991). One of the strangest species might be *Nymphograptus velatus* Elles & Wood, 1908 (Figure 11.11E), a taxon with an extremely wide and irregularly developed lacinia. Unfortunately, the species is poorly known from few shale specimens, and neither the thecal construction nor the development of the lacinia is well understood.

Unfortunately, most genera of the Lasiograptidae are found on shale surfaces as flattened and strongly distorted tubaria. It is often impossible to disentangle the many rods and lists and to understand the construction of these peculiar graptolites (Figure 11.11E, M). As the sicula and the first theca are preserved in some taxa, it is established that the Lasiograptidae are closely related to the Orthograptinae (Mitchell 1987; Bates & Kirk 1991; Štorch et al. 2011). Mitchell et al. (2007a, p. 337) identified *Hallograptus mucronatus* Hall with a pattern A astogeny as the earliest taxon of the Lasiograptidae. The clade is not well supported as the proximal development of many of the highly reticulate, derived taxa is virtually unknown and cannot be compared with that of the basal taxa of the group due to the lack of data.

Family Climacograptidae

The Climacograptidae Frech, 1897 is one of the difficult groups of biserial graptolites, and its evolutionary relationships are controversial. Štorch et al. (2011) defined the clade as the superfamily Climacograptoidea and included the genus *Archiclimacograptus* as a paraphyletic basal taxon. Maletz (2011b) suggested an alternative interpretation of the climacograptids, based on constructional details of the tubaria, as he noted the lack of proximal end spines on the first thecal pair in *Haddingograptus, Pseudoclimacograptus* and most of the derived climacograptids. Therefore, he interpreted the climacograptids to be derived from a neograptid ancestor, possibly of

the genus *Undulograptus*, and a secondary derivation of the proximal end spines in *Climaocgraptus bicornis* and related species. The strong zigzag median septum and the intrathecal folding in early climacograptids (Figure 11.12A, E) are still present in *Pseudoclimacograptus scharenbergi* (Figures 11.12F, 11.13B), but are lost in derived taxa. The zigzag median septum has long been regarded as one of the characteristics of the Climacograptidae, but it was known mainly from *Pseudoclimacograptus scharenbergi*. It is now clear that many early climacograptids and even the genera *Archiclimacograptus* and *Haddingograptus* (Figure 11.13A; Plate 12E) possess this feature (Maletz 1997b). These genera can be differentiated by their proximal development types, but it is nearly impossible to separate distal fragments.

Derived climacograptids show a straight median septum and short thecae, as typified by *Climacograptus bicornis* (Figure 11.12H, J). A similar development is already present in the much older *Proclimacograptus* (Figure 11.12G), but the proximal development differs considerably from *Climacograptus bicornis* (cf. Mitchell 1987) or *Climacograptus cruciformis* (Figure 11.13E–F; Plate 10E). The derived climacograptids are quite variable in the development of proximal spines and additional features on the proximal end. The genus *Climacograptus* (Figure 11.12H, J) bears a slightly cocked sicula with a virgellar spine and subapertural to mesial spines on the first thecal pair, while the closely related *Diplacanthograptus* (Figure 11.12K) possesses a virgellar spine and a spine on the first theca only. Thecal spines are lacking in *Styracograptus* (Figure 11.12I), except for the prominent virgellar spine. This extreme variation in the number and position of apertural spines is quite unusual for any group of graptolites and may, therefore, indicate that the proximal spines, except for the virgella, represent a secondary development.

A typical development in the younger taxa of the Climacograptidae is the reduction of the prosicula. It is replaced by two longitudinal rods in *Climacograptus* (Figure 11.13E–F) and *Diplacanthograptus*, while *Styracograptus tubuliferus* bears only a single prosicula rod (see Mitchell 1987: *Climacograptus* sp. cf. *C. caudatus*).

Figure 11.12 The Climacograptidae. (A) *Undulograptus formosus* (Mu & Lee, 1958). (B) *Undulograptus clabavensis* Bouček, 1973. (C) *Oelandograptus oelandicus* (Bulman, 1963). (D) *Haddingograptus eurystoma* (Jaanusson, 1960), reverse view. (E) *Haddingograptus oliveri* (Bouček, 1973). (F) *Pseudoclimacograptus scharenbergi* (Lapworth, 1876). (G) *Proclimacograptus angustatus* (Ekström, 1937). (H, J) *Climacograptus bicornis* (adapted with permission from Riva & Kettner, 1989, with permission from Edinburgh University Press and John F. Riva). (I) *Styracograptus tubuliferus* (Lapworth, 1876), reverse view (based on Mitchell 1987). (K) *Diplacanthograptus spiniferus* (Ruedemann, 1908), reverse view (adapted from Mitchell 1987, with permission from The Palaeontological Association). (L) *Appendispinograptus venustus* (Hsü, 1959), showing complex parasicular development, Wufeng Formation, China. (M) *Appendispinograptus longispinus* (Hall, 1902), flattened (adapted from Riva 1974, with permission from The Palaeontological Association and John F. Riva). (N) *Climacograptus hastatus* (Hall, 1902), flattened specimen with long parasicula (adapted with permission from Riva & Kettner, 1989, with permission from Edinburgh University Press and John F. Riva). Illustrations are largely reconstructions, based on various sources, not to scale.

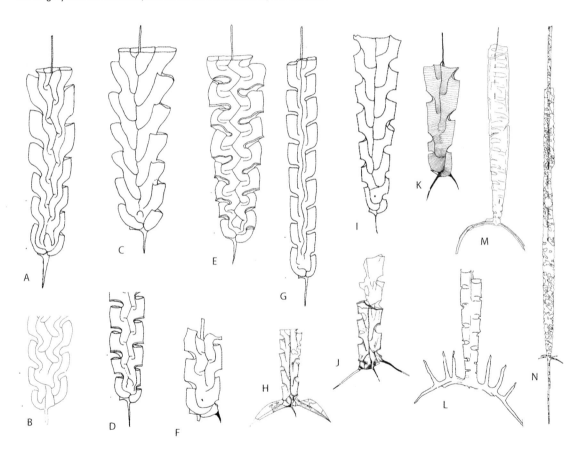

The Climacograptidae are also well known for a number of extrathecal developments on the proximal end. Extensive membranes are typically found on the upper sides of the proximal spines in mature specimens of *Climacograptus bicornis* (Riva 1974, 1976) (Figure 11.12H). They can reach up to the sixth or seventh thecal pair in gerontic specimens. It is unclear whether they occlude the thecal apertures, as these features are found only in flattened material on shale surfaces.

The presence of parasicular tubes and other features initially interpreted as basal membranes is typical for the genus *Appendispinograptus* (Mitchell et al. 2007b; Loxton et al. 2011)

Figure 11.13 (A) *Haddingograptus eurystoma* (Jaanusson, 1960), reverse view, showing zigzag median septum and intrathecal folding, Elnes Formation, Slemmestad, Norway. (B) *Pseudoclimacograptus scharenbergi* (Lapworth, 1876), showing zigzag median septum and intrathecal folding (based on Bulman 1945, pl. 8, Fig. 6). (C) *Proclimacograptus angustatus* Ekström, 1937, showing slightly undulating median septum, no intrathecal folds, Elnes Formation, Slemmestad, Norway. (D) *Appendispinograptus supernus*? (Elles & Wood, 1906), NIGP 139881, internal cast of proximal end showing sicula (S) with parasiculae, sub-scalariform view, Wufeng Formation, Guizhou, China. (E–F) *Climacograptus cruciformis* VandenBerg, 1990, Viola Limestone Formation, Oklahoma, showing prosicula reduced to two rods uniting distally to form the nema (photos provided by D. Goldman). Bar indicates 1 mm in each photo.

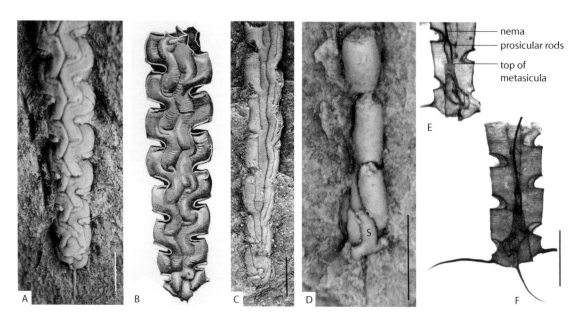

(Figure 11.12 L–M). Extensive heart-shaped, apparently planar membranes may surround the proximal ends of *Appendispinograptus leptothecalis* (Mu & Ge, in Fu, 1982), but no constructional details are available (Loxton et al. 2011). Tubular parasiculae and parathecae are present in other species of the genus *Appendispinograptus* (Mitchell et al. 2007b). The identification of the features as tubular relies on the evidence from a chemically isolated proximal end of *Appendispinograptus supernus* with paired tubes originating from the aperture of the sicula and growing along the apertural spines of th1^1 and th1^2 (Loxton et al. 2011). A single relief specimen of *Appendispinograptus supernus*(?) from the Upper Ordovician of China (Figure 11.13D) shows the sicula with two parasicular tubes growing horizontally away from the sicular aperture (Mitchell et al. 2007b). The typical complex erect structures

on the elongated proximal end spines of *Appendispinograptus venustus* (Hsü 1959) (Figure 11.12 L) can also be regarded as late stage tubular developments. Mitchell et al. (2007b) termed them secondary parasiculae and suggested that they may influence the hydrodynamic behaviour and feeding of the colony.

Family Dicranograptidae

In terms of colony shape, the family Dicranograptidae (Figure 11.14; Plate 11) is one of the most diverse families of the Ordovician axonophorans. With the evolution of secondarily two-stiped colonies in *Dicellograptus* (Figure 11.14D; Plate 10A) and *Dicranograptus* (Figure 1.1B, 11.14I) as an important innovative step in the Middle Ordovician, this group starts to dominate the

Figure 11.14 The Dicranograptidae. (A) *Levisograptus sinicus* (Mu & Lee, 1958). (B) *Dicaulograptus hystrix* (Bulman, 1932b). (C) *Levisograptus dicellograptoides* (Maletz, 1998b), reconstruction. (D) *Dicellograptus* sp., reconstruction. (E) *Jiangxigraptus alabamensis* (Ruedemann, 1908). (F) *Jiangxigraptus* sp. (adapted from Bulman 1970a, with permission from Paleontological Institute, and also from Bulman 1970b). (G) *Dicellograptus caduceus* Lapworth, 1876 (after VandenBerg & Cooper 1992, Fig. 9 V). (H) *Tangyagraptus typicus* Mu, 1963 (from Mu 1963, Fig. 3e). (I) *Dicranograptus clingani* Carruthers, 1868, reconstruction. (J) *Neodicellograptus dicranograptoides* Mu and Wang in Wang & Jin, 1977, reconstruction (JM). Illustrations not to scale.

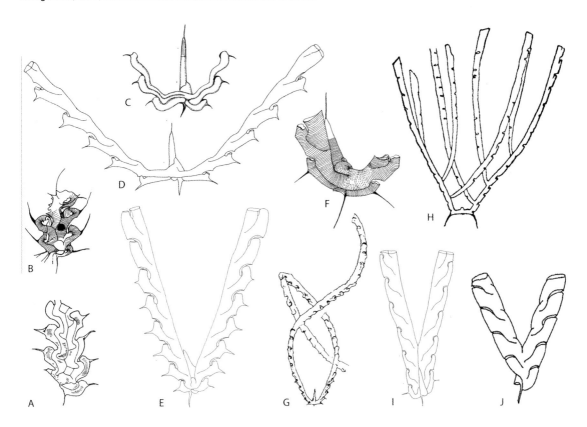

Middle to Upper Ordovician graptolite faunas and evolves into numerous and often difficult to differentiate species. Melchin and Mitchell (1991) regarded the Dicranograptidae as one of the important groups in their DDO (Dicranograptidae–Diplograptidae–Orthograptidae) fauna, dominating the late Ordovician until the Hirnantian extinction event.

In a further step, the introduction of cladial branching in various lineages of the Dicranograptidae produces secondary multiramous colonies and mimics in their colony shapes the long-extinct Dichograptidae. Only two little steps, and the impression of the Ordovician graptolite

faunas changes dramatically! The faunas dominated by biserial axonophorans suddenly include numerous members with multiramous colonies, recalling the Lower to Middle Ordovician Dichograptina, but differ in many details of the proximal development and thecal style from their older homeomorphs.

The start of this evolutionary change is still uncertain. Maletz (1998b) suggested an origin of the dicranograptids through *Levisograptus sinicus* (Mu & Lee, 1958), a biserial with apertural spines on all thecae (Figure 11.14A), similar to the situation in the earliest two-stiped dicellograptid-like species *Levisograptus dicellograptoides* (Maletz, 1998)

Figure 11.15 Torsion in the Dicranograptidae. (A) *Dicellograptus* with left-handed torsion showing axial angle (adapted from Williams 1981, Fig. 1, with permission from Cambridge University Press). (B) *Dicellograptus bispiralis* (Ruedemann, 1947) (adapted from Bulman 1964, Fig. 3, Geological Society of London). (C) *Nemagraptus gracilis* (Hall, 1847), OSU 32962 (based on Finney 1985, Fig. 19-1). (D) *Dicranograptus zigzag* Lapworth, 1876, reconstruction showing independent spiralling of stipes (adapted from Williams 1981, Fig. 4, with permission from Cambridge University Press). (E) *Dicellograptus caduceus* Lapworth, 1876, reconstruction (after Bulman 1964, Fig. 3. Adapted from Bulman 1970a, with permission from the Paleontological Institute, and from Bulman 1970b, with permission from Société Belge de Géologie de Paléontologie et d'Hydrologie.). Illustrations not to scale.

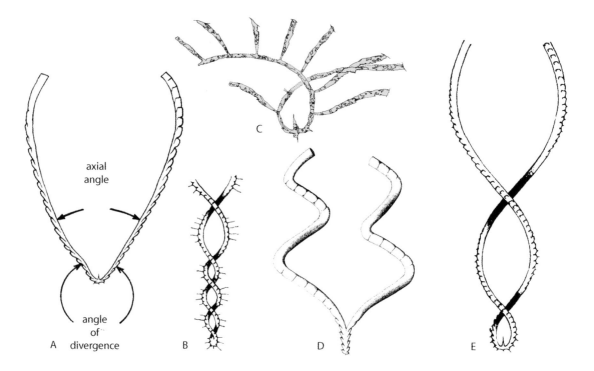

(Figure 11.14C) and in *Undulograptus* sp. (Kraft & Kraft, 2003) from the lower Darriwilian of North America and the Czech Republic. The cladistic analysis of Mitchell et al. (2007a) followed a similar direction and included the biserial, dipleural *Dicaulograptus hystrix* (Figure 11.14B) as a basal member of Dicranograptidae, but did not consider the early two-stiped species of the genus *Levisograptus* as ancestral, but interpreted them as an evolutionary sideline and dead end. Thus, in the Mitchell et al. (2007a) interpretation, a secondary two-stiped colony shape evolved more than once from the biserial axonophorans. For a long time interval in the Darriwilian, the dicranograptids are either extremely rare or not present, and their early evolution is not documented through the fossil record.

The specimens of *Dicellograptus* and *Dicranograptus* are often coiled to a greater or lesser extent (Williams 1981), forming a three-dimensional spiral (Figures 11.14G, 11.15; Plate 11A). Thus, the axial angle between the two stipes (Figure 11.15A) may vary considerably based on the preservation. Specimens of the same species may show left-handed or right-handed torsion, but the direction of torsion is constant in many species. Williams (1981) recognized right-hand torsion in *Dicellograptus complanatus complanatus* and left-hand torsion in *Dicellograptus complanatus complexus*. Torsion is easily visible in some species like *Dicellograptus bispiralis* (Figure 11.15B) and *Dicranograptus zigzag* (see Elles & Wood 1904), but more difficult to recognize in others with a more open spiral and larger colonies. The torsion in *Dicranograptus clingani* and *Dicranograptus clingani resicis* can be seen through the change in the appearance of the thecal apertures even in flattened specimens (Williams & Bruton 1983; Maletz 1995). The torsion of the stipes could lead to two stipes encircling each other as in *Dicellograptus bispiralis* (Hall), but the example of *Dicranograptus zigzag* Lapworth, 1876 shows two independent stipes (Williams 1981) (Figure 11.15D).

A number of *Dicellograptus*-type species evolved thecal cladia either individually or in pairs during the Sandbian and Katian. The best known is the genus *Tangyagraptus* (Figure 11.14H), a reclined dicellograptid with cladia found in the Upper Ordovician Wufeng Shale of China (Mu 1963). Another genus with cladial branching is *Syndyograptus* Ruedemann, 1908 from the Sandbian of the Normanskill Shale of New York State. It differs from *Tangyagraptus* through the possession of paired cladial branches, but details of its development are uncertain.

The members of the genus *Dicellograptus* have more recently been revised, and a number of new genera were established (e.g. Mu et al. 2002). Of these, *Jiangxigraptus* Yu & Fang, 1966, with its often strongly leaning sicula (Figure 11.14E–F), appears to be quite characteristic and easily distinguishable from *Dicellograptus* with its straight sicula (Figure 11.14D). Nevertheless, the differentiation is far from being firmly established.

A spiralling of the somewhat reclined stipes can be seen in the multiramous *Nemagraptus gracilis* (Hall) (Figure 6.8I). A group of slender two-stiped to multiramous dicranograptids is united in the subfamily Nemagraptinae based on the proximal development (Finney 1985; Mitchell 1987). The species typically show a derived pattern A astogeny (Mitchell 1987) with a high origin of th1[1] and a freely pending apertural part of the sicula (Figure 11.16). Initially, the species appeared to be two-stiped, as in *Nemagraptus subtilis* Hadding, 1913 and *Nemagraptus linmassiae* Finney, 1985 (Figure 11.16A), but multiramous taxa became more common.

Nemagraptus gracilis (Figures 6.8I, 11.15C) is one of the most recognizable and also most important graptolites for the biostratigraphy of the Upper Ordovician. Its first appearance has been used to define the base of the Sandbian Stage in the Fågelsång section of Scania, Sweden (Bergström et al. 2000, 2006). The species can be found worldwide in a short time interval where it is common and easily recognized (Finney & Bergström 1986; Brussa et al. 2007). Based on the development of the cladial branching (Finney 1985), multiramous nemagraptids can be differentiated easily and are included in the genera *Nemagraptus* and *Pleurograptus*. The simple thecal style of *Nemagraptus* and all derived taxa initially was difficult to compare with biserial graptolites. Therefore, Bulman (1955, 1970a) included the Dicranograptidae and Nemagraptidae in the Didymograptina. Mitchell (1987) referred the Dicranograptidae (including the Nemagraptinae as a subfamily) to the Diplograptina, and thus established the modern notion that the Dicranograptidae represent a group of highly variable species even developing a secondary two-stiped colony, sometimes with additional cladial branching.

Nemagraptus linmassiae Finney, 1985 may represent one of the earliest taxa of the Nemagraptinae. It is the only species of the clade in which intrathecal folds and high thecal overlap are present (Figure 11.16A). These characteristics may be regarded as remnants indicating the ancestral condition found in dicellograptids. Unfortunately, this species is found only at a single locality and its biostratigraphical range is unknown. A single stipe fragment may indicate the development of cladia on distal stipes in this species (Finney 1985, Fig. 26-12).

The Neograptina

Štorch et al. (2011, p. 368) defined the Neograptina (Plates 13–14), based on a cladistic analysis, as the "total clade comprising all species sharing a more recent common ancestor with *Monograptus priodon* than with *Diplograptus pristis* (i.e., the species on the branches arising from the right side of node 1 in Fig. 6 [of Štorch et al. 2011] and all their descendants", and defined the Normalograptidae as a paraphyletic stem group. Maletz (2014a) used the two superfamilies Monograptoidea and Retiolitoidea, but did not assign the Normalograptidae and Neodiplograptidae to a superfamily and also

Figure 11.16 The Nemagraptinae. (A) *Nemagraptus linmassiae* Finney, 1985, holotype. (B) *Amphigraptus* sp. (C) *Nemagraptus gracilis* (Hall, 1847), proximal end in reverse view, showing aperturally free sicula. (D) Proximal and distal thecae of *Nemagraptus gracilis*. (E) Proximal and distal thecae of *Nemagraptus linmassiae*, compare with shorter thecal overlap of thecae in *Nemagraptus gracilis*. Illustrations not to scale.

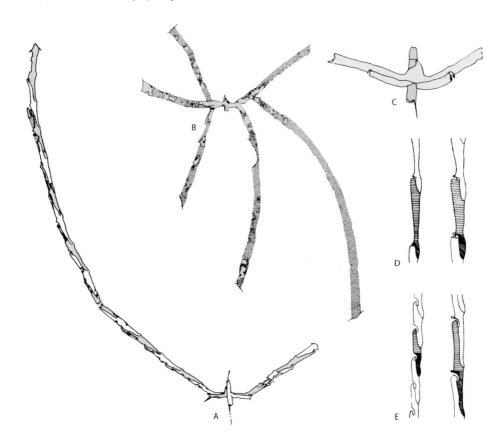

redefined the Neodiplograptidae to include the Petalolithinae. The author restricted the Retiolitoidea to the family Retiolitidae.

The Neograptina initially were a group of relatively inconspicuous biserial graptolites in the Middle Ordovician and may easily be overlooked, especially as they were usually not the most common faunal elements. The early Darriwilian taxa *Undulograptus* and *Skanegraptus* appear to be restricted biogeographically to low latitude regions (Goldman et al. 2011), where *Undulograptus* is widely distributed in China (Mu & Lee 1958), Baltica (Maletz & Ahlberg 2011a, b) and Central Europe (Bulman 1963; Bouček 1973). *Skanegraptus* is known from the Krapperup drillcore of southern Sweden (Maletz

2011c) and has been found only once. It may represent one of the early members with a derived proximal development pattern resembling the pattern H astogeny, but differs considerably from the derived members of *Normalograptus* in the mid to late Darriwilian, exemplified by *Normalograptus antiquus* (Ge, in Ge et al. 1990).

The Neograptina experienced their first radiation during the Middle and Upper Ordovician (Sadler et al. 2011), but never reached a higher diversity during this time interval (Figure 11.10). A dramatic increase in diversity can only be seen in the Ordovician/Silurian boundary interval with the origination of numerous new genera (Figure 11.17). The normalograptids were most common in the cold-water regions of Baltica and

Figure 11.17 The Neograptina radiation (based on Melchin et al. 2011, Fig. 7). Note that the Retiolitoidea is defined differently in Maletz (2014a). HME, Hirnantian mass extinction interval.

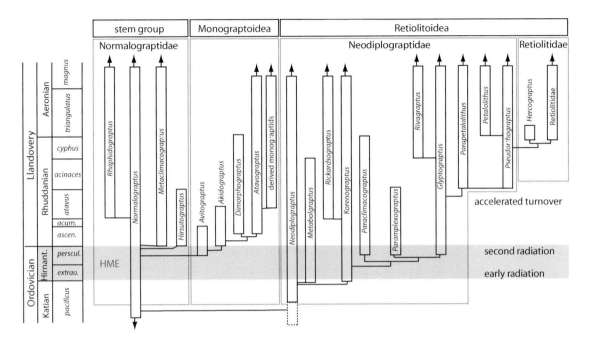

peri-Gondwana, and a distinct faunal gradient existed across the Iapetus Ocean at least from the Katian onwards (Zalasiewicz et al. 1995). *Normalograptus* specimens increasingly dominated the fauna at Whitland, south Wales, on the low latitude side of the Iapetus Ocean, and Zalasiewicz et al. (1995) differentiated a "*Normalograptus* proliferation interval" in the uppermost Katian. In the same time interval, the graptolite fauna of the Hartfell Shales of Scotland is much more diverse, and both intervals are difficult to correlate based on the graptolite faunas.

A major problem in this general picture appeared when a detailed investigation of Upper Ordovician "normalograptids" revealed that many taxa previously identified as *Normalograptus* are actually referable to its climacograptid homeomorph *Styracograptus* (Goldman et al. 2011; Štorch et al. 2011). Thus, the taxonomy of the biserials of the Upper Ordovician had to be revised considerably to sort out the climacograptids of the DDO-fauna from the N-fauna normalograptids (cf. Melchin & Mitchell 1991). Goldman et al. (2011) described the complex biogeographical history of

the normalograptids and differentiated five evolutionary phases during the Middle and Upper Ordovician. They postulated an evolutionary origin in the early to mid-Darriwilian high latitudes and a global spread in the later Darriwilian. A phase of retreat from the low latitudes started in the early Katian, probably slightly later in Laurentia, and was followed by a complete extirpation from all low latitudes in the mid-Katian. A reinvasion of *Normalograptus* in low latitudes was followed by the ecological and evolutionary replacement of the DDO-fauna during the latest Katian to early Hirnantian.

In the latest Ordovician, the recovery of the Neograptina in the aftermath of the Hirnantian extinction led to one of the most fascinating increases in diversity of the graptolites at all time. From a few surviving *Normalograptus* species (Melchin et al. 2011; Štorch et al. 2011), a whole array of biserial colony shapes evolved and also the Monograptidae emerged during this time of change. The Normalograptidae, as the only family-level taxon surviving the Hirnantian extinction, provided a number of new genus-level taxa, but

Figure 11.18 Ordovician and Silurian Normalograptidae. (A) *Normalograptus brevis* (Elles & Wood, 1906), Scania, Sweden. (B) *Normalograptus scalaris* (Hisinger, 1837), Dalarna, Sweden. (C) *Rhaphidograptus toernquisti* (Elles & Wood, 1906), Röstånga drill core, Scania, Sweden. (D) *Pseudoglyptograptus vas* Bulman & Rickards, 1968, Röstånga drill core, Scania, Sweden. (E) *Metaclimacograptus* sp., reverse view, Röstånga drill core, Scania, Sweden. (F) *Metaclimacograptus undulatus* (Kurck, 1882), two proximal ends in obverse views, Röstånga drill core, Scania, Sweden. Scale indicated by a 1 mm long bar in each photo.

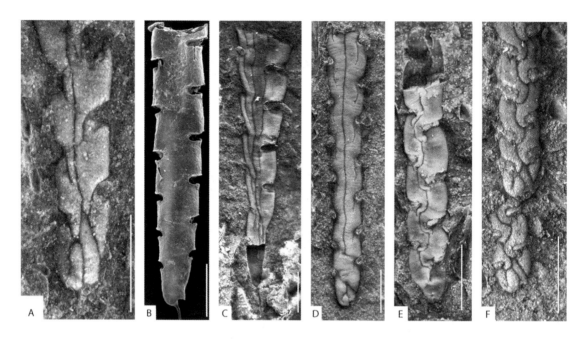

their diversity was quickly outshone by the diversity of the Retiolitidae and Monograptidae (Sadler et al. 2011; Maletz 2014a). Interestingly, Melchin et al. (2011) recognized a radiation of the Neograptina in three phases, starting in the basal Hirnantian during the first phase of the Hirnantian mass extinction (HME). A second phase can be seen in the post-glacial, latest Hirnantian, during the second phase of the HME, including the early Silurian recovery interval. An interval of accelerated turnover then followed in the mid-Rhuddanian (Figure 11.17).

The taxonomy of the biserial Neograptina is complex, and many constructional details are only available from chemically isolated material or from full relief specimens (Mitchell 1987; Melchin 1998). The species invariably show a pattern H astogeny or a derived one (Melchin et al. 2011; Štorch et al. 2011) with a rounded or pointed proximal end bearing the virgella as the only prox-imal apertural spine. Numerous new genus-level taxa have been described in recent literature since better-preserved material is available (e.g. Bulman & Rickards 1968; Koren & Rickards 1996), providing the constructional details for a precise interpretation of the structure and, thus, evolutionary relationships.

During the Lower Silurian, the simple normalograptid construction with short thecal overlap, mostly geniculate thecae and a straight median septum (Figure 11.18A) changes to more complex thecal styles, and recalls the constructional details known from the Ordovician Diplograptina. The median septum may be straight, sinuous, or strongly undulating to zigzag shaped (Figure 11.18C–F). Genicular hoods become larger and cover the thecal apertures in *Metaclimacograptus* (Figure 11.18 F). The median septum may be delayed in other species or even lacking completely in the Silurian Petalolithinae

Figure 11.19 Examples of the Neodiplograptidae. (A) *Metabolograptus persculptus* (Elles & Wood, 1907), Röstanga, Sweden. (B) *Parapetalolithus* sp., Röstanga, Sweden. (C) *Rivagraptus bellulus* (Törnquist, 1890), obverse view, showing alternating thecae and no median septum, Röstanga, Sweden. (D) *Rivagraptus bellulus* (Törnquist, 1890), Dalarna, Sweden, proximal end showing apertural spines on thecae. (E) *Paraclimacograptus innotatus* (Nicholson, 1869), Southern Urals, Russia (from Koren & Rickards 2004, reproduced with permission from The Palaeontological Association). (F) *Hirsutograptus* sp. cf. *H. villosus* Koren & Rickards, 1996, Arctic Canada (infrared photo by Jason Loxton). Scale indicated by 1 mm long bar in each photo.

(Figure 11.9D) and Retiolitidae (see Chapter 12). A very interesting Llandovery metaclimacograptid is the genus *Neodicellograptus* (Figure 11.14J) in which the two stipes are separated and the shape of the colony resembles that of a *Dicellograptus* species (Wang & Jin 1977; Melchin 1998). In genera like *Rhaphidograptus* (Figure 11.18C), *Agetograptus* and *Dimorphograptoides*, a short uniserial part of the colony is developed, similar to the development in *Dimorphograptus*, but a closer phylogenetic relationship may not be expressed (see Melchin et al. 2011).

A considerable diversity and disparity can be seen in the taxa of the recently erected Neodiplograptidae (Melchin et al. 2011), used here in the concept of Maletz (2014a). Early members are very similar to the Norma-lograptidae. Its basal member is *Neodiplograptus*

Legrand, 1987 with a pattern H astogeny and a characteristic thecal gradient from strongly geniculate thecae proximally, to thecae with a higher inclination of the ventral wall and strong widening of the colony distally, even though *Metabolograptus* Obut and Sennikov, 1985 (Figure 11.19A) appears even more like a typical *Normalograptus*. The group also includes the petalolithids with their often elongated and outwards inclined, non-geniculate thecae (Figure 11.19B). The median septum is absent at least on the reverse side, but is present on the obverse side. Variously spined taxa like *Rivagraptus* with paired apertural spines (Figure 11.19C–D), or *Hirsutograptus* with paired genicular spines and additional spines on the sicular aperture (Figure 11.19F), are common.

Outlook

We believe that we understand the biserial axonophorans quite well, but a number of questions are still open to discussion. New material has revealed the general patterns of diversification of Ordovician and Silurian clades, but the early diversification of the Diplograptina is still a mystery. How did the differentiation of the late pseudisograptid descendants *Exigraptus* and *Apiograptus* lead to the early diversification of *Levisograptus* and related taxa? How did the Monograptidae originate from the biserials? The loss of the second stipe is an easy explanation, but not sufficient to explain the patterns we see from the dimorphograptids in Chapter 13. More work on the Dapingian/Darriwilian and Ordovician/Silurian boundary intervals will be needed, as well as good luck in finding faunas that help us to answer our questions.

12

The Retiolitid Graptolites

Jörg Maletz, Denis E. B. Bates, Anna Kozłowska and Alfred C. Lenz

If the siculozooid in *Holoretiolites* had become sexually mature the budding of the last six blastozooids could have been omitted altogether and the sclerotized framework could have disappeared. But the holoretiolite stock need not have become extinct. (Kirk 1978, p. 546)

This remark by Nancy Kirk provides her personal view on graptolite evolution and extinction, but does not capture what we think we know now. It originates from the notion that the retiolitids, this strange and atypical group of graptolites, in which only a meshwork of rods is normally preserved, represent the answer to the graptolite extinction. Losing more and more of the housing construction and the number of zooids may be a way to eventually become a "naked" organism that is impossible to find in the fossil record, or be compared with the fossil graptolite remains if it is found in the modern seas. Thus, according to Kirk, we would be unable to see the connection if we accidentally come across a modern, extant planktic graptolite.

Naturally, Kirk only considered the extinction of the retiolitids here, but her hypothesis also extended to the extinction of the Monograptidae in the Early Devonian, and with this the extinction of all planktic graptolites. Shedding the tubarium could have been a good idea for the evolving graptolites, reacting to a changing environment, but we are pretty sure now that this was not the case, and we may have to look for another reason to explain the extinction of the planktic graptolites. This also means we have to modify our ideas on the Retiolitidae, their tubarium construction and evolutionary history.

Graptolite Paleobiology, First Edition. Jörg Maletz.
© 2017 John Wiley & Sons Ltd. Published 2017 by John Wiley & Sons Ltd.

Tubarium Reduction?

The retiolitids were long regarded as unusual graptolites in which the tubarium is reduced to a meshwork of lists and this meshwork is even today considered to be the defining characteristic of this group (Bulman 1970a; Melchin et al. 2011). The retiolitid meshwork has been termed the clathrium and reticulum, the clathrium being composed of strong regular lists making the major framework of the tubarium, and the reticulum of much finer, fairly irregular lists. Until recently, the retiolitid graptolites have been seen as the result of a reduced development, with the loss of the fusellum, leaving us with this meshwork of bars. This thinking changed only with the advent of modern research tools, the SEM (scanning electron microscope) and TEM (transmission electron microscope) and their use in the study of chemically isolated, well-preserved specimens. Suddenly, we realized that there are different ways of achieving the same or at least similar-looking results in the construction of graptolite tubaria (Figure 12.1). The reduction of the thecal walls is

only one of them. It is now clear through the detailed investigations of Bates and Kirk (1992, 1997) that the meshwork of the Retiolitidae includes a new development – the ancora umbrella and ancora sleeve. This is formed as a mantle outside the thecal-bearing parts, a kind of secondary cover and not the result of a reduction of the fusellum. Interestingly, this was recognized 125 years ago by Holm (1890) and Törnquist (1890) working with three-dimensionally well-preserved retiolitids from Sweden that retained almost complete fusellar thecae (see Figure 12.3), but these discoveries were "forgotten" or overlooked through the years.

In the Retiolitidae, the tubarium includes two membrane-based constructions: the original thecal framework inside, formed from the fusellum, and a second layer, the ancora sleeve (Figure 12.1B) on the outside, probably also constructed of fuselli (Lenz & Thorsteinsson 1997), as indicated by the presence of seamed lists (Bates 1987). The precise origin and secretion of the ancora sleeve membrane is unknown, as the preserved remains have so far not shown any indication of the mode of

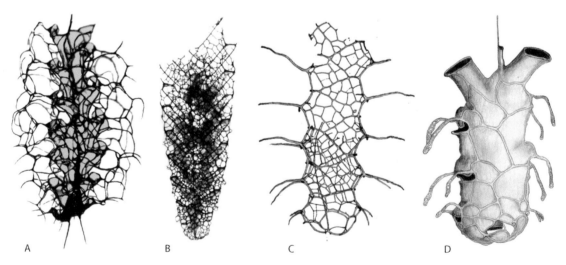

Figure 12.1 (A) *Pipiograptus* sp., a lasiograptid species with the thecal series in the centre and the lacinia outside, thecal row outline of tubarium digitally enhanced. (B) *Retiolites geinitzianus* Barrande, 1850, showing partially preserved thecal fusellum covered by the ancora sleeve. (C) *Spinograptus spinosus* Wood, 1900, showing ancora sleeve with sparse reticulum (modified from Maletz 2010b, Fig. 5A). (D) *Spinograptus tubothecalis* Kozłowska, Dobrowolska & Bates, 2013, reconstruction (based on Kozłowska et al. 2013, Fig. 5), showing hypothetical development of ancora sleeve membranes. Illustrations not to scale.

secretion. Thus, it cannot be explained by the model of fusellar or cortical secretion of the Pterobranchia.

The original wall material of the thecate tubarium and the ancora sleeve is rarely found in the Retiolitidae, and the remains usually consist of the reticulum and clathrium only. A few specimens of Silurian retiolitids have been discovered in which these membranes have survived intact (Lenz & Melchin 1987a; Lenz 1994a, b; Kozłowska-Dawidziuk 1997). Thus, retiolitids have been misunderstood for a long time. It was earlier assumed that, beginning with a particular group of "normal" diplograptid (biserial) ancestors, a reduction of the fusellum ultimately led to an organism that retained only cable-like lists (Bates et al. 2005). With the recognition of the ancora sleeve development, it is clear that the retiolitids gained rather than lost structure, making them more complex than their biserial ancestors by adding a new construction to their tubaria.

Many members of the Lasiograptidae (Chapter 11), often identified in the past as "archiretiolitids" or more formally the subfamily "Archiretiolitinae" (Bulman 1955, 1970a), form a secondary development, a lacinia (Figure 12.1A), outside the normal tubarium, by the complex branching of apertural and genicular spines (Bates & Kirk 1987, 1991). Their outer meshwork of lists and bars is completely different and independently developed from the lists of the Silurian retiolitids (see Plate 14), as we see in a precise analysis of the construction of these features. A lacinia is also typical of some derived glossograptids (Maletz & Mitchell 1996), but is again independently derived from the lacinia of the Lasiograptidae. In earlier Lasiograptidae, a lacinia is not present and the whole colony development is based on the remains of the thecal framework after the attenuation of the fusellum. This is clearly seen in the construction of the tubarium of *Orthoretiolites hami* (Whittington 1954; Bates & Kirk 1991). In this genus, the tubarium is based on the thickening and development of a number of lists, outlining the thecae, and a reduction of the thecal walls (Plate 14D). A reduction of the thickness of the thecal walls and an addition of thecal lists is also present in the early Middle Ordovician Abrograptidae (Chapter 10). Maletz (2014a)

referred the Abrograptidae to the Sinograpta, but did not discuss the Upper Ordovician biserial taxa of the group (*Abrograptus*, *Metabrograptus*, *Parabrograptus*), as they are not known from isolated material.

Retiolitid Origins

Initially, the retiolitid graptolites and other constructionally similar graptolite taxa were just lumped together as the Retiolitidae, because of the apparent lack of thecal walls and the presence of the meshwork of lists, masking the evolutionary relationships of "retiolitid" graptolites. The recognition of the Retiolitidae as a possibly monophyletic group through the development of the ancora sleeve, and a number of cladistic analyses (Lenz & Melchin 1997; Bates et al. 2005; Kozłowska et al. 2009) with exclusion of other, phylogenetically unrelated groups, led to the question of their origins. While it is clear that the retiolitids originated from an ancorate neograptine ancestor (Kozłowska-Dawidziuk et al. 2003; Melchin et al. 2011), the taxonomic differentiation and definition of the Retiolitidae has been much discussed. Kozłowska-Dawidziuk et al. (2003) used the presence of an ancora umbrella as the defining feature of their superfamily Retiolitoidea, but Melchin et al. (2011) employed a more inclusive definition and referred numerous neograptines without any indication of the ancora development, such as the family Neodiplograptidae, to the Retiolitoidea. However, both agree in the evolutionary origin of the retiolitid stock from a neograptine ancestor through the addition of the ancora sleeve. Maletz (2014a) preferred the concept of the Retiolitoidea of Kozłowska-Dawidziuk et al. (2003) and excluded the Neodiplograptidae from the Retiolitoidea.

The presence of an ancora umbrella with four prongs originating from the end of the virgella in *Petalolithus* and *Pseudorthograptus* (Bates & Kirk 1992) (Plate 14B) may be seen as the initiation of the ancora sleeve development. The re-curved extensions (the secondary ribs of Bates & Kirk 1992) of the ancora grow upwards, with vertical lists extending well above the level of the sicular

Figure 12.2 The ancora umbrella in the Petalolithinae. (A) *Petalolithus minor* (Elles, 1897), sicula with four-pronged ancora (from Bates & Kirk 1992, Fig. 32). (B) *Pseudorthograptus inopinatus* (Bouček, 1944), showing fusellum of sicula and thecae and the circular ancora umbrella; note the paired apertural spines on first theca. (C). *Pseudorthograptus* cf. *obuti* Koren & Rickards, 1996 with spiral ancora (first illustrated in Bates & Kirk 1984, pl. 5). (D) *Hercograptus introversus* Melchin, 1999, holotype, showing preservation of fusellum (photo by M.J. Melchin). Figures not to scale.

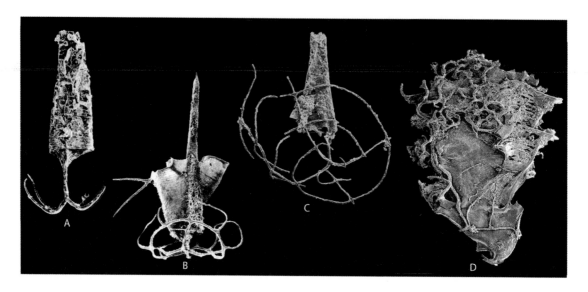

aperture in some species, although at this stage they are not connected to the thecae in *Petalolithus* (Figure 12.2A) or *Pseudorthograptus*. The development recalls the construction of the ancora umbrella in the derived retiolitids, especially as there appears to be a membrane connecting the branches, as is clear from the seamed lists on the four prongs (Bates & Kirk 1992) and remnants of a fusellum on some specimens.

Ancora Umbrella and Ancora Sleeve

The secondary wall construction of the Retiolitidae includes the membranes covering the proximal end and the lateral walls of the retiolitid colony. The early stage development starts with the formation of the ancora umbrella (Figure 12.2), a rounded, often umbrella-shaped structure, an extension of the virgella below the sicular aperture. In all retiolitids, the ancora umbrella is based on a four-pronged construction on the sicular aperture, the ancora hub, with additional branch-

ings or secondary lists formed on the surface of the ancora umbrella membrane. The ancora umbrella can be shallow and formed from regular or irregular meshes (Figure 12.2B), or deep with a distinct spiral structure (Figure 12.2C). This differentiation into the two types appears early in the evolution of the group, but little material of early retiolitid graptolites is available to follow the evolutionary relationships of these types. Bates and Kirk (1992) used the differences in the development of the ancora umbrella in the Retiolitidae to suggest a possible polyphyletic origin of the ancora sleeve. More recent works suggest a monophyletic origin (e.g. Lenz & Melchin 1997; Kozłowska-Dawidziuk et al. 2003).

The species of the genus *Pseudorthograptus* possess fully preserved thecae, but also have an ancora umbrella and often an indication of a partially developed ancora sleeve (Plate 14B). Membranes are seen as vague outlines in some *Pseudorthograptus* species (Koren & Rickards 1996), but the exact development of clathrial and reticular lists is uncertain for most taxa.

The details of the ancora umbrella have rarely been explored and descriptions exist for few species. Bates and Kirk (1991, 1992, 1997) described the development of the individual meshes of the ancora umbrella for a number of retiolitids and petalolithids. Later, Dobrowolska (2013) compared the ancora umbrella development in a number of Upper Wenlock and Ludlow plectograptine retiolitids, and recognized consistent differences to separate the genera, based on the number and arrangement of the individual ancora umbrella meshes.

The ancora sleeve walls start from the rim of the ancora umbrella and enclose the thecal framework of the colony. The ancora sleeve is attached to the thecal walls at the pleural lists and lateral apertural lists in the Plectograptinae. The ancora sleeve appears to be differently anchored and more complex in the Retiolitinae (see Bates & Kirk 1997), but this may in part be due to the development of the reticulum from the inside or outside of the ancora sleeve. In the Retiolitinae and rare Plectograptinae, orifices in the ancora sleeve membranes, called stomata, can be present on the lateral walls, complicating the construction.

Reticulum and Clathrium

The attenuation of the thecal walls, the fusellum, in the Retiolitidae is associated with the development of a secondary and more important structure, a meshwork of bars or lists, often identified as the clathrium and reticulum (Bulman 1955, 1970a). These form lists on the surface of the thecal fusellum or the membranes of the ancora sleeve (Figure 12.1C–D). The differentiation of the clathrium and reticulum is not straightforward and often difficult. The reticulum is generally regarded as the delicate, often irregular network of lists on the thecal walls and the ancora sleeve, while the clathrium forms the coarser "skeletal" lists, outlining the thecae and supporting the fusellum and the ancora sleeve membranes (Bates et al. 2005). A complex terminology exists for the identification of the ancora sleeve lists (Bates et al. 2005) (Figure 12.3A), but is not entirely consistent, and naming varies between the main groups of retiolitids.

The presence of seams on the clathrium and reticulum (Figure 12.4) indicates their original deposition on a membrane surface. Kozłowska-Dawidziuk (2001) discussed the position of these seams on the ancora sleeve wall in the genera *Cometograptus* and *Plectograptus*. She recognized that *Cometograptus* has the seams on the outside, indicating secretion of the lists from the inside of the ancora sleeve wall. In *Plectograptus*, the seams are on the inside, thus the secretion of the reticulum must have been taken place on the outside of the membranes. The general distribution pattern of the lists on the inside or outside of the retiolitid colonies and the reason for these differences is unknown. Kozłowska-Dawidziuk and Lenz (2001) indicated that there are two main groups, one earlier with outward-facing seams and a second one with inward-facing seams, but a few taxa may have both types of seams, such as *Stomatograptus* (see Bates & Kirk 1997).

The reticulate lists of the Retiolitidae are formed from the precise organization and concentration of the cortical bandages (e.g. Bates 1990) on the surface of the thecal and ancora sleeve membranes (Figure 12.4). The secretion differs from the cortical bandages found in other graptolites through their development in the form of distinct linear or cable-like features. The bandages are not laid down irregularly and randomly as the sometimes excessive thickening of the thecal walls in many benthic, but also a few planktic taxa (e.g. Bates et al. 2011). A distinct micro-ornamentation can be seen on the list surfaces of the retiolitids, and two types can be differentiated. These ornamentations have been used to differentiate the Retiolitinae and Plectograptinae (Lenz & Melchin 1987b). The list surfaces of the Retiolitinae are either smooth or show a parallel striation. In the Plectograptinae, a characteristic pustular ornamentation is present (Figure 12.4C). These differences as reliable indicators in the differentiation of the retiolitine from plectograptine retiolitids have recently been questioned (Melchin et al. 2011), and more research is necessary to trace the origin and distribution of these features.

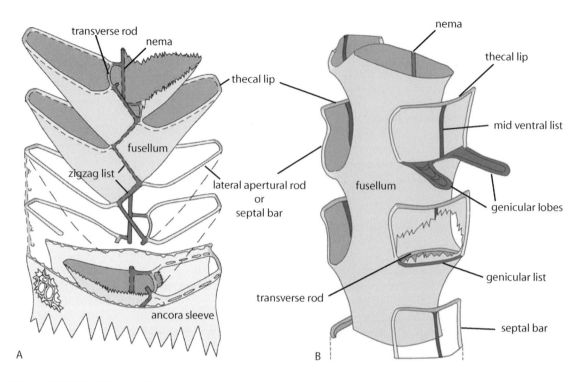

Figure 12.3 (A) *Retiolites* sp., showing connection between thecal framework and ancora sleeve (based on Bates et al. 2005, Fig. 5A, D). (B). Thecal development in *Spinograptus*, ancora sleeve not shown (based on Bates et al., 2005, Fig. 5D).

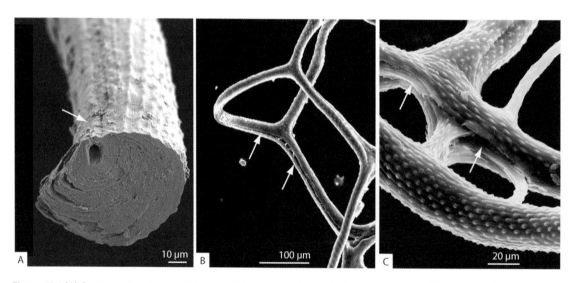

Figure 12.4 (A) Cross-section through thickened list with seam (arrow), showing the cortical bandages of a plectograptine retiolitid. (B) *Paraplectograptus* sp., lists with seams (arrows) indicating the presence of a membrane. (C) *Quattuorgraptus muenchi* (Eisenack, 1951), pustular ornamentation of plectograptine retiolitid.

Early Ancora Sleeve Development

The origin of the ancora umbrella and ancora sleeve membranes can be traced back to the Petalolithinae, a subfamily of the family Neodiplograptidae (Maletz 2014a). The presence of an ancora is well known in some *Petalolithus* species, but it was Koren and Rickards (1996) who established the genus *Parapetalolithus* for species without an ancora. Kozłowska-Dawidziuk et al. (2003) suggested that all ancorate taxa should be included in the superfamily Retiolitoidea. Melchin et al. (2011) further enlarged the Retiolitoidea to include numerous non-ancorate neograptines based on a cladistic analysis of the Silurian members of the Neograptina, and defined the Retiolitoidea as a sister group of the Monograptoidea. The ancora umbrella appears to be fairly simple in most taxa of the genus *Petalolithus*, and in a very few it reaches up to the thecal aperture of the first theca in *Petalolithus folium* as illustrated in Koren and Rickards (1996) (Figure 12.5A). The authors also illustrated a specimen of *Petalolithus ovatoelongatus* showing an ancora umbrella with multiple branchings and possibly preserving a number of ancora sleeve lists (Figure 12.5C). None of the specimens shows a reduction or attenuation of the fusellum of the thecae, and a membrane indicating the presence of a true ancora sleeve is not preserved.

A single specimen of *Pseudorthograptus* sp. C with extensive ancora sleeve lists covering the

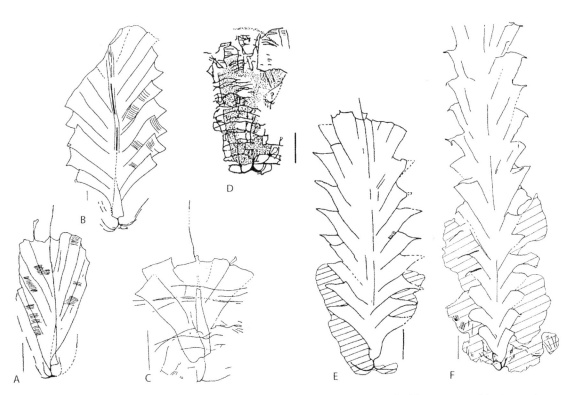

Figure 12.5 Early ancora sleeve development. (A) *Petalolithus folium* (Hisinger, 1837) with ancora reaching to aperture of th1. (B) *Petalolithus ovatoelongatus* (Kurck, 1882) with short ancora. (C) *Petalolithus ovatoelongatus* (Kurck, 1882), showing ancora sleeve lists in proximal end. (D) *Pseudorthograptus?* sp. C, specimen with preserved thecal walls and ancora sleeve. (E) *Pseudorthograptus mutabilis* (Elles & Wood, 1907), showing thecal outlines, ancora and proximal ancora sleeve membrane. (F) *Pseudorthograptus obuti* Rickards & Koren, 1974, showing extensive proximal membranes. All from Koren & Rickards (1996), reproduced with permission from The Palaeontological Association.

colony was found in the southern Ural Mountains (Koren & Rickards 1996). The specimen shows the complex development of the ancora sleeve lists and fully preserved thecae. The outline of the thecal fusellum is visible through the mesh of the ancora sleeve (Figure 12.5D). The specimen indicates that the development of an ancora sleeve and the reduction of the thecal walls may have happened independently in the ancestors of the Retiolitidae.

The most unusual petalolithine is the genus *Hercograptus* (Melchin 1999), a strange graptolite known from a few chemically isolated specimens found in the early Silurian of Arctic Canada (Figure 12.2D). It shows a combination of features connecting the petalolithines with the retiolitids, but is unlikely to be in the direct line of evolution due to its unusual tubarium construction. A strong mid-ventral list can be seen on the thecae, but the thecal apertures are quite unusual with slight dorsal isolation and ventral extensions with widened lateral lappets preserving thecal increments in a similar fashion to those in *Pseudoretiolites*.

The Retiolitinae

The Retiolitinae (cf. Bulman 1970a; Bates at al. 2005; Maletz 2014a) are the earlier of the two groups of the Retiolitidae of Bouček and Münch (1952). Robust members are common in the Llandovery to mid-Wenlock, easily differentiated into such genera as *Retiolites* (Figure 12.6A, C–D), *Pseudoretiolites* (Figure 12.6I), *Pseudoplegmatograptus* (Figure 12.6J) and *Stomatograptus* (Figure 12.6B, E–H). Bates and Kirk (1992, 1997) described these taxa from chemically isolated material, and their construction is understood in great detail. Characteristically, the ancora umbrella development is quite different among the individual genera and difficult to compare.

The ancora umbrella is very complex and deep with a strong spiral development (Figure 12.7C) in *Pseudoretiolites* (Lenz & Melchin 1987b; Bates & Kirk 1992). *Pseudoretiolites* also bears a distinct construction of zigzag lists on the ventral thecal wall extensions (Figure 12.6I). *Retiolites* has a much simpler ancora umbrella (Figure 12.7E;

Plate 14C). In *Retiolites angustidens* (Figure 12.7 F) the ancora umbrella is very shallow with a considerable number of meshes showing a vague spiral arrangement and moderate-sized lateral openings (Bates & Kirk 1997). *Stomatograptus* has a moderately deep ancora umbrella with hexagonal meshes and no indications of a spiral arrangement. A shallow ancora umbrella with few meshes and a vague spiral arrangement is also known from *Pseudoplegmatograptus* (Bates & Kirk 1992).

Some lesser-known members of this group are *Rotaretiolites* and *Eiseligraptus*. The genus *Rotaretiolites* Bates & Kirk, 1992 was found in a few specimens in the Llandovery of Dalarna, Sweden, and was subsequently recognized also in Arctic Canada (Kozłowska-Dawidziuk & Lenz 2001). The construction of the colony has a number of unusual features and the tubarium completely lacks an ancora sleeve. The ancora umbrella is shallow and small, with only four meshes (Figure 12.7B). It is connected to the clathrium by two lists, of which one can be identified as a mid-ventral list of th1^1, while the other is a junction list connected to the mid-ventral list of th1^2 (Figure 12.7A). An unusual taxon, *Eiseligraptus*, found only in the Silurian of Germany (Hundt 1965), has a coarse, fairly unordered mesh and bears a conspicuous and robust, multibranched nematularium not seen in any other retiolitid (Figure 12.6 K).

The Plectograptinae

Bouček and Münch (1952) differentiated the Plectograptinae as the younger of the two groups of retiolitids in the Silurian, because the authors recognized an interval without retiolitids, the *Monograptus firmus/Monograptus riccartonensis* interval in central Europe. They also noted the small size of most of the younger taxa and the reduction and even lack of a reticulum in these. The temporal differentiation is not applicable any more, as many retiolitids from intermediate levels have since been discovered, filling earlier gaps (see Kozłowska-Dawidziuk 2004). Lenz and Melchin (1997) initially recognized the ornamentation on

Figure 12.6 (A, C–D) *Retiolites geinitzianus* Barrande, 1850. (A) Distal fragment, from Tullberg (1883). (C) Cross-section, from Holm (1890). (D) Incomplete proximal end, from Holm (1890). (B, E–H) *Stomatograptus toernquisti* Tullberg, 1883 (= *Stomatograptus grandis* Barrande, 1850). (B) Proximal end, from Tullberg (1883). (E) Fragment, from Tullberg (1883). (F) Cross-section showing stomata, after Bates and Kirk (1997). (G) Fragment showing ventral thecal wall and reticulum with stomata, from Holm (1890). (H) Fragment. (I) *Pseudoretiolites perlatus* (Nicholson, 1868b), from Štorch (1998b). (J) *Pseudoplegmatograptus obesus* (Lapworth, 1877), from Blumenstengel et al. (2006). (K) *Eiseligraptus cystifer* Hundt, 1959, Weinberg Hohenleuben, Thuringia, Germany. Illustrations not to scale.

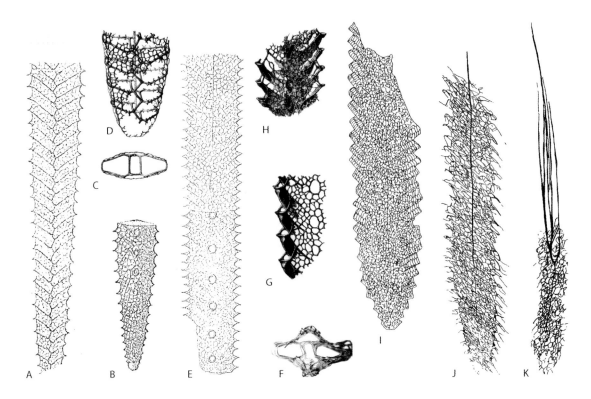

the cortical bandages as a useful indicator for the separation of the two groups. In general, the plectograptine retiolitids are small, often with clearly restricted growth, and eventually their colonies included only a handful of zooids. Additionally, many of the Upper Silurian taxa have appendices (see Figures 12.8, 12.9).

Plectograptine retiolitids are often common in glacial boulders of Scandinavian origin, and many species have been found for the first time in these transported materials in northern Germany and Poland (e.g. Münch 1931; Eisenack 1951; Kozłowska-Dawidziuk 1995; Maletz 2008, 2010b). Numerous specimens are also found in Arctic Canada (e.g. Lenz 1993; Lenz & Kozłowska-Dawidziuk 2004) and in Polish drill cores (e.g. Kozłowska-Dawidziuk 1990, 1995, 1997). The material is often excellently preserved in full relief in limestones, and can easily be isolated and investigated with a SEM. Prior to this, investigations were carried out using high-powered, normal light microscopes, and illustrations were produced as line drawings, some of them excellent, as in Holm (1890), Wiman (1896a) and Eisenack (1951) (see also Figure 12.8).

The development of the ancora sleeve lists is quite variable in the Plectograptinae (Figure 12.9), and species range from those with a very dense reticulum to those with a very few – or no – lists. *Neogothograptus balticus* has barely any reticu-

Figure 12.7 The ancora development in the Retiolitinae. (A) *Rotaretiolites exutus* Bates & Kirk, 1992, reconstruction. (B) *Rotaretiolites exutus*, ancora umbrella from outside (based on Bates & Kirk 1992, Fig. 88). (C) *Pseudoretiolites perlatus* (Nicholson, 1868b), spiral ancora umbrella, side view (based on Bates & Kirk, 1992, Fig. 122: identified as *Pseudoretiolites* sp. cf. *P. decurtatus* Bouček & Münch, 1944). (D) *Pseudoretiolites* sp., clearly spiralled ancora umbrella from below (based on Bates & Kirk 1992, Fig. 167). (E) *Retiolites geinitzianus* Barrande, 1850, proximal end in side view, showing shallow ancora umbrella with hexagonal meshes (based on Bates & Kirk 1986a, Fig. 28, reproduced with permission from The Royal Society). (F) *Retiolites angustidens* (Elles & Wood, 1908), reconstructed ancora umbrella from below (based on Bates & Kirk 1997, Fig. 91). Illustrations not to scale.

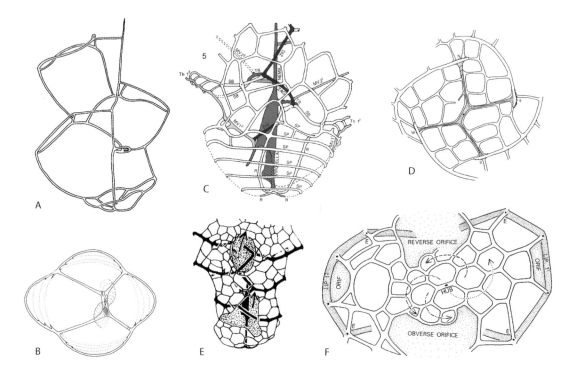

lum (Figure 12.9A) and looks quite different from the closely related *Neogothograptus eximinassa* (Figure 12.9B), in which a dense reticulum covers the surface of the ancora sleeve and masks the fairly simple clathrial construction of the colony. Reticular lists are often formed at later stages throughout the colony's astogeny, and juveniles do not bear a reticulum (Maletz 2008). Reticular lists are restricted to the lateral ancora sleeve walls of the specimens and are not found on the ventral thecal walls.

The Plectograptinae bear geniculate thecae, and the genicula are often adorned with spines, paired lobes or genicular hoods (Figure 12.9G–K). The largest and the most variable genicular processes occur in gothograptids, in which extensive hoods may extend proximally covering the apertures and part of the theca, as in *Gothograptus nassa* Holm, 1890 (Kozłowska-Dawidziuk 2004) (Figure 12.9I). Its characteristic genicular hood is covered by densely packed cortical bandages. Reticulated hoods are present in *Gothograptus kozlowskii* Kozłowska-Dawidziuk, 1990 and in *Neogothograptus reticulatus* Kozłowska, Lenz & Melchin, 2009 (Figure 12.9H). Paired genicular extensions separate *Plectograptus robustus* from *Plectograptus mobergi* (Figure 12.8). *Neogothograptus* can also develop relatively huge genicular elaborations, as is seen in *Neogothograptus alatifromis* Lenz

Figure 12.8 Biostratigraphy of plectograptine graptolites from German glacial boulders, specimens based on line drawings by Hermann Jaeger (modified from Maletz 2008).

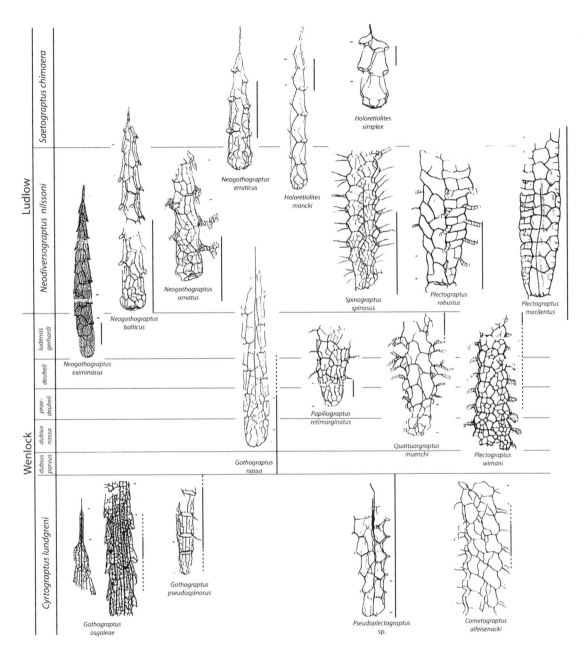

& Kozłowska, 2004 (Figure 12.9G), while the development of the clathrium is unchanged. These paired processes may be quite variably formed as fusellar or microfusellar developments or as massive spines, as in *Spinograptus spinosus* (Wood, 1900) (Figure 12.9 F, K), or *Spinograptus latespinosus* (Kozłowska-Dawidziuk 2004) (Figure 12.9 J).

Figure 12.9 Examples of the Plectograptinae. (A) *Neogothograptus balticus* (Eisenack, 1951). (B) *Neogothograptus eximinassa* Maletz, 2008. (C) *Gothograptus nassa* (Wiman, 1895), proximal end of long specimen. (D) *Papiliograptus* sp. (*P. regimarginatus* in Maletz, 2010b, Fig. 2). (E) *Holoretiolites erraticus* (Eisenack, 1951). (F) *Spinograptus spinosus* (Wood, 1900). (G) *Neogothograptus alatiformis* Lenz & Kozłowska-Dawidziuk, 2004. (H) *Neogothograptus reticulatus* Kozłowska et al., 2009. (I) *Gothograptus nassa* (Holm, 1890). (J) *Spinograptus latespinosus* Kozłowska-Dawidziuk, 1997. (K) *Spinograptus spinosus* (Wood, 1900). SEM photos of material from north German glacial boulders (Maletz 2008, 2010b), Poland (Kozłowska-Dawidziuk 1997) and Arctic Canada (Lenz & Kozłowska-Dawidziuk 2004). The scale indicates 1 mm in (A–G) and 200 µm in (H–K).

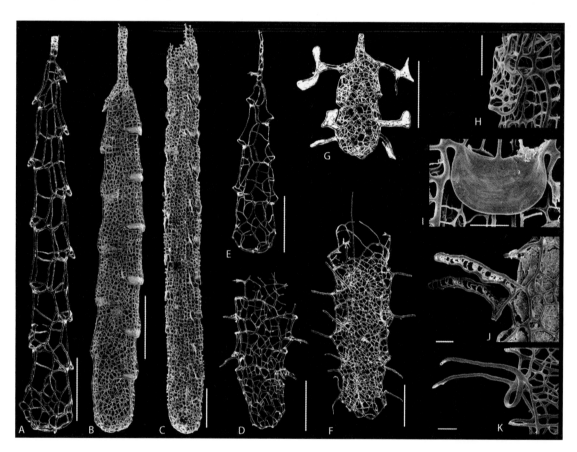

Appendix and the Retiolitid Extinction

The most conspicuous characteristic of many plectograptines is the appendix (Figure 12.9A, B, E). The appendix is a narrow tube at the tip of the colony, assumed to be the housing of the last formed zooid and indicating that the colony ceased to grow. The appendix appeared in the upper Wenlock and became common in most of the taxa during the Ludlow. It is formed differently from the normal thecae, and thus may indicate some change in the anatomy of the graptolite zooid forming it. The appendix may be short and simple, but can reach a considerable length. In *Holoretiolites erraticus*, the appendix can be as long as the main body of the colony (Maletz 2008, Fig. 13G). The nema may be incorporated into the appendix, but in other taxa it is free inside. The appendix is outlined by a few stout lists in

Figure 12.10 Comparison of large (A) and small retiolitid colonies (B, C). (A) *Stomatograptus* sp, fragment of distal part of tubarium with four pairs of thecae. (B) *Holoretiolites helenaewitoldi* Kozłowska-Dawidziuk, 2004, finite tubarium with four pairs of thecae. (C) *Plectodinemagraptus gracilis* Kozłowska-Dawidziuk, 1995 with strongly reduced ancora sleeve. 1 mm scale for all specimens.

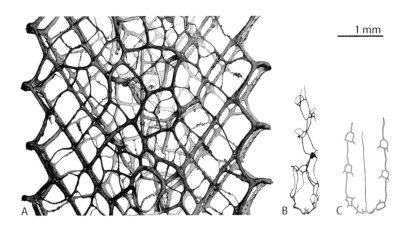

Holoretiolites erraticus (Figure 12.9E), but in *Neogothograptus eximinassa* it includes a dense mesh of numerous reticular lists (Figure 12.9B). In *Spinograptus tubothecalis*, a diminutive appendix can be seen between the distalmost two thecae, forming an unusual development (Figure 12.1D).

The retiolitids at the brink of their final extinction are characterized by a reduction in tubarium length and the development of the appendix (Figure 12.10). The earlier retiolitids from the Aeronian to the Sheinwoodian generally have larger tubaria, containing sometimes up to 80 pairs of thecae and a relatively short sicula, and thus probably had unlimited colony growth (Kozłowska-Dawidziuk 2004). The later retiolitids, by contrast, are represented by small tubaria with a long sicula and a limited number of thecae, reduced to four thecae in *Holoretiolites helenaewitoldi* (Figure 12.10B) and only two in *Neogothograptus alatiformis* Lenz & Kozłowska-Dawidziuk, 2004 (Figure 12.9G). The small colonies replaced the large ones in the Homerian, probably in relation to environmental changes following the mass extinction of graptolites at the end of the lower Homerian (Jaeger 1991; Lenz et al. 2006).

An additional differentiation can be seen in the variable development of the meshwork of the tubarium. There is a considerable reduction in the skeletal elements in the younger taxa, but the development of reticular lists and reticulation of thecal processes is quite variable even within individual genera (e.g. *Neogothograptus*: Figure 12.9A–B). *Plectodinemagraptus gracilis* Kozłowska-Dawidziuk, 1995 from the *hemiaversus/aversus* and *leintwardinensis* biozones of the lower Ludfordian (Upper Ludlow, Silurian) of Poland is the youngest known retiolitid (Figure 12.10C). The tubarium of this species is highly reduced with few clathrial lists and lacks a reticulum (Kozłowska & Bates 2014). The colony basically consists of a reduced ancora umbrella with attached mid-ventral lists and loops outlining the thecal openings.

Outlook

Still the question of why the graptolites developed the complex meshwork of lists of the retiolitid tubaria remains, and we are not much closer to an answer than we were more than a hundred years ago when the first retiolitids were described. However, we have made considerable improvements in our understanding of graptolites in general through the investigation of the Retiolitidae, such as differentiation of the fusellum of the thecae and the outer walls formed from the ancora sleeve and reticulum. From a paleontological standpoint, we need to investigate in more detail the origin of the ancora umbrella and sleeve developments and the precise connection of the ancora sleeve to the thecal framework. We have a good grip on the evolutionary history within the Retiolitidae, and we can even use their evolutionary changes for biostratigraphical purposes, but the organisms themselves and the origins of their tubarium construction are still an enigma.

13

The Monograptids

Jörg Maletz

Even though the Monograptidae only appear in the early Silurian, their general tubarium style is long known, and constructionally comparable taxa with a single thecal row originating from the sicula already appear in the Floian (Early Ordovician) with the genus *Azygograptus*. These early, single-stiped taxa were not successful in the long run, thus the Monograptidae must have made something differently and obviously much better. More than a hundred described genera spanning a time interval of at least 40 million years certainly indicate their success. During the time of their existence they created an astonishingly wide array of colony shapes, some of these not explored by earlier graptolites, and with this they demonstrated an unrivalled mastery of tubarium architecture. It would be wrong to call their architectural designs minimalistic, but starting from simple tubes, they definitely deserved their success.

The monograptids again and again played with the near-endless possibilities of tubarium construction, and mastered the challenges related to the limitations of fusellar construction. There were, however, features we are familiar with through earlier graptolites that the monograptids did not master. For example, no monograptid explored fusellum reduction or the development of something like an extra-tubarial meshwork construction as seen in the lacinia of the Ordovician Lasiograptidae or the ancora sleeve development of the Silurian Retiolitidae. Monograptids limited their mastery of tubarium architecture to the modification of thecal tubes and the addition of apertural modifications instead.

Figure 13.1 The monograptid tubarium. (A) *Atavograptus ceryx* Rickards & Hutt, 1970, reconstruction. (B) *Azygograptus validus* Törnquist, 1901, reconstruction. Arrows in (A, B) indicate direction of stipe growth. (C) *Sinodiversograptus lientanensis* (Mu, 1948), multiramous streptograptid with numerous cladial branches (adapted from Loydell 1990, Fig. 1, which was adapted from Mu & Chen 1962). (D) Resorption foramen for th1 in amplexograptid sicula. (E–G) Sinus and lacuna stages of primary foramen for th1 in monograptids. (D–G) adapted from Bulman 1970a, with permission from The Paleontological Institute, and also from Bulman 1970b. Illustrations not to scale.

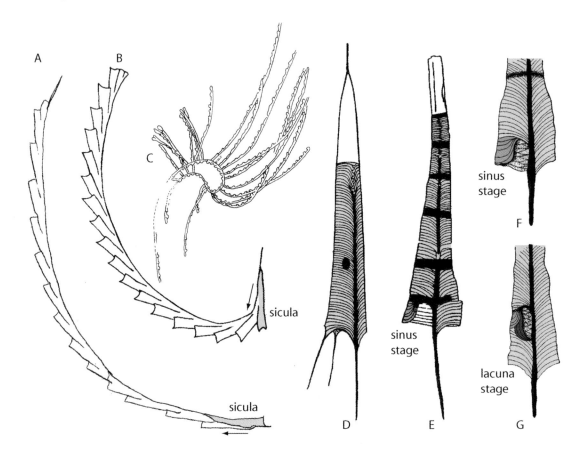

Monograptid Construction

There are a few characteristics that define a truly monograptid tubarium: the sicula with its aperture pointing in the opposite direction to the apertures of the thecae and the stipe growth (Figure 13.1A); the presence of a single stipe or one with a variable number of cladial secondary branches (Figure 13.1C), typically expressed in the genus *Cyrtograptus*, but also present in a variety of further genera; and the foramen for the development of the first theca. The construction of the monograptid colony initially strongly resembles

the Floian to Dapingian *Azygograptus* (Figure 13.1B) and *Jishougraptus*, but the direction of growth of the stipe is quite different. In the monograptids, the stipe grows in the opposite direction to the sicular aperture, while in *Azygograptus* the stipe follows at least initially the direction of the sicular aperture, but the stipe may bend dorsally afterwards.

Differences in the details of tubarium formation can be seen in the construction of the foramen or porus for the development of the first theca. It was generally thought to be through a sinus and lacuna development (Figure 13.1E–G) since the original description of this feature by Eisenack (1942) from isolated material of *Pristiograptus frequens* (Jaekel, 1889), and supported by observations in numerous monograptid specimens (e.g. Walker 1953; Urbanek 1958), but the construction was already illustrated by Münch (1928: Fig. 6). The development differs considerably from the resorption foramen in the diplograptids (Figure 13.1D) and in all earlier graptolites, in which a hole was punched into the wall of the sicula. The supposedly universally distributed sinus and lacinia construction in monograptids is now recognized as a feature restricted to later monograptids. The original type of foramen or porus appears to be a resorption foramen even in the monograptids. An early type of modification of the resorption foramen is visible in a number of *Monoclimacis*? species (Melchin & Koren 2001), but its further distribution is uncertain, and a transition from a resorption porus to the sinus and lacuna stages during the Llandovery can be observed (Lukasik & Melchin 1994, 1997; Melchin & Koren 2001; Dawson & Melchin 2007). The pre-porus fuselli are unchanged, but the fuselli involved in the formation of the porus are slightly deflected and the sicula bulges outwards at this point. A number of truncated fuselli can be seen as a result of the resorption of the thecal wall by the first post-sicular zooid, and a primary porus has not been detected.

Based on the tubarium construction of the monograptids, the origin of the clade should be recognized easily. However, this is not the case, and various scenarios have to be explored. The uni-biserial taxa of the Dimorphograptidae were long considered as ancestral to the Monograptidae. As these were thought to appear much later in the stratigraphic record, this origin was considered unlikely, and Hutt et al. (1972) considered a diphyletic origin of the Monograptidae, while Rickards et al. (1977), based on new records of earliest monograptid specimens, discussed a monophyletic origin and early differentiation of the monograptids. Mitchell (1987, Fig. 17) indicated a possible origin of the Monograptinae through a *Glyptograptus*-type ancestor with pattern I astogeny, but did not discuss the proximal development or transition in detail. The recent cladistic analysis of Melchin et al. (2011, Fig. 3) again indicated the origin of the Monograptidae from a neograptid biserial through a dimorphograptid ancestor, and suggested a direct

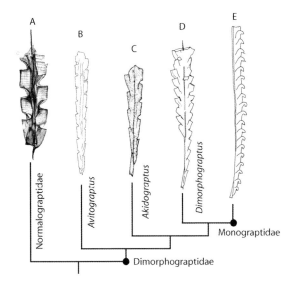

Figure 13.2 Diagram showing the concept of the Dimorphograptidae based on the analysis of Melchin et al. (2011, Fig. 2). (A) *Normalograptus brevis*. (B) *Avitograptus avitus*. (C) *Akidograptus ascensus*. (D) *Dimorphograptus swanstoni*. (E) *Monograptus priodon*. Graptolite specimens from various sources, not to scale.

ancestor–descendant relationship of the dimorphograptids and monograptids (Figure 13.2).

Family Dimorphograptidae

The origin of the Dimorphograptidae may be seen in a form similar to *Avitograptus avitus* (Figure 13.2) with an elongated, slender proximal end and a "normal" pattern J proximal development. Melchin et al. (2011), in their cladistic analysis, identified this lineage leading through the genera *Akidograptus* and *Parakidograptus* (Figure 13.3A) to *Dimorphograptus*, and finally to a monograptid colony shape (Figure 13.2). Melchin et al. (2011) defined the Dimorphograptidae as a paraphyletic family with a pattern J astogeny and a fully biserial to proximally uniserial tubarium. Details are not available, since chemically isolated material and relief specimens showing the proximal and distal thecal development are rare. The group originates in the Hirnantian, latest Ordovician, and ranges into the Rhuddanian (Llandovery) with a small number of genera.

Figure 13.3 (A) *Parakidograptus acuminatus* (Nicholson, 1867b), LO 1284t, reverse view, low relief, latex cast, Tomarp, Scania (Törnquist 1897, pl. 2, Fig. 7). (B) *Dimorphograptus* sp. cf. *D. swanstoni* Lapworth, LO 476t, latex cast, showing delay of median septum on reverse side, Bollerup, Scania (Kurck 1882, Figs 5, 6). (C) *Rhaphidograptus toernquisti* (Elles & Wood, 1906), on slab with LO 1456 T (Törnquist 1899, p. 2, Fig. 1: *Monograptus incommodus*, holotype), latex cast, Röstanga, Scania, Sweden. (D–E) *Agetograptus* sp., based on Bulman (1970a, Fig. 61: *Dimorphograptus* sp.). Scale indicated by 1 mm long bar in each photo.

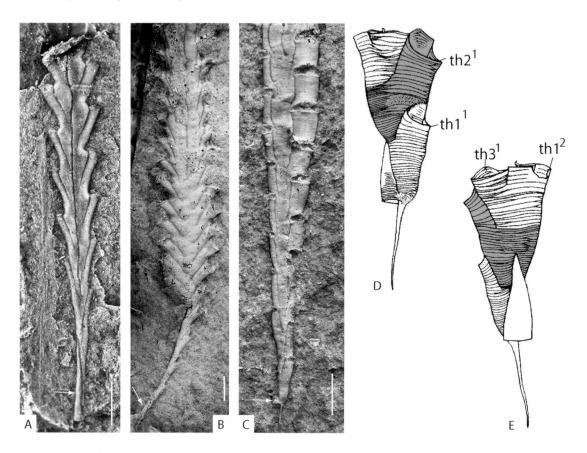

Unfortunately, the typical loss of the second stipe in the proximal end (Figure 13.3B) is not restricted to the Dimorphograptidae. Li (1987) discussed the possibly independent origin of a number of "dimorphograptid" genera, an idea supported by later research, showing the position of species with a proximally uniserial tubarium, lacking at least th1² in various positions on the cladistic tree of the Silurian biserials (Melchin et al. 2011). Thus, a loss of the second stipe, possibly through a redirection of th1² (Melchin 1998; Melchin et al. 2011), may have led independently to dimorphograptid colony shapes in several lineages. A typical example is *Agetograptus* (Figure 13.3D–E) with its redirected and elongated th1², forming a proximally uniserial colony. A similar development can be seen in *Rhaphidograptus toernquisti* (Elles & Wood, 1906). The species has a slender uniserial proximal end and widens considerably distally (Figure 13.3C). Melchin et al. (2011) suggested an origin of the genus from a *Normalograptus*-type ancestor in the Rhuddanian.

Monograptid Thecal Styles

The thecal styles of the monograptids are quite variable, but the underlying concept is simple and clear. Initially, simple widening tubes with straight apertures and moderate thecal overlap are

Figure 13.4 Important thecal styles in Monograptidae. (A) *Pristiograptus dubius* (Suess, 1851), glacial boulder, northern Germany. (B) *Heisograptus micropoma* (Jaekel, 1889), glacial boulder, northern Germany. (C–D) *Monograptus priodon* (Bronn, 1835). (E–F) *Oktavites spiralis* (Geinitz, 1842). (G–H) *Lapworthograptus grayae* (Lapworth, 1876). (I–J) *Neomonograptus* sp. cf. *N. praehercynicus* (Jaeger, 1959), adapted from Jaeger (1959). (C–H) from Dalarna, Sweden, based on Bulman (1932a). Illustrations not to scale.

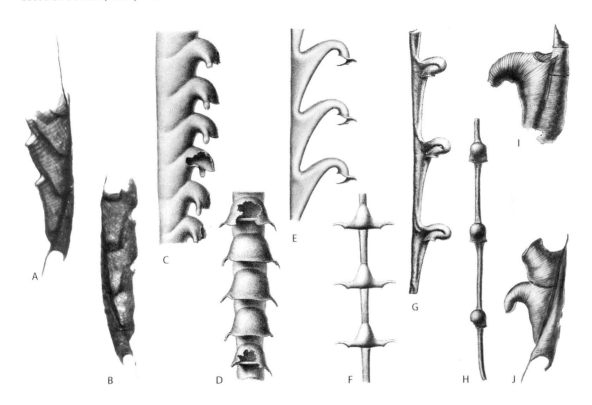

present in *Atavograptus* (Figure 13.1A), the oldest monograptid genus (Hutt & Rickards, 1970; Lukasik & Melchin, 1994), but quickly the thecal apertures became modified. Variously shaped hoods, lateral lobes and genicular developments became the standard of early monograptids (Lukasik & Melchin 1997; Koren & Bjerreskov 1997). The early diversification of the monograptids in the Llandovery also saw the evolution of thecal isolation, paired lateral apertural spines, and variously hooked and hooded thecae (Figure 13.4). Another group of monograptids with straight tubaria kept the simple thecal style with moderate thecal inclination and unmodified thecal apertures, and evolved into the *Pristiograptus* stock (Figure 13.4A–B). They probably had their ancestors in the *Pristiograptus variabilis* group (see Loydell 1993).

The most important new development in monograptid thecae is the hooked theca, in which the distal part of the dorsal wall is free and does not form part of the interthecal septum of the next theca. The distal, apertural part of the theca, thus, is a recurved tubular outgrowth. A typical example can be seen in the genus *Monograptus*, where the hook is conspicuous in lateral view (Figure 13.4C). The thecae retain a considerable thecal overlap, but aperturally they bend back towards the sicula, forming a hook of variable size. Often a slight rounded apertural lip is formed with two lateral apertural spines or lobes, often only visible in ventral view (Figure 13.4D). This development can lead to completely isolated metathecae without thecal overlap, lateral expansion, strong lateral spatulate processes or spines, and conspicuous dorsal hoods as is seen in the thecae of *Oktavites*

Figure 13.5 Biform monograptids. (A–B) *Pernerograptus revolutus* Kurck, LO 475 t, proximal and distal part of single specimen. (C) *Pernerograptus difformis* Törnquist, LO 1470 T, two fragments. None of the specimens show the extreme proximal end with the sicula. All specimens from the Llandovery, Silurian, Sweden, adapted from Hutt (1974b), with permission from The Palaeontological Association. 1 mm scale is for all specimens.

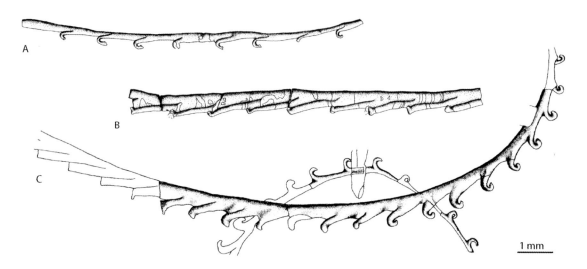

spiralis (Figure 13.4E–F). A considerable asymmetry in the thecae and a lateral twisting of the thecal apertures may also be present in this genus (Lenz & Melchin 1989; Loydell & Nestor 2006).

An extreme isolation of the thecae can be seen in the genus *Rastrites* and in related taxa. Here the metathecae are completely isolated and tubular. The thecal apertures are slightly hooked and sometimes bear indications of lateral spines or spatulate processes. Early *Rastrites* species bear short thecae, but during the evolution of the genus the thecae increase in length and in *Rastrites maximus* they may reach a length of 12 mm (Štorch & Loydell 1992) and form some of the longest thecae known in graptolites.

Typical of many early monograptids is the development of biform thecae (Figure 13.5) or species showing thecal gradients in which the proximal thecae are different from the distal thecae. Hutt (1974b) described the biform thecae in the *Pernerograptus revolutus* group and illustrated a number of well-preserved specimens demonstrating this feature. Bulman (1955, 1970a) and Urbanek (1973) used *Monograptus argenteus* as an example for the development of biform thecae. In many species, the proximal thecae are very slen-

der and shallowly inclined, often with negligible overlap. The thecal apertures are strongly hooked (Figure 13.5A) and may show additional apertural modifications. Along a gradient to the distal end, the thecae become less and less hooked and increase their thecal overlap. The distal thecae appear to be simple with straight, outwards inclined thecal apertures (Figure 13.5C). The gradient is seen along the proximal 20 to 50 thecae and is difficult to recognize when only small tubarium fragments are available. Distal thecae may also show introverted thecal apertures and variably developed geniculae. The thecal apertures can be laterally expanded and adorned with lateral lobes, as in "*Monograptus*" *argenteus* (Loydell & Maletz 2009).

Tubarium Shapes

The tubarium of the monograptids is highly variable in shape (Figure 13.6) and ranges from completely straight to dorsally and ventrally coiled, planar colonies to the three-dimensionally coiled colonies of *Spirograptus* or *Torquigraptus*. Dorsally

Figure 13.6 Tubarium shapes in the Monograptidae. (A) *Pristiograptus*, straight. (B) *Coronograptus*, dorsally coiled. (C) *Streptograptus*, ventrally coiled, fish-hook shape. (D) *Cochlograptus*, ventrally coiled. (E) *Rastrites*, dorsally coiled. (F) *Spirograptus*, 3D-spiral. (G) *Abiesgraptus*, paired cladia. (H) *Cyrtograptus*, dorsally coiled, two generations of cladia. (I) *Cyrtograptus*, dorsally coiled, with single cladium. Reconstructions not to scale.

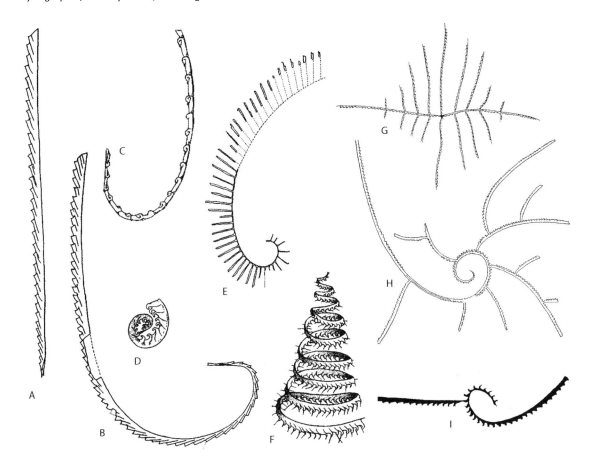

coiled planar colonies like *Rastrites, Coronograptus, Lituigraptus* and others bear the thecal rows on the outside of the coil (Figure 13.6B, E), while in ventrally coiled genera, the thecae are on the inside as in the neocucullograptines (*Bohemograptus, Polonograptus*), *Cochlograptus, Testograptus* (Figure 13.6D) and in many streptograptids (Figure 13.6C).

The three-dimensionally coiled colonies of *Spirograptus, Oktavites, Torquigraptus* and others appear to have the thecae originally on the outside of the spiral, but a twisting of the thecae or at least the thecal apertures is recognizable in many of these taxa (Figure 13.6F).

In the Llandovery, species with a more or less straight, slender proximal end and a distinct bend in the middle of the colony also appear with the *Monograptus inopinus* and *Monograptus limatulus* group, while the species of the genus *Streptograptus* and related taxa show a fish-hook shaped (Figure 13.6C) to sigmoidally bent colony.

Cladia

The growth of secondary branches or cladia from thecal apertures was rare in Ordovician graptolites, but was a standard modification of monograptid colonies in the Silurian and Lower Devonian. Secondary branches were branches originating from the thecal apertures and were formed differently from normal stipes. They appeared first in the Llandovery genus *Diversograptus* (Manck 1923; Rickards 1973), in which irregularly developed sicular and possibly thecal cladia are commonly found, and very disorganized, irregular colony shapes resulted from this development. The exact construction of these cladia has not been investigated, as well-preserved specimens do not exist.

The genus *Sinodiversograptus* from the *Spirograptus turriculatus* Biozone of Australia, China and North America (Loydell 1990) bore a sicular cladium and numerous thecal cladia, and formed a colony reminiscent of the genus *Cyrtograptus*. Loydell (1990) suggested a thecal style of hooked thecae like in *Paradiversograptus runcinatus* (Lapworth, 1876) and not of a streptograptid type, even though he termed the initial stage in the colony formation the streptograptid stage. The hook-like tubarium shape also recalls the typical shape of streptograptids (see Loydell 1993). As all known specimens are flattened, the details of the thecal style are uncertain. Specimens of *Streptograptus sartorius* (Törnquist, 1881) from the *Spirograptus turriculatus* Zone of Sweden show cladial branching (Rickards 1973) and may indicate that streptograptids were able to create cladial branches and could have been related to *Sinodiversograptus*.

Very regular cladia were developed in the Mid-Silurian cyrtograptids (Figure 13.6H–I). Initially, species with a single cladium existed, but were quickly replaced by species with multiple cladia. The colonies could be extremely large and bore numerous cladial stipes on a single main stipe (e.g. Wang et al. 2011: *Cyrtograptus robustus*) (Plate 16A). The cladia were usually developed at regular distances on the main stipe of the colony. Secondary cladia on the first-order cladia (Figure 13.6H) were formed in some species, but were uncommon and restricted to a few species.

Multiple cladia occur also in the Upper Silurian to Lower Devonian linograptines. While the genus *Linograptus* may bear several sicular cladia, paired sicular and thecal cladia are common in the genus *Abiesgraptus* (Figure 13.6G), producing a colony resembling the Ordovician multiramous taxa.

Urbanek (1963) described and illustrated the cladial development in *Neodiversograptus* and *Linograptus* in some detail, as an example of the cladial development in the Monograptidae.

Llandovery Diversification

The diversity of the monograptids in the Silurian shows a number of peaks (Sadler et al. 2011), of which the lower Telychian one is the highest (Figure 11.10). During this interval the diversity and disparity of the Silurian monograptids attained an all-time high. Numerous genera can be differentiated easily due to differences in thecal style and colony shapes. Even though a number of smaller extinction events diminished the diversity, the Aeronian and Telychian included the peak diversity of the monograptids, and never again was the group this successful.

It is still difficult to understand the taxonomy and evolutionary history of many Llandovery monograptids, but a number of groups may be differentiated easily. The most recognizable includes the genus *Coronograptus* (Figure 13.6B) with its strongly elongated sicula and slender, highly overlapping thecae. *Atavograptus*, *Huttagraptus*, *Lagarograptus* and *Pribylograptus* are among the best-known taxa from the Rhuddanian and Aeronian time intervals (Figure 13.7). Lukasik and Melchin (1997) described a number of species of these genera from chemically isolated material collected in Arctic Canada and discussed the evolutionary relationships of this early stock of monograptids. They possess a straight to dorsally curved, slender tubarium with low inclined thecae and moderate overlap. The thecae often, but not invariably, show moderate dimorphism with a gradual change in thecal style from the proximal to the distal end. Similar faunas are also present in the southern Urals and on Bornholm (Koren & Bjerreskov 1997).

Rastritid Monograptids

The monograptids of the Llandovery are largely slender species with straight to variably coiled and curved tubaria, and often a distinct thecal gra-

Figure 13.7 The early differentiation of the monographtids in the Rhuddanian, lower Silurian (based on Lukasik & Melchin 1997; Koren & Bjerreskov 1997).

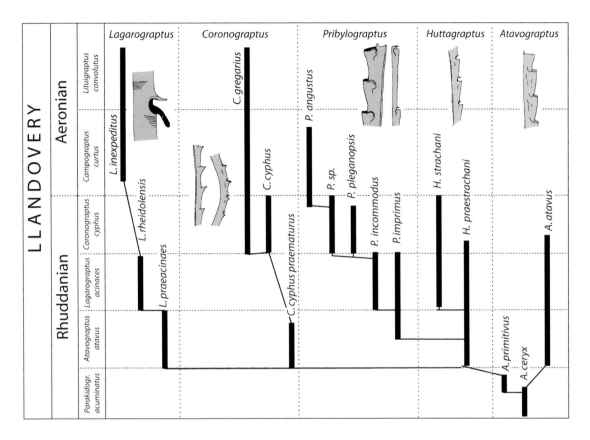

dient. The precise evolutionary relationships of these taxa are unknown, and a cladistic analysis does not exist. In general the species show a dorsally curved tubarium with the thecae on the outside of the curve. After the initial differentiation of a number of genera in the Rhuddanian (see Figure 13.7), the Aeronian tops the situation with the evolution of even more strange groups with the genera *Demirastrites* and *Rastrites*.

The "triangulate monographtids" of Sudbury (1958) (Figure 7.5) or the species of the genus *Demirastrites* (Eisel 1912; Přibyl & Münch 1942), as these taxa are also identified, dominated many faunas of the Aeronian time interval and developed into a plethora of difficult-to-separate species. Of these, *Demirastrites triangulatus* (Harkness, 1851) may be the most well known and biostratigraphically useful taxon (see Figure 15.2). *Demirastrites* species often

bear a strongly coiled proximal end with a small, inconspicuous sicula and a nearly straight distal end, with robust, triangulate thecae. Sudbury (1958) suggested an origin of the group from a *Pernerograptus revolutus* type ancestor, based on the dorsally curved colony shape and the modification of the thecae.

The genus *Demirastrites* was also inferred to be ancestral to the genus *Rastrites* with its increasingly long, isolated metathecae. The *Rastrites* shales of Scandinavia (Törnquist 1899) were named after the typical genus found in this succession. *Rastrites* originated in the Aeronian and evolved in the Telychian into species showing the largest recorded thecae of all graptolites (see Törnquist 1907; Přibyl 1941; Schauer 1967; Štorch & Loydell 1992). They are typified by a small and inconspicuous sicula and a very slender dorsal common canal from which the metathecae origi-

Figure 13.8 (A) *Demirastrites denticulatus magnificus* Přibyl & Münch, 1942, Münch collection, Thuringia (from Blumenstengel et al. 2006). (B) *Demirastrites pectinatus* (Richter, 1853), proximal end showing sicula and rastritiform first theca. (C) *Rastrites geinitzii* (from Loydell et al. 2004). (D) *Lituigraptus convolutus* (Hisinger, 1837), Thuringia (Blumenstengel et al. 2006), proximal end with rastritiform thecae. (E) *Demirastrites pectinatus* (Richter, 1853), Thuringia (Schauer, 1971). (F) *Demirastrites triangulatus* (Harkness, 1851), Thuringia (Schauer 1971, pl. 26). (E–F) adapted from Schauer (1971) with permission from TU Bergakademie Freiberg. (G) *Rastrites linnaei* Barrande, 1850, Bohemia (adapted from Štorch & Loydell 1992 with permission from E. Schweizerbart'sche Verlagsbuchhandlung). Scale indicated by 1 mm long bar close to each specimen.

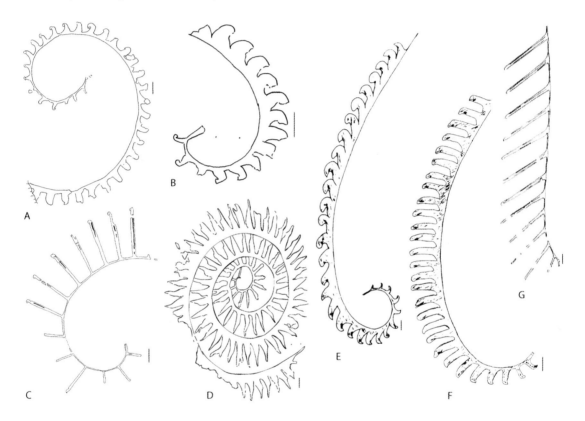

nate at an acute angle (Figures 13.8G, 13.9A). The tubular metathecae are much wider than the very slender prothecae and the dorsally positioned nema is often impossible to recognize in these species.

The genus *Lituigraptus* Ni, 1978 (Figures 13.8D, 13.9B) may be closely related to the rastritids. It possesses a distinct thecal gradient from long, proximal rastritid thecae to more triangulate, massive distal thecae, but also shows stronger apertural modifications as is seen in the strong, laterally expanded, crescendic thecal apertures of proximal thecae in *Lituigraptus convolutus*, changing into ventrally or latero-ventrally expanded paired horns, apparently with some asymmetry expressed in different dimen-

sions of the lateral horns (Loydell & Maletz 2009). A similar development is also seen in *Demirastrites*(?) *muenchi* Přibyl, 1942 (Figure 13.9E–F), in which a distinct torsion of the symmetrically developed, laterally expanded horns can be noted even in flattened specimens.

Streptograptids

The streptograptids (Figures 13.10, 13.11) are here used as an informal group of Llandovery to Wenlock monograptids with a typical development of the thecal aperture and the possession of cupulae in most

Figure 13.9 (A) *Rastrites* sp., Osmundsberget, showing sicula and bases of two thecae, apertures lacking.
(B) *Lituigraptus convolutus* (Hisinger), isolated theca with lateral horns. (C) *Rastrites geinitzi* (Törnquist, 1907), fragment,
LO 2033 t. (D) *Demirastrites raitzhainiensis* (Eisel, 1899), LO 2071 t, showing coiled proximal end and straight distal part of
colony. (E–F) *Demirastrites muenchi* (Přibyl, 1942), Eisel material, showing lateral expansion of thecal apertures.
(G) *Rastrites abreviatus* Lapworth, 1876, LO 2057 t, showing hooked thecal apertures in relief. Scale indicated by 1 mm
long bar in each photo.

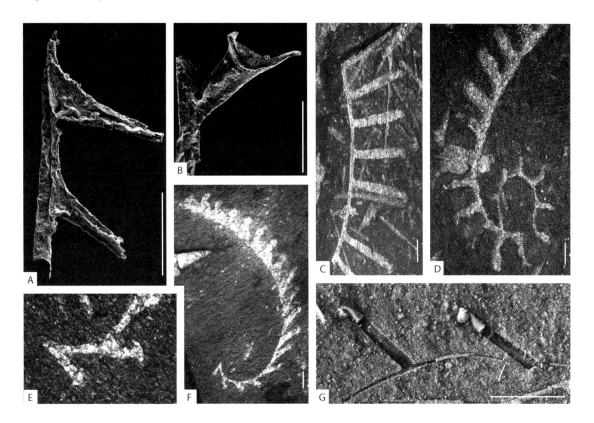

species (Loydell & Maletz 2004). Cupulae are lacking in a few very slender members of the group (e.g. *Streptograptus ansulosus*: Figure 13.11A), but these species still possess the typical mctathccae of streptograptids with their distinct hooked and introverted apertures provided with an upturned nozzle (Loydell & Maletz 2004). The origin of the group is uncertain, but an early possible member of the group may be seen in *Pseudomonoclimacis sidjachenkoi* (Obut & Sobolevskaya, 1965) and related species from the *Lituigraptus convolutus* Biozone of Arctic Canada (Figures 13.10A–B, 13.11C). The species has a dorsally curved or coiled colony with dramatically changing thecal styles in a proximal–distal gradient, showing increasing thecal overlap and reduction of

the apertural modifications. The proximal thecae bear recoiled, introverted thecal tubes with a typical streptograptid aperture, with lateral lobes and a dorsal nozzle (Figure 13.10A). Typical also is the development of the cupulae (Figure 13.10), the paired protuberances at the thecal origins, as in the genera *Streptograptus* and *Mediograptus* (see Loydell et al. 1997, pl. 1, Figs. 10, 12). The cupulae in *Mediograptus*, however, are poorly known since chemically isolated material is extremely rare and the development of the cupulae appears to be reduced, while the thecal apertures are highly complex.

Streptograptids are either nearly straight with a slight S-shaped curvature, or fish-hook shaped (Figure 13.11). A strong dorsal curvature can be noted

Figure 13.10 Tubarium features in streptograptids. (A–B) *Pseudomonoclimacis sidjachenkoi* (Obut & Sobolevskaya, 1965), Arctic Canada, proximal end with partially preserved sicula and four thecal apertures (B) and distal end (A) showing the gradual change in the shape of the thecal apertures. (C) *Mediograptus flittoni* Loydell & Cave, 1996, Arctic Canada, ventral view, showing considerable lateral widening and poorly developed nozzle. (D) *Streptograptus* sp., Dalarna, Sweden, specimen without cupulae. (E–G) *Streptograptus sartorius* (Törnquist, 1881), Solberga, Dalarna, Sweden. The characteristic paired cupulae are indicated by arrows. The scale is indicated by a 1 mm long bar in each photo.

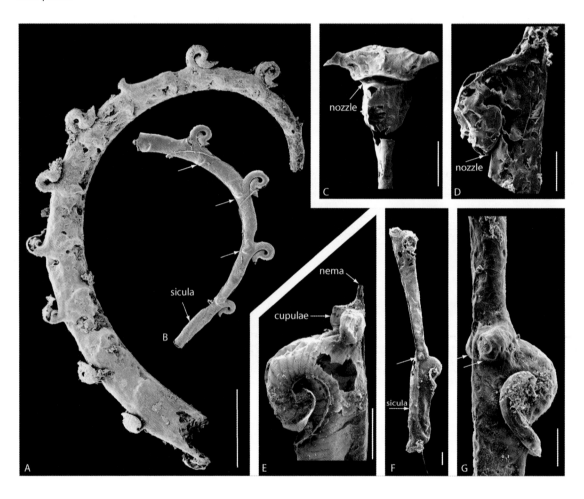

in *Paramonoclimacis* and related taxa. The remains of this ventral curvature are even visible in the extreme proximal parts of the fish-hook shaped younger taxa of the *Streptograptus plumosus* group (Figure 13.11E) and in the genus *Pseudostreptograptus*. In these taxa, the colonies are ventrally curved with the thecae on the inside of the curve and the distal ends are nearly straight. The species of the genus *Mediograptus* are straight to slightly curved dorsally in the proximal end, and are straight distally (cf. Bouček & Přibyl 1951; Štorch 1994). They are common in the Sheinwoodian, Wenlock, in many regions, but may not survive this interval (Štorch 1995). The differentiation of poorly preserved material of *Mediograptus* and *Streptograptus* is difficult, and it may be expected that younger material described as *Streptograptus* or even *Monograptus* (e.g. Jaeger 1991: *Monograptus serexiguus*; Teller

Figure 13.11 Evolutionary relationships of early streptograptids. (A) *Streptograptus ansulosus*. (B) *Streptograptus crispus*. (C) *Paramonoclimacis*. (D) *Streptograptus pseudobecki*. (E) *Streptograptus plumosus*. (F) *Pseudostreptograptus williamsi*. Illustrations of graptolites not to scale.

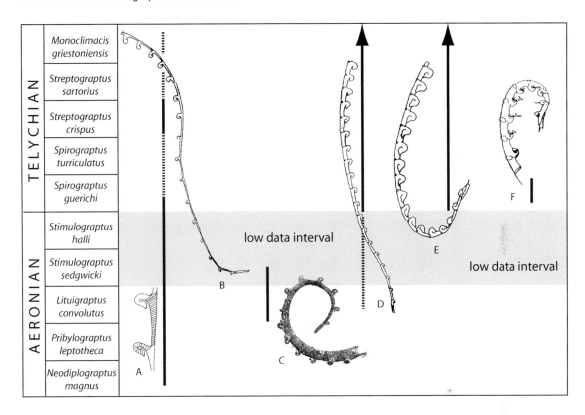

1986: *Monograptus flexuosus, Monograptus antennularius*) may belong to *Mediograptus* instead. The lateral expansion of the thecal apertures in *Mediograptus* specimens (Figure 13.10C) may be impossible to see in laterally preserved specimens, in which the hooked thecal apertures are visible (Loydell & Cave 1996; Loydell et al. 1997). A considerable lateral expansion of the thecal aperture is also present, however, in some *Streptograptus* species (Loydell & Maletz 2004; Loydell & Nestor 2006).

Spirograptus, Cyrtograptus and their Relatives

One of the most conspicuous and common groups of the monograptids in the later Llandovery to Wenlock includes the planispiral to three-dimen-

sionally coiled colonies of the genera *Spirograptus, Oktavites* and related forms. A clear phylogenetic relationship of these genera has not been demonstrated, but similarities in their thecal development and colony shapes suggest a phylogenetic connection, and a connection between these has been used to infer the evolutionary relationships of the cladial taxa (*Cyrtograptus*) with the non-cladial ones (see below). *Spirograptus turriculatus* (Barrande, 1850) may be one of the most iconic graptolites and is known to every paleontologist (Plate 16). Melchin and Lenz (1986) first described isolated material of *Spirograptus turriculatus* and identified *Monograptus sedgwicki* (Portlock, 1843) as a possible ancestor of the *Spirograptus* clade. Loydell et al. (1993) discussed the taxonomy and biostratigraphy of the species of this genus and illustrated chemically isolated specimens of *Spirograptus guerichi* and *Spirograptus turricula-*

Figure 13.12 (A) *Oktavites contortus* (Perner, 1897), Dalarna, Sweden, thecal aperture. (B) *Oktavites contortus* (Perner, 1897), Dalarna, Sweden, median fragment with six thecae. (C) *Oktavites contortus* (Perner, 1897), Dalarna, Sweden, proximal theca. (D) *Spirograptus turriculatus* (Barrande, 1850), proximal end with three thecae, Dalarna, Sweden. (E) *Oktavites spiralis* (Geinitz, 1842), Arctic Canada, single thecal aperture. (F–G) *Torquigraptus denticulatus* (Törnquist, 1899), Dalarna, Sweden, showing torsion of thecal apertures towards the reverse side (G). The scale is indicated by a 0.5 mm long bar in each photo.

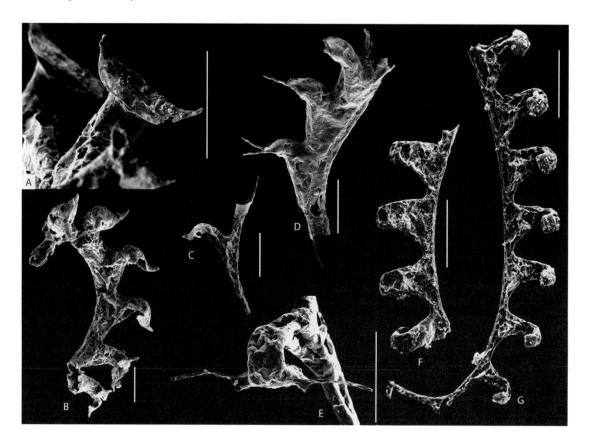

tus. Lenz and Melchin (1989) illustrated isolated material of *Oktavites spiralis* (Geinitz) from the Silurian of Arctic Canada, showing the complexity of the thecal apertures and the asymmetry of the lateral spines or rostral processes.

The thecae of this loosely assembled group are generally isolated and without thecal overlap, often triangulate and more slender in the proximal parts of the colonies. The thecal apertures are hooked and bear either lateral spines or are widened laterally with variably developed modifications. The thecal apertures are adorned with paired lateral spines in *Spirograptus guerichi*, but are strongly asymmetrically developed in *Spirograptus*

turriculatus (Melchin & Lenz 1986; Loydell et al. 1993). In *Oktavites spiralis*, paired apertural thecal spines are associated with ventral lateral lobes (Figure 13.12E), while in *Oktavites contortus* the lateral extensions of the thecal apertures do not bear spines (Figure 13.12A). Considerable variation can be seen in the thecal apertures of the cyrtograptids (Teller 1976, 1994; Lenz et al. 2012). The thecae may bear paired lateral spines on hooked thecal apertures, but distally often the thecae are straight and provided with short lateral lobes or even simple straight apertures.

The phylogenetic connection between the spirally coiled *Oktavites* and *Spirograptus* species and the multiramous cyrtograptids with their variably developed cladial branches is still speculative, even though some thecal characteristics can easily be compared. An origin from a spirograptid ancestor (e.g. Bouček 1933; Deng 1986) was considered for at least some cyrtograptids. Based on details of the thecal construction, Lenz and Melchin (1989) considered the evolutionary origins of the cyrtograptids as possibly polyphyletic, following the ideas of Rickards et al. (1977). Fu (1994) followed a similar suggestion and discussed four lineages leading to cyrtograptid graptolites. Urbanek and Teller (1997) also repeated the view of a polyphyletic origin for the cyrtograptids, but no taxonomic implications were suggested. They suggested an evolution of the cyrtograptids along at least five lineages, but their interpretation was based on the general shape of the tubarium, and constructional aspects of the thecae were not considered.

A general pattern of cladial development is uncertain, but it appears that early species show only a single cladium and multiramous taxa with numerous cladia, and eventually species with secondary cladia on the primary cladial stipes appeared. Thorsteinson (1955) first recognized the actual cladial growth in Cyrtograptus, which has been verified from isolated material of several species (Teller 1994; Zhang 1994; Lenz & Kozłowska-Dawidziuk 2001; Lenz et al. 2012). Huo et al. (1986) tried to describe the growth of the main stipes of the Cyrtograptus sakmaricus group as a logarithmic spiral, and used the curvature of the stipe as a measure of the phylogenetic relationship within the clade.

The relationship of the spirally to planispiral coiled Torquigraptus (Loydell 1993) to the Spirograptus group is uncertain. Torquigraptus has a highly variable tubarium shape, but most species bear a very slender proximal end with a short and inconspicuous sicula. Distally, the thecae grow larger until they reach a final thecal size after the secretion of a variable number of thecae. All Torquigraptus species share the distinct torsion of their thecal apertures towards the reverse side of the tubarium (Figure 13.12F–G). The thecal apertures are often laterally widened and may bear distinct paired lateral apertural lobes (Štorch 1998c), but apertural spines have not been noted in the genus.

Pristiograptus Clade

The *Pristiograptus* clade describes a large group of mainly straight monograptids with a fairly simple thecal style (Figure 13.13C, F). Early species of the genus *Pristiograptus* appear in the *halli* Biozone (Loydell 1993) or even earlier, but the origins of the clade is not well known. They are similar to *Pristiograptus variabilis* (Perner 1897), with a short sicula and a slender, straight stipe. The thecae show low inclination and overlap. A distinct dorsal curvature of the proximal end can be seen in *Pristiograptus renaudi* (Philippot, 1950), but is rare. A first major diversification of the *Pristiograptus* lineage occurred in the Sheinwoodian to early Homerian (Wenlock) in the *Pristiograptus dubius* group (Figure 7.3), but ended with the *Lundgreni* extinction event, after which only a single member of the group survived (Jaeger 1991; Urbanek et al. 2012). This bottleneck led to another explosive diversification and re-establishment of the monograptids in the upper Silurian. Urbanek et al. (2012, p. 589–590) speculated on the reason for the survival of the *Pristiograptus dubius* lineage, and suggested that the "eurybiotic nature of its adaptations and the lack of specialization" was the main cause, following Jaeger (1991), who called *Pristiograptus dubius* a "generalist". Numerous species and subspecies of *Pristiograptus* have been established (e.g. Přibyl 1943; Radzevičius 2003, 2006, 2007; Urbanek et al. 2012) from shale specimens and also from chemically isolated material showing important constructional features and documenting the evolutionary changes in great detail.

The most important information on the Wenlock to Ludlow monograptids originates from glacial erratic boulders (Figure 13.13) found in northern Germany and Poland (e.g. Urbanek 1958; Jaeger 1959). They include diverse graptolite faunas from the Homerian *Cyrtograptus lundgreni* Biozone to the Ludfordian *Bohemograptus cornutus/praecornutus* Biozone and thus cover a considerable Middle Silurian time interval. The glacial boulder of this "greenish-grey graptolite rock" or "Grünlich-Graues Graptolithengestein", as it was called in Germany based on its colour and main fossil content (Heidenhain 1869; Haupt 1878; Jaekel

Figure 13.13 Isolated graptolites from the Gorstian and Ludfordian. (A) *Saetograptus chimaera* (Barrande, 1850), partly isolated specimens in glacial boulder, Germany. (B) *Saetograptus chimaera*, juvenile. (C) *Pristiograptus frequens* (Jaekel, 1889). (D) *Saetograptus chimaera*, longer specimen. (E) *Saetograptus leintwardinensis* (Lapworth, 1880), bleached to show fuselli. (F) *Pristiograptus frequens*, bleached. (G) *Saetograptus leintwardinensis* (Lapworth, 1880), long specimen in oblique view. Scale indicated by 1 mm long bar in each photo.

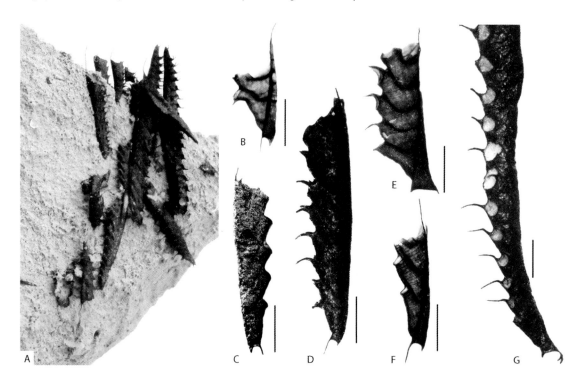

1889), originated from the glacial erosion and southward transport of the sediments of the Silurian Colonus Shale Trough of southern Scandinavia. The graptolite faunas can easily be extracted from the limestones by using acid techniques, as Gümbel (1878) did.

The largely clastic sediments of the Colonus Shale basin include numerous carbonate concretions that are more robust and are preserved through the glacial transport. They are now found in many moraine deposits south of their original depositional region, where they have been collected because of the presence of beautifully preserved fossils. These rocks are not only a source for Silurian monograptids, but also yielded some of the most spectacular retiolitid graptolites ever found (Münch 1931; Eisenack 1951; Maletz 2008, 2010b).

Aftermath of the *Lundgreni* Extinction

The *Lundgreni* Extinction Event represents one of the most profound extinction intervals in graptolite history (e.g. Koren 1987, 1991a, b; Jaeger 1991; Porębska et al. 2004). Early post-*Lundgreni* Extinction graptolite faunas are of low diversity, and the survival of only two lineages, *Pristiograptus dubius* and the plectgraptine retiolitids, was postulated (Jaeger 1991; Lenz 1993; Koren 1994a; Štorch 1995; Lenz et al. 2006; Kozłowska 2015), but this does not fit with the diversity curve provided by Cooper et al. (2014), indicating a distinct extinction event, but not the near-extinction of all planktic graptolites (see also Figure 7.4). Numerous papers deal with this extinction event and discuss the paleontological, sedimentological and geochemical aspects of the interval (Porębska et al. 2004; Lenz et al. 2006; Noble et al. 2012).

Jaeger (1991) discussed the "Great Crisis" in some detail and illustrated the situation as seen in the Silurian succession of Thuringia, Germany. He indicated the presence of an interval with *Pristiograptus dubius parvus* (Ulst, 1974), followed by the *Pristiograptus dubius/Gothograptus nassa* Interregnum and the *Pristiograptus praedeubeli* Biozone. The graptolite faunas of the recovery interval, the *Gothograptus nassa* to *Colonograptus praedeubeli/deubeli* Biozone interval, have been described subsequently, and we have learned a lot about the faunal composition and evolution of these graptolite successions (Lenz 1994c, 1995; Gutiérrez-Marco et al. 1996; Lenz et al. 2012). The faunas clearly show the derivation from a pristiograptid ancestor evolving into numerous straight to variably dorsally curved colony shapes (Koren 1993, 1994b). While one group evolves into the *Lobograptus-Bohemograptus* lineage of the cucullograptines with its mainly slender tubaria, a second lineage leads to the genera *Colonograptus* and *Saetograptus* with their pristiograptid-like thecae adorned with paired lateral lobes or spines (Figure 13.13).

The *Pristiograptus* lineage must be regarded as the main feeder for the robust, straight monograptids dominating the Upper Silurian and Lower Devonian. *Pristiograptus dubius parvus* and *Pristiograptus praedeubeli* appear to be the early members of this evolving clade, quickly leading to the robust members of the genera *Colonograptus* and *Saetograptus*. *Saetograptus* particularly dominated the higher part of the Gorstian and the early Ludfordian, before they died out during the *Leintwardinensis* extinction event (Štorch et al. 2014). A number of *Saetograptus* species, starting with *Saetograptus varians*, can be seen to form paired lateral apertural spines, initially on the first few thecal pairs, but eventually all thecae are spined as in *Saetograptus chimaera* and *Saetograptus leintwardinensis*. The aperture of the sicula bears a short virgella and develops variously shaped additions on the antivirgellar side. Initially, the antivirgellar side is adorned with a pointed rutellum, but in *Saetograptus leintwardinensis* a forked spine is present, recurved over the

sicular aperture (Maletz 1997c) (Figure 13.13G). In other species, the dorsal projection evolves into a long spine, as in *Saetograptus clavulus* (Perner, 1899) (Štorch et al. 2014) and *Saetograptus argentinus robustus* (Maletz et al., 2002).

Cucullograptinae and Neocucullograptinae

Urbanek (1966, 1970) described the Cucullograptinae and Neocucullograptinae in great detail based on chemically isolated specimens from Polish drill cores and glacial boulder material. Based on the details of the tubarium construction, he was able to infer their phylogenetic relationships in some detail. The Cucullograptinae possibly evolved from an ancestor close to *"Monograptus" idoneus* (Koren, 1992) in the basal Gorstian (Rickards & Wright 1999), but went extinct during the *Leintwardinensis* extinction event (Štorch et al. 2014). Many three-dimensionally preserved species of the Cucullograptinae (Urbanek 1966) provide information on the general construction of their slender, straight to slightly curved colonies and the thecal apertural complexities. The thecae are slender, with low inclination and thecal overlap. The species bear thecal apertures with a special, often asymmetrically formed apparatus, based on two lateral lobes. These lobes develop differently and often a strong asymmetry can be seen in the construction. One lobe may develop into a large hood, covering the thecal aperture, while the second one is strongly reduced. The hood may form a lateral extension or lobe that appears like a spine in flattened specimens (*Cucullograptus aversus rostratus* Urbanek, 1960). The apertural apparatus is formed from the addition of normal fusellar tissue, as can be seen in chemically isolated material (Urbanek 1966). The thecal development may be consistently present or show a gradient along the colony, with increasing complexity towards the distal end.

The Neocucullograptinae originate after the *Leintwardinensis* extinction event from a *Bohemograptus*-type ancestor. They are similar in their tubarium development to the Cucullograptinae, but their apertural apparatus is formed from microfusellar additions instead of normal fusellar tissues.

Figure 13.14 Neocucullograptidae. (A) *Egregiograptus egregius* Urbanek, proximal end (adapted from Urbanek 1970 under the terms of the Creative Commons Attribution Licence 4.0 - CC-BY-4.0). (B–D) *Neolobograptus inexpectatus supernus* Urbanek, 1970, proximal end and distal thecal style. (E) *Bohemograptus cornutus* Urbanek, showing lateral apertural modifications of first theca. (F) *Bohemograptus tenuis* (Bouček, 1936), small colony. (G) *Bohemograptus cornutus* Urbanek, large colony. (H) *Egregiograptus dimitrii* Koren & Suyarkova, 2004. (F–H: Adapted from Koren and Suyarkova (2004) with permission from Taylor & Francis. (I) *Korenea sherwini* Rickards et al., 1995 (adapted from Rickards et al. 1995, reproduced with permission from Australasian Palaeontologists). Illustrations not to scale.

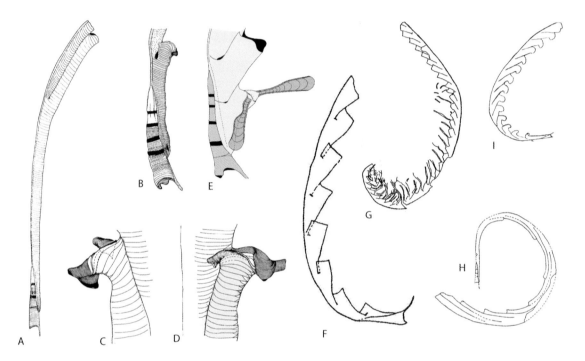

All Neocucullograptinae show a ventrally curved tubarium with considerable variation in the thecal length. The thecal shape and overlap varies from low with moderate inclination in the genus *Bohemograptus*, to high overlap and extreme length in the genus *Egregiograptus*. Many taxa have been found in the southern Tien Shan (Koren & Suyarkova 1997, 1998, 2004), providing information for a new interpretation of the taxonomy of the Neocucullograptinae and a very precise biostratigraphy for the late Ludlow time interval.

Bohemograptus cornutus Urbanek, 1970 represents a special development of the thecal apertures. Paired lateral lobes develop into large membranes formed from fusellar tissues (Figure 13.14E). The initial parts of these ventro-lateral lappets are secreted as microfusellar tissue, but soon normal fuselli can be seen to construct the considerably

widening processes. These processes with their thickened rims appear to be constructed in a similar fashion to the scopulae of the archiretiolitids (see Bates & Kirk 1991: *Orthoretiolites hami*).

Koren (1993) termed the extinction of the Neocucullograptinae in the Upper Ludlow the *Podoliensis* Event, but it is otherwise also known as the *Kozlowskii* Event (Štorch 1995). It led to the extinction of most graptolite lineages, except for the *Pristiograptus* lineage and the genus *Linograptus*.

Linograptinae

The Linograptinae represent a small group of taxa characterized by the successive evolution of multiple cladia (Urbanek 1997b), generating colony

Figure 13.15 The evolution of the Linograptinae (after Urbanek 1997b). Biostratigraphy modified from Loydell (2012).

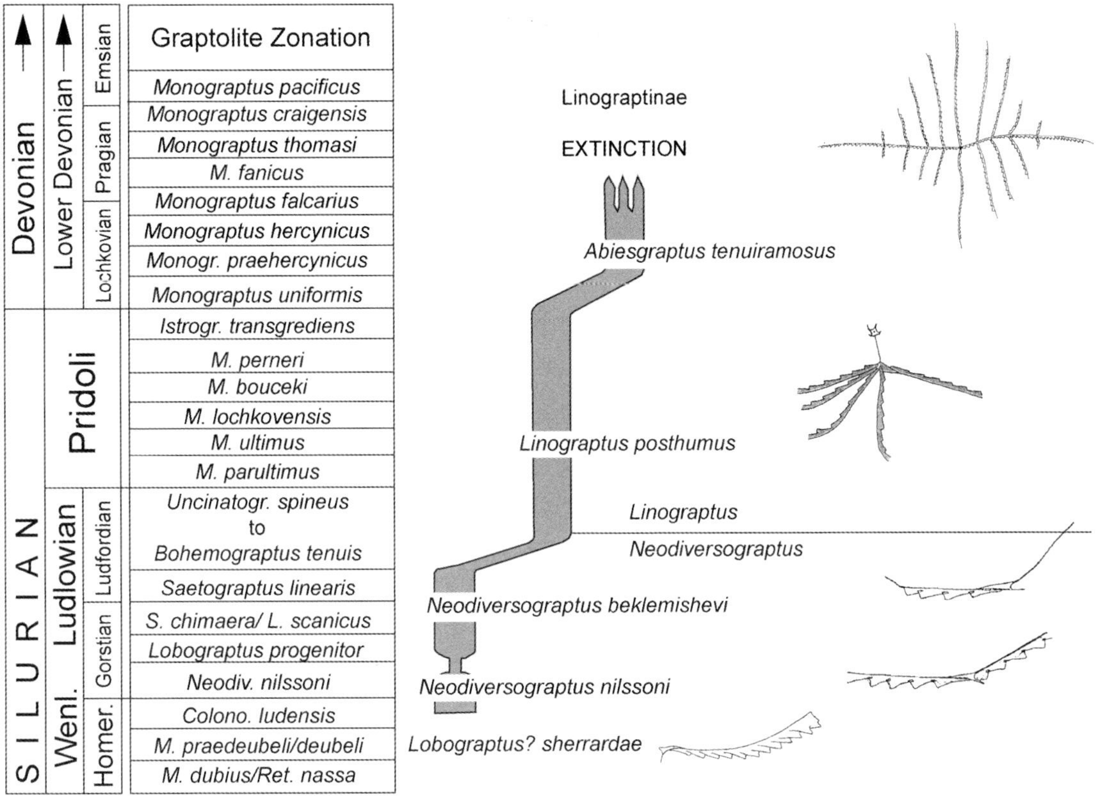

			Graptolite Zonation
Devonian	Lower Devonian	Emsian	*Monograptus pacificus*
			Monograptus craigensis
		Pragian	*Monograptus thomasi*
			M. fanicus
		Lochkovian	*Monograptus falcarius*
			Monograptus hercynicus
			Monogr. praehercynicus
			Monograptus uniformis
S I L U R I A N	Pridoli		*Istrogr. transgrediens*
			M. perneri
			M. bouceki
			M. lochkovensis
			M. ultimus
			M. parultimus
	Ludlowian	Ludfordian	*Uncinatogr. spineus* to *Bohemograptus tenuis*
			Saetograptus linearis
		Gorstian	*S. chimaera/ L. scanicus*
			Lobograptus progenitor
			Neodiv. nilssoni
	Wenl.	Homer.	*Colono. ludensis*
			M. praedeubeli/deubeli
			M. dubius/Ret. nassa

Linograptinae

EXTINCTION

Abiesgraptus tenuiramosus

Linograptus posthumus

Linograptus

Neodiversograptus

Neodiversograptus beklemishevi

Neodiversograptus nilssoni

Lobograptus? sherrardae

shapes and sizes not seen since the multiramous Ordovician Dichograptidae. Large specimens of the early Devonian *Abiesgraptus multiramosus* Hundt, 1935 may have covered an area close to a square metre, and thus represent one of the largest known graptolite colonies. Urbanek (1997b) interpreted the evolution of the lineage in great detail as an example of anagenetic evolution of a single lineage, and recognized four chronospecies (Figure 13.15).

All Linograptinae share a simple thecal style, with low thecal inclination and straight to slightly lobate thecal apertures. The origin of the Linograptinae may be traced to the genus *Lobograptus* and especially to *Lobograptus? sherrardae* (Sherwin, 1975) from the late Homerian, Silurian (Koren & Urbanek 1994). The evolution continued through the genus *Neodiversograptus* with its bipolar tubarium, formed by the addition of a single sicular cladium. Multiple sicular cladia

can be seen in the genus *Linograptus*, forming a multiramous tubarium in which all secondary stipes originate from the sicula. The development of thecal cladia can be found in specimens in the early Pridoli to early Lochkovian time interval, before the cladial development was stabilized and the paired sicula and thecal cladia evolved in the Lochkovian *Abiesgraptus*. Rarely, specimens of *Abiesgraptus* may develop secondary cladia (Hundt 1939: *Gangliograptus*).

A number of important changes in the sicula and the proximal end can be followed from *Neodiversograptus* to *Linograptus* (Figure 13.15). The earliest species *Neodiversograptus nilssoni* has a short sicula in which the prosicula makes up about a quarter of the whole sicular length. The most important new characteristic is the development of a bilobed apertural process that developed into a ventro-lateral spine on the dorsal side of the

Figure 13.16 The proximal end of *Neodiversograptus* and *Linograptus*. (A) *Neodiversograptus nilssoni* Lapworth, 1876, proximal end. (B) *Neodiversograptus nilssoni* Lapworth, 1876, with sicular cladium. (C) *Neodiversograptus beklemishevi* Urbanek, 1963, proximal end with sicular cladium. (D) *Linograptus posthumus* (Richter, 1875), proximal end with single sicular cladium and virgellarium. Illustrations adapted from Urbanek (1997b) with permission from Instytut Paleobiologii, and adapted from Loydell (2012).

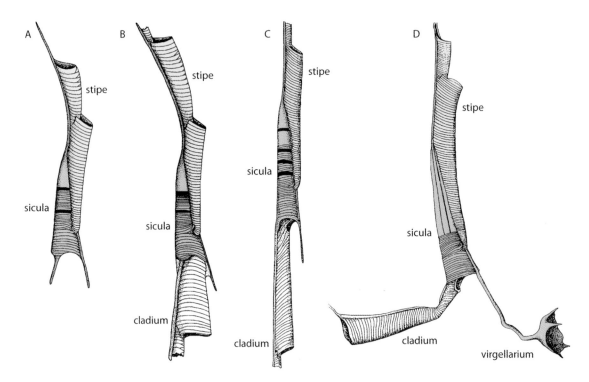

apertures. In dipolar tubaria, a sicular cladium is formed, covering most of the sicular aperture and growing along the elongating ventro-lateral spine (Figure 13.16.B). In the slightly younger *Neodiversograptus beklemishevi* Urbanek, 1963 (Figure 13.16C), the origin of the cladial stipe is more slender and a conspicuous opening for the sicular zooid is visible. Otherwise the development is similar to that of *Neodiversograptus nilssoni*. A change in development can be seen in the proximal end of *Linograptus posthumus* (Figure 13.16D). The sicula shows a much shorter metasicula and a relatively large prosicula. Sicular annuli are lacking in *Linograptus* or are reduced to a single annulus at the prosicular/metasicular boundary (see Urbanek 1963). Another new development is the typical nematularium (Figure 13.16D) that is restricted to the genus *Linograptus*.

Kozlowskii Event

Another major impact on graptolite diversity was exerted by the *Kozlowskii* Extinction Event (Figure 13.17), correlatable with the Lau Event (Jeppsson 1998; Calner 2005; Melchin et al. 2012). The Lau Event represents a strong positive carbon isotope excursion, but was originally identified as a conodont extinction event (Jeppsson 1998). Slavík and Carls (2012) discussed the problems of the international correlation of this important time interval in Silurian geology.

Štorch (1995) described the *Kozlowskii* extinction in some detail and regarded the *Pristiograptus fragmentalis* Biozone as the crisis interval, while the *Neocolonograptus parultimus* to *Neocolonograptus ultimus* interval represents the recovery interval after the event (Figure 13.15).

Figure 13.17 The *Kozlowskii* Extinction Event in the mid Ludlow (based on Štorch 1995, Fig. 10). Graptolite illustrations from various sources, not to scale.

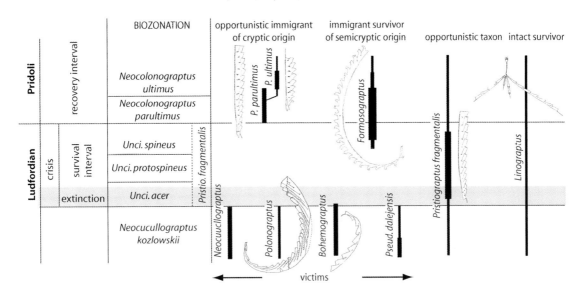

The whole clade of the neocucullograptids died out at the top of the *kozlowskii* Biozone, together with *Paramonoclimacis dalejensis*, while the *Pristiograptus dubius* lineage survived in the species *Pristiograptus fragmentalis* through the interval. Štorch (1995) regarded the species as an opportunistic taxon, representing the most common species during the crisis interval. Otherwise, only *Linograptus* survived the crisis without any apparent damage or fluctuation in its occurrence.

The recovery after the *Kozlowskii* event in the upper Ludfordian has been the focus of a number of studies, and considerable new information has been assembled in the last two decades. Urbanek (1997a) described and illustrated the graptolite faunas from the late Ludfordian and early Pridoli from chemically isolated material. The specimens from the Mielnik-1 drill core provided the most complete and best-preserved graptolite succession through this time interval. A number of ingressions of the long-ranging *Pristiograptus dubius* group (Urbanek 1997a, Fig. 3) can be recognized in this interval of relatively low-diversity faunas, largely dominated by straight monograptids with variably but moderately developed paired lateral apertural lobes, also including the *Uncinatograptus*

lineage, in which species with paired lateral spines (Chapter 7: Figure 7.7) evolved again as homeomorphs of the earlier *Saeograptus* species of the Gorstian. Although not as well preserved, the Late Ludfordian to Pridoli graptolite faunas of the Turkestan-Alai mountains in southern Tien Shan (Koren & Suyarkova 1997) supported the biostratigraphical interpretation of Urbanek (1997a). The succession provided an important insight into the faunal composition and distribution of these faunas in a wider paleogeographical context. Unfortunately, the graptolites of this time interval are often difficult to identify, and Urbanek (1997a, p. 89) complained about the poor and inadequate taxonomy published so far, which according to him led to confusion with respect to species identification and biostratigraphical interpretation.

Jaeger (in Kriz et al. 1986) and Jaeger (1991) described the Pridoli graptolites from chemically isolated material and provided information to complete our understanding of these faunas. The Pridoli (Figure 13.18) was established as the latest time interval to be recognized as a Silurian Series and was studied in great detail in the type area of the Czech Republic. The base of the Pridoli is defined by the *Neocolonograptus parultimus*

Figure 13.18 The Pridoli graptolite biostratigraphy (based on Loydell 2012). Graptolite illustrations based on various resources, not to scale.

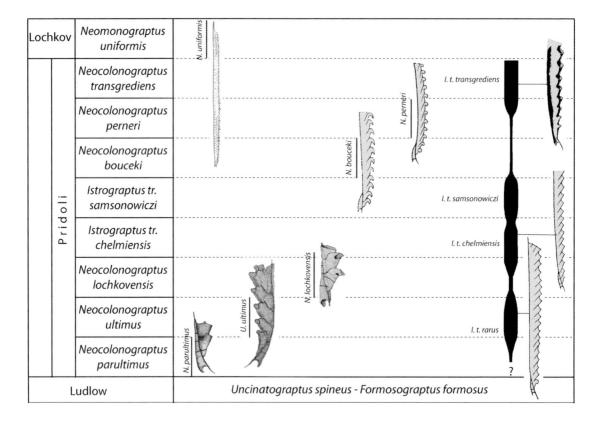

Biozone and the interval ends with the *Istrograptus transgrediens* Biozone (Kriz et al. 1986). The graptolite faunas of the Pridoli are generally of low diversity and simple development, and typically occur as monospecific or nearly monospecific assemblages in southern Tien Shan (Koren & Suyarkova 1997). Perner (1899) already described a number of the typical species of the Pridoli from the succession of the Prague Basin in the Czech Republic, but the poor preservation and difficult correlation for a long time prevented a better understanding of the interval, even though it is widely distributed and can, for example, be found in Arctic Canada (Lenz 1990), Poland (Teller 1964) and Germany (Kraatz 1958). Teller (1997a, b) discussed the Pridoli succession from the East European Platform of Poland and provided important information on the biostratigraphical ranges and identification of the graptolite faunas. He dif-

ferentiated the *Istrograptius transgrediens* lineage into four chronospecies (Teller 1997b, p. 73), based on the number of proximal thecae with "beak-like apertures". These beak-like apertures are formed by paired lateral lobes on the thecal apertures, as is seen in the SEM photos of chemically isolated specimens in his publication.

Final Extinction

It is thanks to Hermann Jaeger's work (Jaeger 1959, 1978, 1988) that we now know about the presence of planktic graptolites in the Lower Devonian and have a precise graptolite biostratigraphy for this time interval (Figure 6.3). Only two lineages made the transition into the Devonian – the multiramous *Linograptus/Abiesgraptus* clade and the

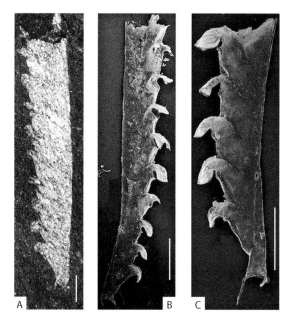

Figure 13.19 Devonian monograptids. (A) *Uncinatograptus uniformis* (Přibyl, 1940), Lower Devonian, Germany, tectonically distorted specimen. (B–C) *Uncinatograptus birchensis* (Berry & Murphy, 1975), Arctic Canada. Scale indicated by 1 mm long bar in each photo.

straight monograptids of the *Uncinatograptus uniformis* lineage – and subsequently evolving species are of a quite simple style with straight to proximally curved tubaria (Figure 13.19). Hooded and hooked thecae can be recognized, and many taxa show a distinct gradation from proximal to distal thecal style, often losing the apertural hood or hook distally. Apertural variations of the sicula in the form of dorsal and ventral lips as well as trumpet-like apertures are common, the latter reminiscent of the trumpet-like sicular shapes in derived *Saetograptus leintwardinensis* or of certain *Bohemograptus* specimens.

Lower Devonian graptolite faunas were of low diversity (Porębska 1984; Koren 1974; Lenz 2013) and most biozones include only one or two species. Jaeger (1978, 1988) discussed the biostrati-graphical distribution of these faunas and also provided information on the paleogeographical distribution. Graptolite faunas appeared to be increasingly restricted to equatorial regions in the Lower Devonian (Berry & Wilde 1990; Goldman et al. 2013), and Koren (1979) suggested that the limited biofacies and biogeographical distribution of the youngest graptolite faunas may have led to their final extinction. The latest planktic graptolites may have been *Uncinatograptus yukonensis* and *Uncinatograptus pacificus* from the Emsian, Lower Devonian, of Arctic Canada. Lenz (2013) called the *Uncinatograptus yukonensis* Biozone the most widely recognized Lower Devonian graptolite biozone and discussed a surprisingly high diversity, based on numerous species described especially from China. He also referred to the extreme intraspecific variation of *Uncinatograptus yukonensis* documented first by Lenz (1988a, b). It may be based possibly on oversplitting of a few highly variable monograptid species.

Outlook

The Monograptidae represent the most diverse graptoloids that ever evolved, and their extinction is still unsolved since the presence, diversity and interaction of planktic and nektic organisms in the early Paleozoic is poorly known. They possessed their highest diversity and disparity in the early Silurian, and slowly, through a number of major and minor extinction events, were reduced to the last few species of limited areal distribution in the early Devonian. For no obvious reason they became extinct and left only a few benthic taxa to carry the torch through the millennia into our modern times. Even though we know a lot about the taxonomy and biostratigraphy of the monograptids, so many details of their evolutionary history and their ecological interactions with other organisms are lacking. Where did they come from? Why did they become extinct? We have no answers yet.

14 Collection, Preparation and Illustration of Graptolites

Denis E. B. Bates and Jörg Maletz

Who collected the first graptolite? Was it a prehistoric human interested in beautiful rocks, or a modern scientist focused on the advancement of paleontology and interest in animal evolution? This is certainly not a question that needs to be asked, as graptolites are too small to be of interest to most collectors of precious stones or fossils. Only a scientifically motivated person would appreciate these tiny remains of ancient organisms. Nevertheless, graptolites are quite attractive, but you have to use some special tools to recognize their beauty, as beauty in graptolites is often hidden in the small scale of the remains and revealed only through investigation with high-powered microscopy.

Scientists have collected graptolites for more than 150 years and quickly discovered their scientific usefulness. Large collections of graptolites have amassed in our museums, and a great deal of curatorial work goes into these collections and their preservation and scientific evaluation. For us as scientists, these collections can provide important information and we can evaluate material without having to do extensive fieldwork and travel. Also, many of the collections come from localities no longer available, as exploration has ceased in those regions, or outcrops and quarries have disappeared. So this material is the only chance to evaluate the local geology. Drill core material is another source of important information from subsurface successions that would not be available otherwise.

Graptolite Paleobiology, First Edition. Jörg Maletz.
© 2017 John Wiley & Sons Ltd. Published 2017 by John Wiley & Sons Ltd.

Our work on fossils only starts with the collection and documentation of the material and the deposition in a fossil collection of a museum or research institution. This part of the work of a scientist has not changed much since the beginning of scientific collecting. Detailed information on the locality and the exact position in a certain section has to be provided with the material, or the material will become scientifically useless. Many old collections exist and are in a good shape as curators care for the material and in modern times make the information available in online catalogues, as was done in the past in printed ones.

If you are a fossil hunter interested in dinosaur bones, you know how important preparation is – it may take years to excavate a dinosaur from the surrounding rocks and free the bones from the last pieces of attached mineral grains. Small fossils like graptolites also need careful preparation, but the tools and methods differ considerably from those used for the large beasts. New methods have been introduced in recent times to prepare graptolites. Only through good preparation and illustration techniques is it possible to showcase the details of our small but precious fossils, and find the data we need for our research.

Collecting Graptolites

Unlike large animals such as the dinosaurs, graptolites are relatively easy to collect and do not require great technical skills. Their size allows, in most cases, several specimens to be found on a single "hand specimen" sized slab (Figure 14.1), and even picked up in weathered and loose pieces. In a sequence of strata, a record should be made of the precise horizon of each specimen collected. The graptolites may vary from horizon to horizon, due to evolutionary or ecological changes. Specimens can be orientated, particularly if they were deposited under the influence of a bottom current (Figure 14.1A). If so, they should be marked with their orientation in the field.

Graptolites might be easy to spot on shale surfaces, but often the contrast with the surrounding sediment is low and the specimens are only visible under special light conditions. Where the rock is broken open to reveal the specimens on each side of the split, both specimens should be collected, and labelled to indicate that they are counterparts of the same specimen. If the rock has been cleaved, it may be difficult to persuade it to split along bedding planes rather than the cleavage planes. Here it is possible to persuade the rock to split along a bedding plane, by holding and hitting it as shown in the diagram (Figure 14.2A). Even though the cleavage may disrupt and deform the graptolites in the rock, they may still be recognizable (Figure 14.2B).

Specimens should be well wrapped in the field to avoid damage, using paper such as newsprint or kitchen paper, and carefully labelled in the bags. If material is collected in bulk, for later splitting or for acid digestion, it need not be wrapped so carefully. Wherever collected, local and national rules on collecting should be obeyed. Note that some countries forbid the export of fossils, and even collecting is prohibited in many famous localities. For example, it is prohibited to collect the Burgess Shale fauna from the famous Walcott Quarry in the Yoho National Park (see Caron & Rudkin 2009), and you need permits for many sections in Canadian Parks, including the famous Green Point section, the GSSP for the base of the Ordovician System in western Newfoundland and an important locality for the investigation of basal Ordovician graptolite faunas (Cooper et al. 1998, 2001). Other countries have similar laws, and it is important before you plan a research project to make sure you are able to get a working permit to collect and export material.

Figure 14.1 Graptolite hand slabs. (A) Current-aligned specimens of *Orthograptus apiculatus* (Elles & Wood, 1907) from Laggan Burn, Ayrshire, Scotland. (B) Hand specimen from Llangammarch Wells, mid Wales. There are several species present, including three synrhabdosomes of *Saetograptus varians* (Wood, 1900).

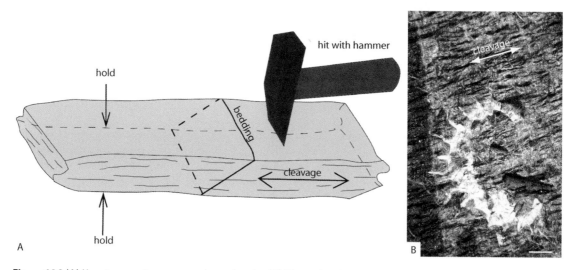

Figure 14.2 (A) How to use a hammer on cleaved rocks. (B) Cleaved rock showing exposed graptolite on tectonically crenulated bedding plane, *Arienigraptus zhejiangensis* (Yu & Fang, 1981), Darriwilian, Peru. Scale indicates 1 mm.

Figure 14.3 Preparation of graptolites. (A) *Arienigraptus geniculatus* (Skevington, 1965), LO 10601t, Lerhamn drillcore, Scania, Sweden, specimen before preparation, coated with ammonium chloride. (B) Same specimen after preparation, uncoated. (C) *Normalograptus scalaris* (Hisinger, 1837), LO 1097t, Tomarp, Scania, Sweden (Törnquist 1893, pl. 1, Figs 7–8), showing polished section of proximal end, filled with pyrite. Scale indicated by 1 mm long bar in each photo.

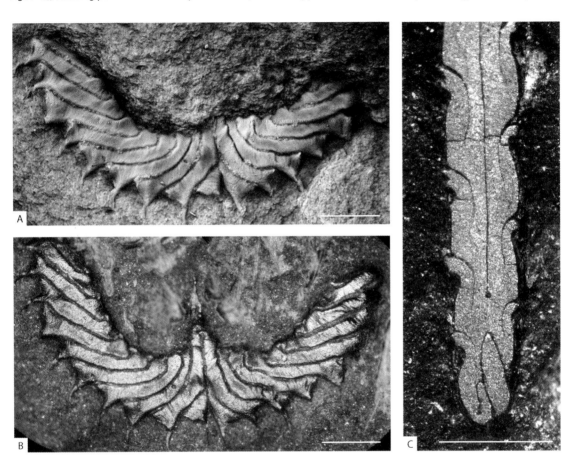

Physical Preparation

Although some specimens are completely exposed after the rock is split, often they are partially buried within the rock, and need to be "developed" to reveal their true extent (Figure 14.3A). The trick is to be able to remove the rock matrix to expose the graptolite, without damaging the specimen (Figure 14.3B). Provided the rock is not too hard, this can be done with either a scalpel, or a steel rod sharpened to a chisel point, mounted on a suitable handle such as a pin vice. A scalpel used on a relatively soft rock can be handed very delicately and precisely, but is useless on very hard and tectonically modified rock.

Hard rock can be removed using a small grinding wheel in a small hand-held drill, or a small engraving tool with a chisel point. However, there is a real danger of damaging the specimen, so such tools have to be used extremely carefully. The rock may not split easily from the graptolite specimen, which may be quickly damaged or even lost.

Care should be taken not to lose any material. Many relief specimens of graptolites tend to flake out of the rock when prepared. These specimens should be kept separate in a small glass bottle or

Figure 14.4 (A–B) *Didymograptus artus* Elles & Wood, 1902, JM 16, original specimen in obverse view, preserved as mould (A) and latex cast (B). (C–D) *Didymograptus artus* Elles & Wood, 1902, JM 17, original specimen in reverse view, preserved as mould (D) and latex cast (C). Both specimens from the Elnes Fomation, Slemmestad, Oslo Region, Norway. Latex casts coated with ammonium chloride to enhance structural details. Scale indicated by 1 mm long bar in each photo.

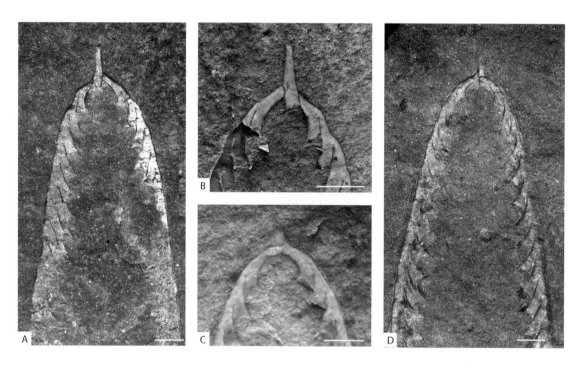

plastic box and labelled carefully. Loosely stored specimens in cardboard boxes easily break and get lost. Varnish has been used to protect graptolites on weathered slabs, but is only a temporary solution. It actually damages more than protects, as the varnish tends to break after a few years and the specimens will be completely destroyed.

Natural moulds of graptolites occur where the material has been infilled by, usually, iron pyrites, the pyrites being oxidized during weathering of the rock, to leave a negative impression on both surfaces (Figure 14.4A, D). A positive can be produced by making a replica, using a flexible material (Figure 14.4B–C). The most widely used material is liquid latex, or silicone rubber. The latex dries clear, but a few drops of Indian ink can be added to give a matt black peel; other colours can be used. The slab should be wetted before adding the latex. This will lower the surface tension of the latex, and prevent air bubbles being trapped on the rock

surface. The liquid latex will seep into every little corner of the mould and you will get a perfect replica. The latex cast method can also be used for cleaning weathered surfaces. Apply a layer of latex on the slab and take it off when dry. It will take away all loose weathered materials and leave a clean surface. A second latex cast will show all the details that are still preserved of the specimen.

Certain preparation methods to add information have also been used on graptolites. Serial sectioning of specimens has been applied to understand the three-dimensional development of species. Graptolite specialists like Holm (1890, 1895), Wiman (1895, 1896a, b, 1897, 1901), Bulman (1944–1947) and Kozłowski (1949) used this method to understand the development of dendroid graptolites, but also applied it to other taxa. The method has been abandoned for graptolite research, as other modern techniques may be applied to reveal their internal construction. Törnquist (1893) used a very

Figure 14.5 Chemically isolated graptolites. (A) *Haddingograptus* sp. cf. *H. oliveri* (Bouček, 1973), BAS 39, specimen in partial relief, western Newfoundland. (B) *Tetragraptus* sp., SK 1C 06, Skattungbyn, Dalarna, juvenile, bleached specimen showing fungus growth (arrows). (C) *Pterograptus elegans* Holm, 1881, ML 1.93.09, Mainland, western Newfoundland, flattened, bleached specimen showing fusellum. (D) *Streptograptus sartorius* (Törnquist, 1881), Sol 01, IR photograph showing fusellar construction of an opaque specimen, Solberga, Dalarna. Scale bar indicates 0.5 mm for each specimen.

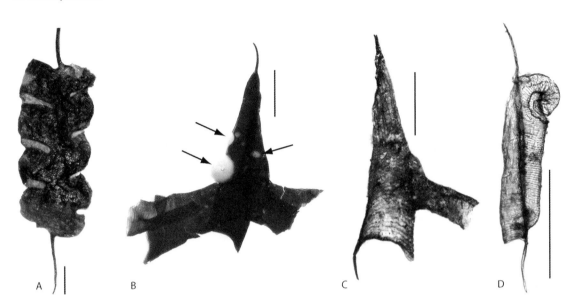

effective, but much simpler method to see internal details of graptolites filled with pyrite. He ground the slabs down to the desired level and polished the surface to investigate the internal details (Figure 14.3C). The method has the obvious disadvantage of being destructive, but as graptolites are often very common, this is usually not a problem. Loydell and Maletz (2009) used a similar method, embedding three-dimensionally preserved, chemically isolated graptolites in epoxy resin, grinding them down and mounting them on glass slides, to investigate the internal construction of *Normalograptus scalaris*. Today, computerized tomography (CT) scans can be used to understand the construction of graptolites (Sutton et al. 2001), but it is more successful in other fossil groups.

Chemical Preparation

The best graptolite material is usually gained from the dissolution of limestones (Figure 14.5) in which graptolites are preserved in full three dimensions. Material in limestone, dolomite or chert is often so well preserved that the graptolites can be extracted by dissolution of the rock, using an appropriate dilute acid. However, if the rock has been tectonized, the graptolites might have suffered from distortion and may break apart. In this case, you will get only small fragments of the graptolite fusellum, unless the specimens are coated with silica (Maletz 2009). Graptolites were first chemically isolated by Gümbel (1878), but it was Holm (1890, 1895) who deserves the credit for his early work on isolated graptolites from the Ordovician and Silurian of Scandinavia, which provided some spectacular new information.

For limestones, dilute hydrochloric or acetic acid can be used, and the specimens are easily liberated as the rock is dissolved. Dolomite needs hydrochloric acid, and chert requires hydrofluoric acid. In all cases, the work should be carried out in a laboratory. Hydrofluoric acid should only be used by professional technicians, under strict conditions to prevent personal damage with the very dangerous liquid. Hydrofluoric acid has rarely been used, but can provide great results from shale samples (e.g. Albani et al. 2001; Maletz 2010a) (Figure 14.5C). Minute specimens can be found that will be impossible to recognize on the shale surfaces. A new method of extracting fossils from clay-rich sediments is the use of the surfactant Rewoquat. Jarochowska et al. (2013) described this method, which provided excellent results with Silurian retiolitids.

Specimens should not be kept in water for any period of time, as fungus will easily grow on the specimens (Figure 14.5B) and destroy the whole collection in a short time, as the material still consists of organic material. If robust enough, specimens can be mounted dry on a glass slide with a well (obtainable from microscope supplies firms), fixed to the bottom of the well with a drop of gum Arabic (obtainable from art supplies firms). More fragile material can be preserved in glycerine, which anyway is preferable, as the specimens may more easily be moved and even prepared for SEM investigation later on. Specimens intended for investigation with the SEM can be cleaned from the glycerine by washing them in warm water or letting them sit in distilled water for about 24 hours.

Once isolated, material can be rendered transparent by the use of a bleaching agent (Figure 14.5C), usually potassium chloride and nitric acid. Again, this should be done in a laboratory under controlled conditions. The specimen should be placed in a watch glass, and the clearing observed under a low-power binocular microscope. The specimens can easily be destroyed by prolonged exposure to the agent.

A different method to get constructional details without bleaching, and thus avoiding damaging the specimens, is the use of infrared (IR) photography (Figure 14.5D). It has a similar effect as bleaching, but does not damage the specimens and therefore is preferable with rare and delicate material. Kraft (1932) and Eisenack (1935) explored the method and used graptolites as examples, but at the time, it was too difficult to become a standard method. Today, with modern computerized techniques and videography, it is much easier to obtain good IR photos. Melchin and Anderson (1998) described the method using a modern IR video camera. The camera can be attached to a micro-scope and the pictures directly transferred to a monitor, allowing direct analysis and storage of the images.

Methods of Illustration

Illustrations are most essential for any scientific publication on fossils. In the past, drawings or engravings were used before photography was sufficiently developed to be used in publications (see also Chapter 15). Some of the illustrations in older scientific publications are unsurpassed in their scientific quality and aesthetics, and may still be regarded as a standard for scientific illustration (Figure 14.6A–B).

For the illustration of graptolites, both drawing and photography are used today – drawing since many graptolites do not show a clear contrast with the adjacent rock surface, either in their relief or in colour, and thus are difficult to photograph. A camera lucida – a drawing mirror – attached to a microscope is normally used to assist and to attain precise outlines. This projects an image of a sheet of paper into the eyepiece of the microscope, superimposing it on the specimen. The image can then be traced, observing both the specimen and the pencil together and adding details as necessary. The result is an interpretation of a specimen (Figure 14.6C), which should be done with the outmost care and precision to avoid too much interpretation, and with this, distortion and possible misinterpretation.

Flattened specimens can be very difficult to photograph, unless there is a good contrast between specimen and matrix. They can be flooded by alcohol, or a 50/50 mixture of alcohol and glycerine, to enhance the contrast between the specimen and the surrounding sediment. Using water instead is possible, as it has the same effect, but is not recommended, as many sediments, especially soft shales, would swell and the specimens may be destroyed. A flat-bed scanner, used at the highest resolution, can be used to photograph relatively flat specimens. It has sufficient depth of focus to produce good results. Even relatively cheap scanners have resolutions of up to 9600 dpi. Place a sheet of clear acetate over the

Figure 14.6 Old and new graptolite illustrations: *Dicaulograptus hystrix* (Bulman, 1932). (A–B) Illustrations by Georg Lilljeval (ca. 1890) for Gerhard Holm (published in Bulman 1932b, pl. 9). (C) Line drawing of specimen (B) from Mitchell (1988, Fig. 11-9). Illustrations not to scale.

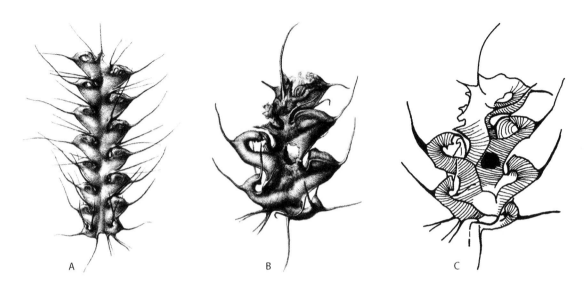

A   B   C

surface of the scanner to avoid scratching it, and cover the area with a black cloth.

Specimens in relief can more easily be photographed (Figure 14.3A, B), preferably after whitening. For this, ammonium chloride is put in a glass tube with a bulb, heated with a Bunsen burner or small camping stove, and "puffed" over the specimen, causing the gas to sublimate on the specimen. This provides details that are impossible to see in the normal uncoated specimens as the whitening enhances the visibility of constructional features (Figures 14.3, 14.4).

Digital cameras have now largely superseded film cameras. Even relatively simple compact cameras allow close-up photography, though "Bridge" cameras and digital SLR cameras allow much higher magnification photographs. A bellows extension can be used for increased magnification. The resulting photographs can be manipulated using a software program such as Adobe Photoshop or Serif PhotoPlus. Contrast and colour can be altered, and blemishes or backgrounds removed.

Stereo pictures can be generated by taking two photographs of the specimen, with rotation of the specimen of 5–10° between shots. However, with the almost two-dimensional nature of even specimens in relief, a stereopair will not usually add appreciably to the information contained in a single picture. An exception to this is in the retiolitids, where the meshwork has considerable complexity (Figure 14.7). Stereo photographs have, however, also been used in a few cases for monograptids (e.g. Lenz & Melchin 2008). These photographs are usually taken using a scanning electron microscope (SEM), an instrument not normally available to the amateur paleontologist.

Permanent Storage

Storage of fossil collections is an important and often ignored issue by scientists, and museum curators can tell some stories. Proper labelling of the slabs with permanent labels, a clear numbering system and all relevant data is essential (Figure 14.8), but is often neglected, and specimens are kept unlabelled in boxes with loose paper labels. Misplacement of labels and/or samples is common during research and can easily go unnoticed. However, samples are rendered useless

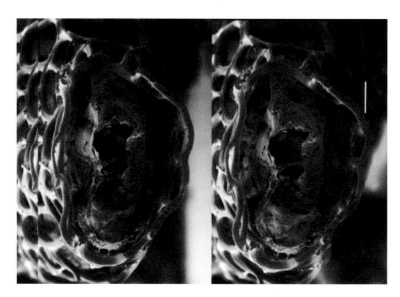

Figure 14.7 Stereopair photographs, taken in the scanning electron microscope (SEM), *Retiolites angustidens* from Gotland, Sweden. Scale bar represents 100 microns.

Figure 14.8 Storage of graptolites. (A) *Pseudophyllograptus angustifolius* (Hall, 1865), isolated, dry specimen on slide, preserved at Swedish Museum of Natural History, Stockholm. (B–D) *Archiclimacograptus wilsoni* (Lapworth, 1876), SMA 19619, shale specimen with labels, preserved at Sedgwick Museum, London. (C) Photograph of the specimen preserved as a pyrite-filled cast, covered with coalified silvery shining fusellum. (D) Specimen photographed after coating with ammonium chloride to show constructional details. Scale in (C–D) represents 1 mm.

when important labels are lost, and locality and collection information is not available any more.

Most graptolite specimens do not require any special storage conditions. However, pyritized material is subject to pyrite decay in the presence of oxygen, breaking down to ferrous sulphate and sulphur dioxide. This is a common and dangerous problem in many fossil collections (Birker & Kaylor 1986; Newman 1998). If water is present, sulphuric acid can also be produced. Pyritic specimens should be stored in sealed containers with a desiccant material.

Pyritic graptolite specimens may become loose on the slabs and easily get lost if stored in open boxes. Loose specimens in boxes or in other small containers may shift and become abraded or even break. Therefore, loose specimens and chemically isolated material are best stored in glycerine, if not mounted in microscope slides for protection. SEM stubs need to be kept in tightly sealed containers so that the delicate specimens do not get coated with dust.

Outlook

The collection of fossils is an essential part of the work of a paleontologist, and surely will remain important as long as paleontologists are working. We work in museums with old collections and understand the problems of curating these and keeping the material available for future generations of researcher. Convincing the general public and the politicians of the value of scientific collections in our natural history museums is a difficult task nowadays, but needs to be on our watch list. So much is still to be discovered, and our museum collections may still yield some astonishing and exciting material – and not just for the graptolite researcher, as is shown by recent findings of new dinosaur species collected more than 75 years ago (e.g. Longrich 2014). An example for the graptolites would be the description of *Hustedograptus bulmani* by Mitchell et al. (2008), based on specimens collected by Gerhard Holm in the late 19th century and first illustrated by Bulman (1932b).

15

History of Graptolite Research

Jörg Maletz

Our modern digital world is so obsessed with visual impressions. Photos, videos, 3D animations, all to provide us with the visual stimulus we want and yearn for. Photos by email from worlds apart, video conferences, and other means of visual contact dominate our world and let us communicate quickly with colleagues on other continents in real time – having virtual face-time has become an obsession to many and we cannot live without our computers and cell phones any more. But it is just a few decades ago when we had only letters and sometimes the phone to communicate. It often took weeks for our letters to reach the addressee and for an answer to reach us. I still remember publishing scientific results – after long exchanges of letters and manuscripts – with people I never met and never knew the face of, a strange idea in our modern world.

If we go into the scientific past, this has been much more common, and as I search for our colleagues from the past, I realize that some of their faces have disappeared completely, along with nearly all knowledge of their personal and scientific achievements. Only the most accomplished and famous names have left their traces and are cherished for their scientific impact. Others left their traces only in the faded scientific papers that survive on old journal pages or theses, and the people behind them are long forgotten. Still, history should tell us that we may not be the first who make a certain discovery, not the first to develop a certain idea, not the first to find an important fossil. As paleontologists we are working with history, geological history, with fossils and with publications written by scientists in the past. We search in old and new publications and extract the information we need, often surprised by the insight of our scientific ancestors.

Graptolite Paleobiology, First Edition. Jörg Maletz.
© 2017 John Wiley & Sons Ltd. Published 2017 by John Wiley & Sons Ltd.

We look through fossil collections in museums and recover so many things that have been seen before, but not necessarily recognized to be of interest. The discovery of the conodont animal is one of these instances, where museum collections, made by scientists and fossil collectors long ago, were preserved and eventually came to light and provided the most important scientific discoveries in decades, long after the collector disappeared and our memories of him or her are gone. This chapter will discuss a few of the scientists working with fossil and extant pterobranchs, hopefully to provide the faces behind so many publications that we cannot live without, and get a glimpse into the scientific past without which we would not be here. It will, understandably, not cover the scientists that are still at work and are acknowledged along with their work in other chapters of this book.

The First Collected Graptolite?

It is not known when graptolites were first collected by a curious person and looked at in detail, most probably not knowing what she or he was looking at. We think it might have been Linnaeus (1735) who noted the name *Graptolithus* in his *Systema Naturae*. He provided at least the name we still use today, but what does this mean? Maybe we have to leave it here and also have to accept that Linnaeus (1751) was the first to publish an illustration of a graptolite in his *Skanska Resa* (Figure 15.1A), as we have no other proof. Here Linnaeus, however, spelled the term *Graptolitus*, but the original *Graptolithus* was accepted subsequently as the correct spelling. It was during a time when curiosity led people to collect many strange and unusual pieces, including rocks, minerals and fossils, in "curiosity cabinets" and private collections (Berg-Madsen & Ebbestad 2013). It also provided us with some of the oldest illustrations of the fossils later identified as graptolites. The illustration of a possible glacial boulder of Silurian age from Stargard, Mecklenburg (northeastern Germany) by Walch (1771, suppl. IVc) (Figure 15.1B) may be regarded as one of the earliest illustrations of a piece of rock with fossil graptolites. Walch (1771) identified the material by the term "orthoceratites", as it was subsequently by many scientists (e.g. Bronn 1835; Geinitz 1842). Walch also coined the term "trilobite" in the same publication, another word well known in paleontology and among fossil collectors. At the time, graptolites were basically unknown, and the few

available descriptions can be found under the heading "fossil plants" (e.g. Brongniart 1828; Bronn 1835; Göppert 1860). Brongniart's (1828) *Fucoides dentatus* and *Fucoides serra* from the Ordovician of Quebec, Canada, for example, have recently been identified as the graptolites *Levisograptus dentatus* (Brongniart, 1828) by Maletz (2011a),

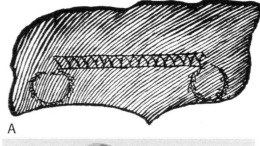

Figure 15.1 (A) Illustration of *Graptolitus*(!) in Linnaeus (1751, p. 147). (B) Illustration of possible graptolites on a piece of rock in Walch (1771, pl. suppl. 4c, Fig. 5), ?glacial boulder, Stargard, Mecklenburg, northern Germany.

and *Tetragraptus serra* (Brongniart, 1828) by Cooper and Fortey (1982). The uncertainty about the relationships of graptolites is clearly reflected in these early works. A number of authors identified the graptolites as a sort of "polyp" in a general sense (Hisinger 1837; Murchison 1839). Quenstedt (1840, p. 274) discussed the "Graptolithi Linn." as "nautileen", but in the text he also suggested an inclusion of the graptolites with the foraminiferans. It was the German zoologist and paleontologist Heinrich Georg Bronn (1800–1862), who coined the term Graptolithina that we still use for our favourite fossil group. In his *Index Palaeontologicus*, Bronn (1849) listed the Graptolithina as a special group of the Anthozoa, but did not discuss or define the taxon.

Tullberg (1882) discussed some of the oldest references to graptolites and indicated that it was probably Magnus (von) Bromell (1727) who first noticed and mentioned graptolites. Tullberg (1882) also discussed the material illustrated by Linnaeus (1751) and identified the locality from which the material originated as a gravel hill near Östra Herrestad in Scania, Sweden. He recognized the graptolite specimens as *Climacograptus scalaris* L. and *Monograptus triangulatus* Harkn., but the whereabouts of Linnaeus' original specimens are unknown. *Climacograptus scalaris* L. is today referred to as *Normalograptus scalaris* (Hisinger, 1837) (e.g. Loydell & Maletz 2009), while *Demirastrites triangulatus* (Harkness, 1851) is an important index fossil for the *Demirastrites triangulatus* Biozone of basal Aeronian, Llandovery age (Loydell 2012). Both species are well known from relief specimens and chemically isolated material (Figure 15.2).

A complex history can be seen in the genus name *Priodon*, now surviving only in the name of the graptolite *Monograptus priodon* (Bronn, 1835). The introduction of the genus name *Priodon*, often referred to Nilsson (see Tullberg 1882; Elles & Wood 1902, p. VII) but apparently used first by Bronn (1835, p. 56), led to some confusion, as the genus name was preoccupied by *Priodon* Cuvier in Quoy and Gaimard, 1825: an actynoperygian fish. Bronn (1835) thus replaced the name *Priodon* with *Lomatoceras*. Hisinger (1837) independently suggested the name *Prionotus* instead, which is a homonym of *Prionotus* Lacépède, 1801 (Osteichtyes, Triglidae). Thus, two names proposed for the same

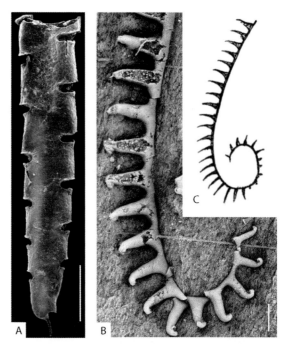

Figure 15.2 (A) *Normalograptus scalaris* (Hisinger, 1837), three-dimensionally preserved, chemically isolated specimen from Kalholn, Dalarna, Sweden (from Maletz 2003, Fig. 2). (B) *Demirastrites triangulatus* (Harkness, 1851) relief specimen from the Llandovery, Lower Silurian, of central Wales (from Palmer & Rickards 1991, pl. 74). (C) *Demirastrites triangulatus* (Harkness, 1851) for comparison (from Harkness 1851). The scale bar indicates 1 mm in each photo.

type of graptolite were recognized as homonyms and needed to be replaced by a new name. The problems finally ended with the suppression of *Lomatoceras* through ICZN (1954: Opinion 198) and the unanimous acceptance of the genus name *Monograptus* Geinitz, 1852, which has been in use ever since. Thus, even at these early times, taxonomy was not easy and many mistakes were documented in the scientific literature.

Foundation of Graptolite Research

In the late 1840s to 1850s, graptolite research was finally gaining more interest and the first detailed taxonomic descriptions appeared in

Figure 15.3 Some of the early paleontologists providing important insight into graptolite research. (A) Joachim Barrande (1799–1883). (B) Hanns Bruno Geinitz (1814–1900). © Senckenberg Naturhistorische Sammlungen Dresden. Reproduced with permission from Senckenberg Naturhistorische Sammlungen Dresden. (C) James Hall (1811–1898). Reproduced with permission from Special Collections and University Archives, University of Iowa. (D) Frederick M'Coy (1817–1899).

Europe (e.g. Barrande 1850; Geinitz 1842, 1852; M'Coy 1850; and many more) and North America (e.g. Hall 1847, 1865). There is no doubt that this was the time when the foundation for graptolite research for the next century was laid down by some extraordinary scientists (Figure 15.3) with a wide range of interest and a precise way of documentation. They produced the earliest monographic descriptions of the fossil group that is now known to be one of the most valuable groups for Paleozoic geology and paleontology. None of these people was interested solely in graptolites, but all regarded the graptolites as one of the fossil groups worth their interest. The descriptions produced during this period show astonishing insight into the construction of the graptolite colonies, and the earliest taxonomic systems were produced. The local interest in geology and paleontology generated the basis for these scientists to look into the graptolite records of many regions.

M'Coy (1850), Harkness (1851) and Salter (1852) described graptolites from the British Isles, while Boeck (1851) and Scharenberg (1851) discussed and illustrated material from the Middle Ordovician of Norway. Their work is still valuable as a basis for information on graptolites and their distribution, even though many of their species cannot be recognized from a modern standpoint. However, as most of their material remained in museum collections until today, we can still gain insight into their understanding of these fossils by comparing the original illustrations with the preserved specimens.

Hall (1847, 1865) (Figure 15.3C) described the Ordovician graptolites of eastern North America in great detail and showed a profound understanding of their nature. He erected many new species and genera from his material, and the illustrations especially in Hall (1865) are much better than previously available illustrations of graptolites. They often show details not generally understood at the time. Unfortunately, his illustrations are also the basis of one of the major misunderstandings in graptolite interpretation: the synrhabdosomes of Ruedemann (1895), based entirely on his reconstruction of *Orthograptus eucharis* (Hall, 1865). Hall reconstructed this species based on the idea that all graptolites have a branched construction

Figure 15.4 (A) *Clonograptus flexilis* (Hall), lectotype, GSC 965c (from Hall, 1865, pl. 10, Fig. 5). (B) *Clonograptus flexilis* (Hall), lectotype (from Lindholm & Maletz 1989, text-fig. 2A). (C) *Orthograptus eucharis* (Hall), reconstruction of species (from Hall 1865, pl. 14, Fig. 9). Boxes added to show the supposed proximal branching in (A) and (C). Illustrations not to scale.

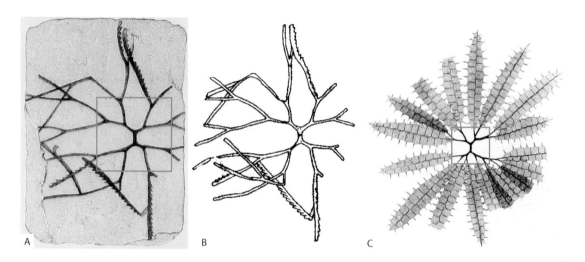

A   B   C

in the centre (Figure 15.4C), devoid of thecae and comparable to the centre part of *Clonograptus flexilis* (see Maletz 2015). It is now common knowledge that all graptolites start from a sicula and all branches are formed from thecal tubes. Thus, this interpretation of James Hall has not been verified and has to be considered an understandable error at the time.

James Hall was one of the excellent paleontologists and observers of his time, and the illustrations of graptolites that were produced under his guidance by R.P. Whitfield (Batten 1987; Blum 1987) were among the most informative published at the time. As Hall's illustrated specimens are largely preserved in museum collections, these can be compared with the illustrations and show the amount of detail and understanding included in his plates. New drawings of the type specimen of *Clonograptus flexilis* by Lindholm and Maletz (1989), for example, did not provide any further details (Figure 15.4A, B). As was usual at the time, Hall (1865) even included fossils that are today identified as trace fossils in the graptolites, as the example of *Oldhamia* shows. Other fossils initially identified as graptolites include *Triplograptus* (Richter, 1871), *Nereograptus* (Geinitz, 1852) and *Protovirgularia* (M'Coy, 1850) – originally

described as graptolites, but later recognized as trace fossils. These cases only indicate that the scientific understanding of graptolites was at an early stage and the concept of graptolites not yet settled. Through our growth of knowledge on graptolites, misidentified taxa were eliminated and referred to other fossil groups.

In Germany, Geinitz (1842) (Figure 15.3B), with his short paper "Ueber Graptolithen", must be regarded as the starting point of graptolite research (Maletz 2001a). Geinitz probably did not see the potential of graptolite research, as might be seen in his taxonomic descriptions, lacking any information on biostratigraphical use (e.g. Geinitz 1852, 1890), even though this was previously explored by Hall (1850). At the same time, Richter (1853, 1871) in Thuringia and Roemer (1855) and Kayser (1878) in the Harz Mountains collected and described the Silurian graptolite faunas of Germany. After a short period of interest, graptolite research disappeared nearly completely in Germany, and only in the early 20th century increased in importance largely through the interest of fossil collectors and amateur scientists like Eisel, Manck and Hundt (Maletz 2001a).

It was Geinitz (1852) who suggested abandoning the genus name *Graptolithus*, as this was in the past used for all graptolites and had become a general term for these fossils. Thus, he compared it with the terms trilobite or ammonite, used as informal labels for a group of fossil organisms. The genus name *Graptolithus* then started to disappear slowly from the scientific literature, but it took a long time to vanish completely. Quite a number of new genus-level names were introduced, while the term *Graptolithus* was eliminated and survived only in the general term "graptolite" for the whole fossil group. The genus *Graptolithus* was officially suppressed only in 1954 (ICZN 1954) and ceased to be used in taxonomy entirely.

M'Coy (1850) (Figure 15.3D) described the graptolites as a group of Silurian Radiata (Zoophyta) and established the family Graptolitidae as the first family group taxon of the Graptolithina Bronn, 1849. M'Coy (1850) differentiated the uniserial taxa as *Graptolites* from the biserial ones, which he called *Diplograpsis* (now *Diplograptus*: Mitchell et al. 2009). This development may be seen as the starting point of graptolite taxonomy: the differentiation of uniserial and biserial graptolites as a seemingly natural way of separating major groups of graptolites. Even though we now know that the story is not that simple, and that biserial graptolites evolved from uniserial ones and later the uniserial species evolved again from biserial ones, this first step was an important one. It shows that the usefulness of various characteristics had to be explored before a stable classification was established and phylogenetic relationships were understood.

Looking back, we might have to regard Joachim Barrande (Figure 15.3A) as the most influential leader of graptolite research, with his insight into the taxonomy of these fossils that were so poorly known at the time. Barrande (1850) for the first time provided a useful terminology for graptolite colonies, based on features he was able to observe and interpret. Barrande differentiated the graptolites, previously referred to a single genus *Graptolithus*, into the subgenera *Monoprion* and *Diprion* and based these on the number of thecal series. He also separated the genera *Rastrites* and *Retiolites* for the first time from the bulk of the graptolite species. His differentiation of genera

was based on the same criteria used by M'Coy (1850) in his taxonomic interpretation, however.

The British Dominance

In the later years of the 19th century, British workers dominated the research on graptolites and produced some of the most influential publications. Harkness, Hopkinson (Figure 15.5E), M'Coy, Lapworth and Nicholson (Figure 15.5A) were among the main players, and their work led to *A Monograph of British Graptolites* (Elles & Wood, 1901–1918), initiated by Charles Lapworth. This compilation dominated graptolite research for a very long time with its detail and usefulness. It covered all planktic graptolite taxa known from the British Isles at the time of research, and is still an invaluable source of information. It was, however, Nicholson's (1872a) monograph that set the stage for this monumental work and has to be regarded as the basis of all monographic work on graptolites in Britain and beyond.

The later part of the 19th century, the interval from 1866–1880 (see Elles & Wood 1902), was characterized by a more comprehensive understanding of graptolites as fossils and key for biostratigraphical interpretations (Nicholson, 1868a) in Britain. The number of described genera increased and the "taxonomic tree" of the graptolites became more complex. Nicholson (1872a, b) and Lapworth (1873a, b) provided important improvements in the understanding of graptolite taxonomy. The basic concepts of differentiation of the genera based on the number of stipes (e.g. *Tetragraptus, Didymograptus, Monograptus*) was established, when Nicholson (1872b) differentiated the Graptolitidae into the Monoprionidae, Diprionidae and Tetraprionidae and referred to them as a subclass of the Hydrozoa.

Nicholson (1872a) and Allman (1872) discussed the relationship of the graptolites to the recently discovered extant pterobranchs, known only through *Rhabdopleura* Allman, in Norman, 1869. Slightly earlier, McCrady (1859) tried to convince the scientific world that graptolites are related to the echinoderms, as he compared the tubarium

Figure 15.5 Founders of graptolite research in Britain. (A) Henry Alleyne Nicholson (1844–1899). (B) *Cephalograptus tubulariformis* (Nicholson, 1867b), holotype, original illustration (from Nicholson 1867b, pl. 7, Fig. 12a). (C) *Cyrtograptus murchisoni* Carruthers (from Hopkinson 1869, Fig. 5a). (D) *Phyllograptus* sp. (from Hopkinson 1869, Fig. 20; based on Hall 1865, pl. 16). (E) John Hopkinson (1844–1919). From Jackson and Alkins (1919), reproduced with permission from Cambridge University Press.

construction with the anatomy of echinoderm larvae, an idea that was quickly abandoned.

One of the most important aspects of graptolite research, the biostratigraphical use, was also one of the aspects strongly promoted during this period. Lapworth (1878) demonstrated the use of graptolites to unravel the complex tectonic situation of the greywacke succession of the Moffat series, a milestone in graptolite research (Fortey 1993). It was James Hall (1850), however, who had already recognized the biostratigraphical use of graptolites or "their value in the identification of strata" as he expressed it in his paper. Today, the biostratigraphical use is probably the most important geological application of graptolite research, and the very precise biozonations of graptolite faunas from the Ordovician to the Lower Devonian in many regions of the world can be regarded as a standard for Paleozoic biostratigraphy (Loydell 2012).

Early Graptolite Research in Scandinavia

During the late 19th century and early 20th century, graptolite research was taken over by Scandinavian scientists, as they discovered the highly productive successions with excellent preservation of faunas in their countries. Holm, Moberg, Törnquist (Figure 15.6A) and Tullberg (Figure 15.6G) dominated the period that may be called the Scandinavian Period in graptolite research. Linnarsson (1871), Törnquist (1879, 1881) and Tullberg (1880, 1882) were among the first to describe graptolites from Ordovician and Silurian successions in Sweden, and provided evidence of the excellent preservation of the Scandinavian graptolite faunas (Figure 15.6B–F). Tullberg unfortunately died at an early age in 1886 and published only a single monograph on Silurian graptolites from Scania, but he also

Figure 15.6 (A) Sven Leonhard Törnquist (1840–1920), ca. 1897, photo provided by Per Ahlberg, Lund. (B–E) *Expansograptus praenuntius* (Törnquist, 1901), LO 1611t and LO 1612t, syntypes (D, E from Törnquist, 1901, pl. 2). (F) *Monograptus priodon* (Bronn, 1835), from Tullberg (1883, pl. 2, Fig. 24). (G) Sven Axel Tullberg (1852–1886).Photo based on Henriksson (1994; SGU Information).

described a number of new Ordovician taxa in several smaller papers. Törnquist (1901, 1904) monographed the graptolite fauna of the Lower Didymograptus Shale (now Tøyen Shale Formation) of Scania and Västergötland, Sweden. Most of the species of Törnquist have never been revised since the original description, but the faunas are now the basis for the stratotype of the Floian Stage of the Ordovician System (Bergström et al. 2004), and their biostratigraphy has been documented in great detail from the sections at Mt. Hunneberg (Egenhoff & Maletz 2007). It was probably the work of Astrid Monsen (1925, 1937) that first gave us an indication of the extremely high biodiversity of the graptolite faunas of the Floian to Dapingian (Lower to Middle Ordovician) time interval from the Oslo Region in southern Norway, and her work is still the basis for graptolite taxonomy of the paleocontinent of Baltica.

Gerhard Holm (1890, 1895) (Figure 15.7A) and Carl Wiman (1893a, b) (Figure 15.7D) described for the first time three-dimensionally preserved graptolites chemically isolated from limestones in some detail, and also showed the growth lines (fusellar construction) of these faunas. In particular, some of the illustrations of Wiman showed enormous detail. The type specimen of *Urbanekograptus retioloides*, even though a fragment (Figure 15.7C), can clearly be identified and compared with the much better and complete specimens illustrated by Urbanek (1959). Holm and Wiman also used serial thin-sections and dissection of isolated specimens (Figure 15.7B) to understand the construction of the graptolite tubaria. Quite a number of monographs on Ordovician and Silurian graptolites from Sweden were produced during the early 20th century and became an important source of information for graptolite research.

Figure 15.7 (A) Gerhard Holm (1853–1926). (B) *Haddingograptus eurystoma* (Jaanusson, 1960), specimen dissected by Holm (from Bulman 1932b, pl. 1, Fig. 30). (C) *Urbanekograptus retioloides* (Wiman, 1895), type specimen, fragment, from Wiman (1895, pl. 9, Fig. 4). (D) Carl Wiman (1867–1944), ca. 1890 (from Svenska män och Kvinnor 8, 1955; provided by Jan-Ove Ebbestad).

  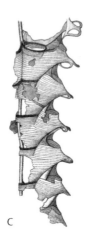 

A   B   C   D

Graptolites from "Down Under"

During the last half of the 19th century, graptolites were also discovered in the successions of Victoria, Australia, by field geologists from the Geological Survey of Victoria, and were published in a number of papers by Frederick M'Coy (see Keble & Benson 1939). M'Coy, who moved to Melbourne in 1854 (Hegarty et al. 2003) to become one of the first four professors at the newly opened university, illustrated a number of specimens in his "Prodromus", and it was clear from very early on that the faunas were very similar to those of Britain and North America. Hall (1892, 1897), M'Coy (1874, 1875) and Pritchard (1892, 1895) provided early descriptions and biostratigraphical information on the Victorian graptolite faunas, leading to the bloom in Australasian graptolite research. In particular Thomas Sergeant Hall (Figure 15.8A) has to be credited for the recognition of graptolites as the key to understanding the complex Ordovician sequence of Victoria. His work was followed by the research of W.J. Harris (Figure 15.8D) and D.E. Thomas (Figure 15.8E), culminating in the revision of the biostratigraphy of the Ordovician succession of Victoria (Harris & Thomas 1938b) and leading to numerous taxo-

nomic papers describing the Ordovician graptolite faunas of the region. Keble and Benson (1939) (Figure 15.8B, C) discussed the early graptolite research of Australia in some detail and differentiated three periods of research from 1856 to 1939. Here they may be termed initial research, biozonation, and systematization of research. The authors complained that most of the early work on Australian graptolites was overlooked by later scientists, largely because of difficulty in access to publications and providing a complete list of graptolite publications available to them.

VandenBerg and Cooper (1992) published a modern overview of the Ordovician graptolite succession of Australasia, including Tasmania and New Zealand, but no overview on the Silurian is available. Even though the Australasian Ordovician succession was regarded as the most complete graptolite succession worldwide, surprisingly little taxonomic work was done in the last half century, except for the faunal revision of the Lower to Middle Ordovician faunas by Morris (1988, unpublished thesis) and the revision of the Bendigonian graptolites of Victoria (Rickards & Chapman 1991), both based entirely on previously published material. Cooper (1973) and Cooper and Ni (1986) described in some detail the isograptids

Figure 15.8 Australasian graptolite specialists and their faunas. (A) Thomas Sergeant Hall (1858–1915). (B) Alexander Robert Keble (1884–1963). (C) William Noel Benson (1885–1957). (D) William John Harris (1886–1957). (E) David Evan Thomas (1902–1978). (F) *Rhabdinopora campanulatum* (Harris & Keble, 1928), holotype, NMV P31903. (G) *Parisograptus subtilis* (Williams & Stevens, 1988), NMVP 324150. (H) *Goniograptus timidus* (Harris & Thomas, 1939), NMVP83313, from paratype slab. (I) *Dicellograptus elegans* (Carruthers, 1868), NMVP 62593A (drawing in VandenBerg & Cooper 1992, Fig. 10G). Photos provided by A.H.M. VandenBerg.

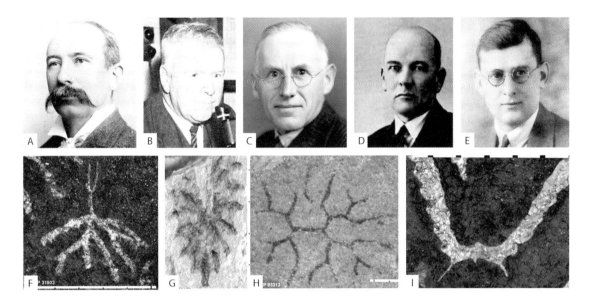

and pseudisograptids of the Dapingian of Australasia, and Cooper (1979b) documented the Ordovician graptolite faunas of the Aorangi Mine area of New Zealand.

The 20th Century Graptolites

The 20th century saw the specialization of research, and with it the detailed documentation of many local graptolite faunas. It brought together some of the most amazing graptolite faunas with some of the most skilled researchers. However, it should not be forgotten that the great compilation of Elles and Wood (1901–1918) set the stage for graptolite research of the 20th century. Gertrude Lillian Elles (1872–1960) and Ethel Mary Reader Shakespear (née Wood) (1871–1946) combined their efforts in the monumental work describing all graptolites known from Britain at the time, and reviewing graptolite literature worldwide. Gertrude Elles was for a long time the leading British graptolite specialist, and many successful graptolite workers spent some time with her at the University of Cambridge, where Oliver Meredith Boone Bulman (1902–1974) also subsequently worked and dominated graptolite research from the 1930s to the 1970s (Rickards 1999). Including the work of Barrie Rickards (1938–2009), who took over from Bulman, Cambridge became the longest lasting and most successful institution for graptolite research and will forever be associated with these names.

Roman Kozłowski (1899–1977) (Figure 15.9B) probably did some of the most fascinating graptolite research. Kozłowski (1938, 1949) dissolved cherts from the lower Ordovician (Tremadocian) of the Holy Cross Mountains of Poland and gained graptolite faunas never seen before and rarely found afterwards (Figure 15.9A). He described several new orders of graptolites (Tuboidea, Camaroidea, Stolonoidea) as well as forms related to the Dendroidea, the Pterobranchia and the unusual Graptoblasti and Acanthastida. All these

Figure 15.9 (A) Original illustration of *Rhipidodendrum samsonowiczi* Kozłowski, 1949 from Kozłowski (1949, pl. 10, Fig. 1). Reproduced with permission from Instytut Paleobiologii PAN. (B) Adam Urbanek (left) and Lech Teller (right) in Urbanek's office in 2004. Roman Kozłowski is seen in the framed picture at the back (photo by Tanya Koren).

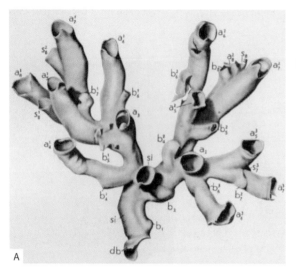

were meticulously described and illustrated from isolated fragments preserved in three dimensions, and he provided a fascinating insight into the benthic graptolites of the Lower Ordovician, their evolution and diversity. He established the Institute of Palaeozoology at Warszaw University in which research was strongly connected to a biological approach. The institute was also the home of Adam Urbanek (1928–2014) and Lech Teller (1928–2013) (Figure 15.9B), providing excellent work on Silurian graptolite faunas, often chemically isolated from drill cores and preserved in three dimensions (e.g. Urbanek 1958, 1997a, b; Teller 1964, 1997b). Urbanek was highly interested in evolutionary studies and the biological affinities of graptolites (e.g. Urbanek 1960, 1986, 2004), and his years in the department mark the time of the most intense and successful graptolite research in Poland.

Graptolites in China and Russia

The earliest graptolite descriptions from China were probably those of Chang (1933, 1938), Hsü (1934) and Sun (1931, 1933, 1935), as indicated by

Mu (1980), but graptolite research in China is most closely connected to Mu An-Tze (Figure 15.10A), or Mu En-Zhi, depending on the Latin spelling of his name. Mu's work started in the 1940s with publications on Upper Ordovician graptolites of the Wufeng Shale (Mu 1946). Early in his career, he published a new graptolite taxonomy (Mu 1950) that was largely ignored by Westerners, but had a strong influence in China. Mu quickly became the leading specialist in graptolite research, as can still be seen in the latest graptolite compilation from China (Mu et al. 2002), prepared largely by Li Ji-Jin (1928–2013) (Figure 15.10C) and published long after Professor Mu died in 1987. Mu published numerous taxonomic papers on graptolites, including the new genus *Sinograptus* and the family Sinograptidae (see Chapter 10), but was also highly interested in graptolite biostratigraphy and evolutionary studies (Mu 1987). Thousands of species have been described from China, and a precise biostratigraphy was established for the graptolite faunal succession of various plates and terranes that form modern China (e.g. Chen et al. 2013). A number of Chinese workers have been active since the 1960s, including Ge Mei-yu, Lin Yao-kun, Li Ji-jin, Ni

Figure 15.10 (A) Mu En-Zhi (1917–1987). (B) *Sinograptus typicalis* Mu, 1957, holotype, one of the most famous "Chinese" graptolites. (C) Li Ji-jin (1928–2013) (photo ca. 1986, provided by John Riva, Québec, Canada).

A B C

Yu-nan and others, enriching graptolite research with data from China. Intense international exchange dramatically modified graptolite research from the 1980s, and many Chinese colleagues were able to attend international conferences and work with colleagues from other countries, enabling the erection of a number of Ordovician stratotype sections in China (see Chapter 6).

Graptolite research in Russia was initiated and dominated for a long time by Aleksandr Obut (1911–1988), who became one of the leading graptolite specialists of the 20th century. He was accompanied by a number of young researchers like Rimma Sobolevskaya, Tanya Koren (1935–2010), N.F. Mikhailova, D.T. Tzai and others, forming a strong graptolite working group (Sennikov et al. 2011) and providing important graptolite information from Russia.

The Pterobranch Connection

It was probably Michael Sars (1868) (Figure 15.11B) who first recognized the strange organism, later known as *Rhabdopleura* Allman, 1869, as new and mentioned it under the name *Haliolophus*.

His son George Ossian Sars (1872) (Figure 15.11C) described the taxon as *Rhabdopleura mirabilis* from the Norwegian coast. When Allman in Norman (1869) introduced the genus *Rhabdopleura* for the first time into the literature, it was not clear that this genus of extant organisms was closely related to the graptolites, and he referred it to the Polyzoa (Bryozoa). A little later, M'Intosh (1882) (Figure 15.11D) found with *Cephalodiscus* a second genus related to the genus *Rhabdopleura*, both anatomically quite similar and both secreting a housing construction, the tubarium, from glands on the head-shield. Thus, a second genus was referred to the class Pterobranchia, previously established by Lankester (1877, 1884) for *Rhabdopleura*. The closer relationships of this group were uncertain, even though Allman (1872) and Nicholson (1872c) had already discussed the possible relationships to the graptolites, but this was only established with certainty more than 100 years later by Mitchell et al. (2013) after a combined cladistic analysis of fossil and extant pterobranchs. Schepotieff (1905) discussed in some detail the possible relationships, but came to a different opinion: the graptolites are related to the ancestors of the modern Pterobranchia, but both groups cannot be

Figure 15.11 Finding and classifying the first pterobranchs. (A) Vladimir Beklemishev (1890–1962). (B) Michael Sars (1805–1869). (C) George Ossian Sars (1837–1927). (D) William Carmichael M'Intosh (1838–1931).

combined into one. Beklemishev (1951a, b) (Figure 15.11A) saw a closer connection and used the class Graptolithoidea to combine the extinct graptolites with the extant pterobranchs.

Graptolite Reconstructions through Time

Initially, graptolites were thought to be extinct and relationships to modern organisms difficult to establish. Thus, ideas on graptolites and their reconstruction might seem strange to us now, but were acceptable and understandable from the standpoint of the time. For a very long time, the phylogenetic relationships of graptolites, based on the poor available data, were discussed controversially, and graptolites were referred to many modern groups before they settled with the Hemichordata (see Maletz 2014a for a review). Hall's (1865) reconstruction of *Orthograptus*

eucharis (Hall 1865, pl. 14, Fig. 9) became the mould on which reconstructions were based for quite some time. This is clear especially after the publications of Ruedemann (1895, 1898) in which the most influential and long-used reconstructions of graptolite colonies were presented. Ruedemann illustrated his synrhabdosomes or "supercolonies" (Figure 4.3) with a basal cyst or pneumatophore, gonangia and numerous interconnected biserial stipes (the tubaria or rhabdosomes of modern taxonomy), but most of the features are not verified from the available material. These reconstructions are still found in many modern textbooks, showing the enormous influence of these interpretations (Maletz 2015). The interpretation of Ruedemann was quickly extended by Frech (1897) for many biserial (Figure 15.12A) and monograptid graptolites, and became a standard for the interpretation of graptolite colonies and lifestyle.

Still, information on the zooidal anatomy of the graptolites did not exist, and the idea that

Figure 15.12 (A) *Petalolithus folium* (Hisinger, 1837), interpretation of colony by Frech (1897, Fig. 132). (B) *Monograptus priodon* (Bronn, 1835), specimen showing "muscle scars" (from Haberfelner 1933, pl. 1, Fig. 2). (C–D) Interpretation of graptolite zooid in two views (from Ulrich & Ruedemann 1931, Figs 9–10).

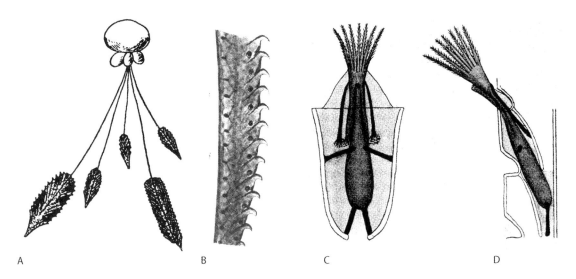

graptolites were some kind of strange hydroid or bryozoan kept its place in the scientific community. Ulrich and Ruedemann (1931) were among the first to speculate on the anatomy of the graptolite zooids and produced a reconstruction of the zooids in their thecal tubes attached with paired retractor and gastral attachment muscles (Figure 15.12C, D). Following this idea, Haberfelner (1933) interpreted the thickened bases of the interthecal septae in many monograptids as the muscle scars of the graptolite zooids (Figure 15.12 B). Only since the 1950s has it been customary to suggest a closer relationship of the graptolites to the extant Pterobranchia and the pterobranch zooids; especially those of the *Rhabdopleura* were regarded as a basis for the reconstruction of graptolite zooids (e.g. Bulman 1945), following the comparison of the graptolites with the pterobranchs by Kozłowski (1938, 1949). Bulman (1970a, Fig. 7) consequently introduced *Rhabdopleura* zooids in his reconstructions for the *Treatise*. The final acceptance was followed by the analysis of the tubarium construction, interpreting the fusellum and cortex as comparable in pterobranchs and graptolites, by Andres (1977), and the iconic reconstruction of part of a biserial graptolite colony (Figure 2.12) by Crowther and Rickards (1977, Fig. 2).

The Kirk Hypothesis – A Controversy

There is one person who deserves a special introduction when we are discussing graptolites and their interpretations. This is Nancy Kirk (1916–2005) (Figure 15.13A), who vehemently argued that the graptolite tubarium was an internal construction covered by a layer of tissue, an extrathecal mantle (Kirk 1978, 1979, 1990) (Figure 15.13B), even though she believed that a relationship to the Pterobranchia might be supported by the evidence and interpreted benthic graptolites as pterobranch-like. She developed ideas on graptolites only later during her scientific career in the 1960s. Wyatt (1997) wrote an interesting account of her life and ideas that is worth reading to get an idea of her scientific impact during this time. "Turning the world of graptolites upside down" is a good expression to describe some of Nancy's ideas. She argued that the planktic graptolites lived with the sicular aperture pointing upwards (Figure 15.13C, D), differently from the accepted orientation (see Bulman 1955, 1970a), due to the way the benthic graptolites evolved into planktic ones by producing lift through coordinated feeding currents. The water flow is directed downwards, leading to the

Figure 15.13 (A) Nancy Kirk (1916–2005). (B) Reconstruction of benthic graptolite sicula (from Kirk 1974, pl. 2 N). Reproduced with permission from the University College of Wales, Aberystwyth. (C–D) Orientation of graptolite colonies preferred by Nancy Kirk (from Bates & Kirk 1986a, reproduced with permission from The Royal Society).

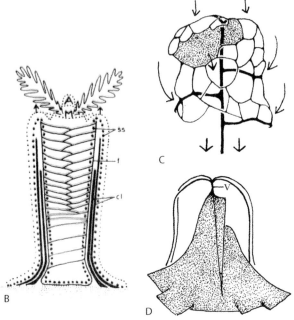

eventual release of the colony into the water column, forming the planktic graptolites. As a next step the bithecae, as supposed cleaning individuals, would be eliminated and the colonies evolved a free-swimming lifestyle: the graptolite automobility model was born.

Nancy Kirk developed her ideas of an extrathecal mantle as she realized that the graptolite constructions are much more complex than she expected at first. Her models of retiolitid graptolites (Plate 14) were among the best produced and have been featured in a number of her publications with Denis Bates to clarify the complex construction of the retiolitid graptolites (Bates & Kirk 1991, 1992, 1997).

Urbanek (1976, 1978) initially supported the idea of an extrathecal mantle, but changed his opinion later on (Urbanek 1986). Bates and Kirk (1984, 1986a, b) further developed the concept and explained that the extrathecal tissue model was the only one explaining all features of the tubarium construction. Rigby (1993b) used information from ultrastructural studies, modern ideas on the pterobranch affinities of the graptolites, phylogenetic studies, as well as graptolite colony models and mathematical analyses of graptoloid shapes, to vehemently reject the Kirk hypothesis.

In any case, the Kirk hypothesis, the extrathecal tissue model, made considerable impact in graptolite research and produced numerous papers trying to reject the idea. Even Rigby (1993b, p. 284) acknowledged that the stimulus of Nancy Kirk's papers "puts us all at her debt." Thus, even though the idea does not have any followers now, it was an important stimulus for graptolite research. As a final result, the British and Irish Graptolite Group (BIGG) dedicated the Palmer and Rickards (1991) graptolite book to Nancy, indicating that while controversial, her idea had found a valuable place in graptolite research.

Outlook

Graptolite research has had its ups and downs with a number of highly influential specialists over the last 150 years. In the 1970s to 1990s, the Graptolite Group of the International Palaeontological Association held a number of conferences, but this ended with the last meeting in Argentina in 2003. New developments in graptolite research since this time include the increase of cladistic methods of investigation and interpretation of graptolite taxonomy and evolution. Graptolite specialists used opportunities to meet at international conferences to discuss relevant topics, but did not have their special meetings. In the last few years, interest in graptolite research is increasing again with the recognition of "hot shales" in the Silurian in North Africa, the Far East and China and the growing demand for oil and gas resources.

Theresa Podhalańska recently discussed the significance of taphonomic research on graptolites to identify zones of increased accumulation of hydrocarbons in Ordovician and Silurian rocks of Poland and the Baltic Region. It is clear that graptolites are an important instrument, in addition to elevated carbon values or increased gamma ray radiation on well logs, to allow the identification of potential source rocks for hydrocarbons, including shale gas. So it seems that – apart from the taxonomic and evolutionary work of paleontologists – graptolites are again regarded as an important part of geological history and are useful for the prosperity of humans on our planet.

References

Albani, R., Bagnoli, G., Maletz, J. and Stouge, S. (2001) Integrated chitinozoan, conodont and graptolite biostratigraphy from the Upper Cape Cormorant Formation (Middle Ordovician), western Newfoundland. *Canadian Journal of Earth Sciences* **38**, 387–409.

Allman, G.J. (1869) *Rhabdopleura normani*, Allman, nov. gen. et sp., in *Shetland final dredging report*. Part II. Crustacea, Tunicata, Polyzoa, Echinodermata, Actinozoa, Hydrozoa and Porifera (ed. A.M. Norman). *Reports of the British Association of the Advancement of Science* **1868**, 311–312.

Allman, G.J. (1872) On the morphology and affinities of graptolites. *Annals and Magazine of Natural History, Fourth Series* **9**, 364–380.

Althausen, M. (1992) Lower Paleozoic (Riphean) metalliferous black shales. *Oil Shale* **9** (3), 194–207.

Andersson, K.A. (1907) Die Pterobranchier der schwedischen Südpolar-Expedition 1901–1903 nebst Bemerkungen über *Rhabdopleura normani* Allman. *Wissenschaftliche Ergebnisse der Schwedischen Südpolarexpedition* **5**, 1–122, 8 pls.

Andres, D. (1961) Die Struktur von Mastigograptiden aus einem ordovizischen Geschiebe Berlins. *Neues Jahrbuch für Geologie und Paläontologie, Monatshefte* **1961** (12), 636–647.

Andres, D. (1977) Graptolithen aus ordovizischen Geschieben und die frühe Stammesgeschichte der Graptolithen. *Paläontologische Zeitschrift* **51** (1/2), 52–93.

Andres, D. (1980) Feinstrukturen und Verwandtschaftsbeziehungen der Graptolithen. *Paläontologische Zeitschrift* **54** (1–2), 129–170.

Antezana, T. (2009) Species-specific patterns of diel migration into the Oxygen Minimum Zone by euphausiids in the Humboldt Current ecosystem. *Progress in Oceanography* **83**, 228–236.

Bapst, D.W., Bullock, P.C., Melchin, M.J., Sheets, H.D. and Mitchell, C.E. (2012) Graptoloid diversity and disparity became decoupled during the Ordovician mass extinction. *PNAS* **109** (9), 3428–3433.

Bardack, D. (1997) Wormlike animals: Enteropeusta, in *Richardson's Guide to the Fossil Fauna of Mazon Creek* (eds C.W. Shabica and A.A. Hay). Northeastern Illinois University, Chicago, pp. 89–92.

Barrande, J. (1850) *Graptolites de la Boheme*. Théophile Haase Fils, Prague.

Barrande, J. (1859) Schreiben an Herrn W. Haidinger, Director der k. k. geologischen Reichsanstalt u.s.w. *Jahrbuch der kaiserlich-königlichen geologischen Reichsanstalt* **10** (4), 479–480.

Barrande, J. (1861) *Défense des Colonies. I. Groupe probatoire comprenant: la colonie Haidinger, la colonie Krejci et la coulée Krejci*. Prague.

Barrande, J. (1862) *Défense des Colonies. II. Incompatibilité entre le système des plis et la réalité des faits matériels*. Prague.

Barrande, J. (1863) Présentation d'un mémoire de M. le Dr. A. de Volborth. *Bulletin de la Societe Géologique de France* **20** (2), 595–598.

Barrande, J. (1865) *Défense des Colonies. III. Etude générale sur nos étages G-H, avec application spéciale sur les environs de Hlubocep près Prague.* Prague.

Barrande, J. (1870) *Défense des Colonies. IV. Description de la Colonie d'Archiac; caractères généraux des colonies dans les bassins siluriens de la Bohême.* Prague.

Barrande, J. (1880) Du maintien de la nomenclature établie par M. Murchison. Exposition univers, internat, de 1878, à Paris. *Congress International de Géologie* **21**, 101–106.

Barrande, J. (1881) *Défense des Colonies. V. Apparition et réapparition en Angleterre et en Ecosse des espèces coloniales siluriennes de la Bohême, d'après les documents anglais les plus authentiques et les plus récents.* Prague.

Bates, D.E.B. (1987) The construction of graptolite rhabdosomes in the light of ultrastructural studies. *Indian Journal of Geology* **59** (1), 1–28.

Bates, D.E.B. (1990) Retiolite nomenclature and relationships. *Journal of the Geological Society of London* **147** (4), 717–723.

Bates, D.E.B. and Kirk, N.H. (1984) Autecology of Silurian graptoloids. *Special Papers in Palaeontology* **32**, 121–139.

Bates, D.E.B. and Kirk, N.H. (1986a) Graptolites, a fossil case history of sessile, colonial animals to automobile superindividuals. *Proceedings of the Royal Society of London* **B228** (1251), 207–224.

Bates, D.E.B. and Kirk, N.H. (1986b) Mode of secretion of graptolite periderm, in normal and retiolite graptolites. *Geological Society Special Publication* **20**, 221–236.

Bates, D.E.B. and Kirk, N.H. (1987) The role of extrathecal tissue in the construction and functioning of some Ordovician and Silurian retiolitid graptoloides. *Bulletin of the Geological Society of Denmark* **35**, 85–102.

Bates, D.E.B. and Kirk, N.H. (1991) The ultrastructure, mode of construction and functioning of Ordovician retiolitid graptolites from the Viola Springs Limestone, Oklahoma. *Modern Geology* **15** (2–3), 131–286.

Bates, D.E.B. and Kirk, N.H. (1992) The ultrastructure, mode of construction and functioning of a number of Llandovery ancorate diplograptid and retiolitid graptolites. *Modern Geology* **17** (1–3), 1–270.

Bates, D.E.B. and Kirk, N.H. (1997) The ultrastructure, construction and functioning of the genera *Stomatograptus* and *Retiolites*, with an appendix on the incremental construction of the rhabdosome in *Petalolithus*, and its comparison with that of the thecal framework in *Retiolites* and *Stomatograptus*. *Institute of Geography and Earth Sciences, University of Aberystwyth Publication* **10**, 1–168.

Bates, D.E.B. and Loydell, D.K. (2000) Parasitism on graptoloid graptolites. *Palaeontology* **43**, 1143–1151.

Bates, D.E.B. and Urbanek, A. (2002) The ultrastructure, development, and systematic position of the graptolite genus *Mastigograptus*. *Acta Palaeontologica Polonica* **47** (3), 445–458.

Bates, D.E.B., Kozłowska, A. and Lenz, A.C. (2005) Silurian retiolitid graptolites: Morphology and evolution. *Acta Palaeontologica Polonica* **50** (4), 705–720.

Bates, D.E.B., Kozłowska, A., Chmielarz, D. and Lenz, A.C. (2011) Excessive thickening of the cortical layer in graptolites. *Proceedings of the Yorkshire Geological Society* **58** (4), 211–222.

Bateson, W. (1885) The later stages in the development of *Balanoglossus kowalevskii*, with a suggestion as to the affinities of the Enteropneusta. *Quarterly Journal of Microscopical Science* **25**, 81–122, pls. 4–9.

Batten, R.L. (1987) Robert Parr Whitfield: Hall's assistant who stayed too long. *Earth Sciences History* **6** (1), 61–71.

Bauert, H. (1994) The Baltic oil shale basin – An overview. *Proceedings 1993 Eastern Oil Shale Symposium*, 411–421. University of Kentucky Institute for Mining and Minerals Research.

Bauert, H. and Kattai, V. (1997) Kukersite oil shale, in *Geology and Mineral Resources of Estonia* (eds A. Raukas and A. Teedumäe). Estonian Academy Publishers, Tallinn, pp. 313–327.

Becker, R.T., Gradstein, F.M. and Hammer, O. (2012) The Devonian Period, in *The Geologic Time Scale 2012* (eds F.M. Gradstein, J.G. Ogg, M.D. Schmitz and G.M. Ogg). Elsevier, Oxford, pp. 559–601.

Beckly, A. and Maletz, J. (1991) The Ordovician graptolites *Azygograptus* and *Jishougraptus* in Scandinavia and Britain. *Palaeontology* **34** (4), 887–925.

Beklemishev, V.N. (1951a) On the systematic structure of animals (Vtorichnorotye (Deuterostomia)), their origin and composition. *Uspekhi Sovremennoy Biologii* **32**, 256–270.

Beklemishev, V.N. (1951b) *Osnovy sravnitelnoi anatomii bespozvonochnykh*. Moskva, Sovetskaia Nauka [Russian Editions in 1951, 1964] **2** volumes, 430, 478 pp. [German Edition: *Grundlagen der vergleichenden Anatomie der Wirbellosen* 1958 (Band 1. Promorphologie); 1960 (Band 2. Organologie). VEB Deutscher Verlag der Wissenschaften, Berlin.] [English edition: 1970. *Principles of Comparative Anatomy of Invertebrates*, Vols 1 and 2. Oliver and Boyd, Edinburgh.]

Bengtson, S. and Urbanek, A. (1986) *Rhabdotubus*, a Middle Cambrian rhabdopleurid hemichordate. *Lethaia* **19** (4), 293–308.

Benton, M. (2007) The Phylocode: Beating a dead horse? *Acta Palaeontologica Polonica* **52** (3), 651–655.

Berg-Madsen, V. and Ebbestad, J.O.R. (2013) The Bromell fossil collection at Uppsala University, Sweden: its history and the people behind it. *GFF* **135** (1), 3–17.

Bergström, S.M. (1980) Conodonts as paleotemperature tools in Ordovician rocks of the Caledonides and adjacent areas in Scandinavia and the British Isles. *Geologiska Föreningens i Stockholm Förhandlingar* **102**, 377–392.

Bergström, S.M. and Cooper, R.A. (1973) *Didymograptus bifidus* and the trans-Atlantic correlation of the Lower Ordovician. *Lethaia* **6**, 313–340.

Bergström, S.M., Finney, S.C., Chen, X., Pålsson, C., Wang, Z.-H. and Grahn, I. (2000) A proposed global boundary stratotype for the base of the Upper Series of the Ordovician System: The Fagelsang section, Scania, southern Sweden. *Episodes* **23** (3), 102–109.

Bergström, S.M., Löfgren, A. and Maletz, J. (2004) The GSSP of the Second (Upper) Stage of the Lower Ordovician Series: Diabasbrotter at Hunneberg, Province of Västergötland, Southwestern Sweden. *Episodes* **27** (4), 265–272.

Bergström, S.M., Finney, S.C., Xu, C., Goldman, D. and Leslie, S.A. (2006) Three new Ordovician global stage names. *Lethaia* **39**, 287–288.

Berner, R.A. (1970) Sedimentary pyrite formation. *American Journal of Science* **268**, 1–23.

Berner, R.A. (1984) Sedimentary pyrite formation: an update. *Geochimica et Cosmochimica Acta* **48**, 605–615.

Berry, W.B.N. (1960a) Graptolite faunas of the Marathon region, west Texas. *University of Texas Publications* **6005**, 1–179.

Berry, W.B.N. (1960b) Correlation of Ordovician graptolite-bearing sequences. *21st International Geological Congress* **7**, 97–107.

Berry, W.B.N. (1967) Comments on correlation of the North American and British Lower Ordovician. *Geological Society of America Bulletin* **78**, 419–427.

Berry, W.B.N. (1968) British and North American Lower Ordovician correlation – reply. *Geological Society of America Bulletin* **79**, 1265–1272.

Berry, W.B.N. and Murphy, M.A. (1975) Silurian and Devonian graptolites of central Nevada. *University of California Publications in Geological Sciences* **110**, 1–109.

Berry, W.B.N. and Norford B.S. (1976) Early late Cambrian dendroid graptolites from the Northern Yukon. *Geological Survey of Canada, Bulletin* **256**, 1–12.

Berry, W.B.N. and Wilde, P. (1978) Progressive ventilation of the Oceans – An explanation for the distribution of the Lower Paleozoic black shales. *American Journal of Science* **278**, 257–275.

Berry, W.B.N. and Wilde, P. (1990) Graptolite biogeography: Implications for palaeogeography and palaeooceanography. *Geological Society, London, Memoirs* **12**, 129–137.

Berry, W.B.N., Wilde, P. and Quinby-Hunt, M. (1987) The oceanic nonsulfidic oxygen minimum zone: a habitat for graptolites? *Geological Society of Denmark Bulletin* **35**, 103–114.

Bertrand, R. (1990) Correlations among the reflectances of vitrinite, chitinozoans, graptolites and scolecodonts. *Organic Geochemistry* **15** (6), 565–574.

Bertrand, R. and Héroux, Y. (1987) Chitinozoan, graptolite, and scolecodont reflectance as an alternative to vitrinate and pyrobitumen reflectance in Ordovician and Silurian strata, Anticosti Island, Quebec, Canada. *American Association of Petroleum Geologists Bulletin* **71** (8), 951–957.

Bigsby, J.J. (1853) On the geology of Quebec and its environs. *Quarterly Journal of the Geological Society, London* **9**, 82–101.

Birker, I. and Kaylor, J. (1986) Pyrite disease: Case studies from the Redpath Museum, in *Proceedings of the 1985 Workshop on the Care and Maintenance of Natural History Collections* (eds J. Waddington and D.M. Rudkin). Life Sciences Miscellaneous Publications, Royal Ontario Museum, 21–27.

Bjerreskov, M. (1978) Discoveries on graptolites by X-ray studies. *Acta Palaeontologica Polonica* **23**, 463–47.

Bjerreskov, M. (1991) Pyrite in Silurian graptolites from Bornholm, Denmark. *Lethaia* **24**, 351–361.

Bjerreskov, M. (1994) Pyrite diagenesis of graptolites from Bornholm, in *Graptolite research Today* (eds. X. Chen, B.-D. Erdtmann and Y. Ni) Nanjing University Press, Nanjing. pp. 217–22.

Blum, A.S. (1987). A better style of art: The illustrations of the Paleontology of New York (by James Hall). *Earth Science History* **6**, 72–85.

Blumenstengel, H., Hansch, W., Heuse, T., Leonhardt, D., Maletz, J., Meisel, S., Samuelsson, J., Sarmiento, G.N., Sehnert, M., Tröger, K.-A., Verniers, J. and Walter, H. (2006) Fauna und Flora im Silur Deutschlands. *Schriftenreihe der Deutschen Geologischen Gesellschaft* **46**, 130–152.

Boeck, C. (1851) *Bemærkninger angaaende graptolitherne*. P.T. Malings Bogtrykkeri, Christiania (Oslo).

Bouček, B. (1933) Monographie der obersilurischen Graptolithen aus der Familie Cyrtograptidae. *Prace geologicko-paleontologickeho ustavu Karlovi university v Praze 1933* (**1**), 1–84, 7 pls.

Bouček, B. (1936) La fauna graptolitique du Ludlovien inférieur de la Bohême. *Rozpravy II, Tridy ceske Akademie* **46**, 137–152.

Bouček, B. (1944) Über einige gedornte Diplograptiden des böhmischen und sächsischen Silurs. *Mitteilungen der tschechischen Akademie der Wissenschaften* **53** (25), 1–6.

Bouček, B. (1957) The dendroid graptolites of the Silurian of Bohemia. *Rozpravy Ustredniho ustavu Geologickeho* **23**, 1–294.

Bouček, B. (1973) *Lower Ordovician Graptolites of Bohemia*. Publishing House of the Czechoslovak Academy of Sciences, Prague.

Bouček, B. and Münch, A. (1944) Die Retioliten des mitteleuropäischen Llandovery und unteren Wenlock. *Academie Tcheque des Sciences, Bulletin International* **44**, 527–579.

Bouček, B. and Münch, A. (1952) Retioliti stredoevropskeho svrchniho wenlocku a ludlowu; the central European retiolites of the upper Wenlock and Ludlow. *Sborník ústredniho ústavu Geologickeho, oddíl paleontologicky* **19**, 1–54 (Czech text), 55–103 (Russian text), 104–151 (English text).

Bouček, B. and Přibyl, A. (1951) On some slender species of the genus *Monograptus* Geinitz, especially of the subgenera *Mediograptus* and *Globosograptus*. *Bulletin international de l'Academie tcheque des Sciences. Extrait du texte czeque publié dans la revue Rozpravy II. tridy Ceske Akademie* **52** (13), 1–32.

Braithwaite, L.F. (1976) Graptolites from the Lower Ordovician Pogonip Group of Western Utah. *The Geological Society of America, Special Paper* **166**, 1–106.

Briggs, D.E.G., Erwin, D.H. and Collier, F.J. (1994) *The Fossils of the Burgess Shale*. Smithsonian Institution Press.

Briggs, D.E.G., Kear, A.J., Baas, M., de Leeuw, J.W. and Rigby, S. (1995) Decay and composition of the hemichordate Rhabdopleura: implications for taphonomy of graptolites. *Lethaia* **28**, 15–23.

Brögger, W.C. (1878) Om Paradoxidesskifrene ved Krekling. *Nyt Magazin för Naturvidenskaberne* **24**, 18–88.

Bromell, von, M. (1727) *Lithographiae Svecanae Specimen II. Telluris Svecanae petrificata lapidesque figuratos exhibens, juxta seriem atque ordinem, quo in Museo Metallico Bromeliano servantur., Sectio prima. De Vegetalibus, fossilibus & Lapidefactis, 306–312. Articulus Tertius, De Lithoxylis, 331–337*. Acta Literaria Sveciae 1727. Upsala.

Brongniart, A. (1828) *Histoire des végétaux fossils*. G. Dufour and Ed. D'Ocagne, Paris.

Bronn, H.G. (1835) *Lethaea Geognostica, Erster Band, das Übergangs-, bis Oolithen-Gebirge enthaltend*. Schweizerbart, Stuttgart.

Bronn, H.G. (1849) *Handbuch der Geschichte der Natur. Dritter Band, Zweite Abtheilung. II. Theil: Organisches Leben (Schluß). Index palaeontologicus oder Ueberblick der bis jetzt bekannten fossilen Organismen*. Schweizerbart, Stuttgart.

Brussa, E.D., Maletz, J., Mitchell, C.E. and Goldman, D. (2007) *Nemagraptus gracilis* (J.

Hall) in Bolivia and Peru. *Acta Palaeontologica Sinica* **46** (Suppl.), 57–63.

Bull, E.E. (1996) Implications of normal and abnormal growth structures in a Scottish Silurian dendroid graptolite. *Palaeontology* **39** (1), 219–240.

Bulman, O.M.B. (1927a) A monograph of British dendroid graptolites. *Palaeontographical Society Monograph* 1–28.

Bulman, O.M.B. (1927b) *Koremagraptus*, a new dendroid graptolite. *Annals and Magazine of Natural History, Series* **9** (19), 344–347.

Bulman, O.M.B. (1928) *Odontocaulis* Lapworth, a synonym of *Callograptus* Hall. *Geological Magazine* **65** (8), 337–342.

Bulman, O.M.B. (1932a) On the graptolites prepared by Holm 2–5. *Arkiv för Zoologi* **24A** (9), 1–29.

Bulman, O.M.B. (1932b) On the graptolites prepared by Holm. 1. Certain Diprionidian graptolites and their development. *Arkiv för Zoologi* **24A** (8), 1–45.

Bulman, O.M.B. (1933) On the graptolites prepared by Holm 6. Structural characters of some *Dictyonema* and *Desmograptus* species from the Ordovician and Silurian rocks of Sweden and the east baltic Region. *Arkiv för Zoologi* **26A** (5), 1–52.

Bulman, O.M.B. (1936) On the graptolites prepared by Holm 7. The graptolite fauna of the Lower Orthoceras limestone of Hälludden, Öland, and its bearing on the evolution of the Lower Ordovician graptolites. *Arkiv för Zoologi* **28A** (17), 1–107.

Bulman, O.M.B. (1938) Graptolithina, in *Handbuch der Palaeozoologie, vol. 2D* (ed. O.H. Schindewolf). Gebrüder Borntraeger, Berlin, pp. 1–92.

Bulman, O.M.B. (1944–47) A Monograph of the Caradoc (Balclatchie) graptolites from limestones in Laggan Burn, Ayrshire. *Monograph of the Palaeontographical Society, London* i–xi, 1–78.

Bulman, O.M.B. (1945) A monograph of the Caradoc (Balclatchie) Graptolites from limestones in Laggan Burn, Ayrshire, Part 1. *Palaeontographical Society Monograph* **98** (430), 1–42.

Bulman, O.M.B. (1947) A monograph of the Caradoc (Balclatchie) Graptolites from limestones in Laggan Burn, Ayrshire, Part 3. *Palaeontographical Society Monograph* **98**, 59–78.

Bulman, O.M.B. (1950). Graptolites from the Dictyonema Shales of Quebec. *Quarterly Journal of the Geological Society, London* **106**, 63–99.

Bulman, O.M.B. (1954) The graptolite fauna of the *Dictyonema* shales of the Oslo region. *Norsk Geologisk Tidsskrift* **33** (1–2), 1–40.

Bulman, O.M.B. (1955) *Graptolithina*, in *Treatise on Invertebrate Paleontology, Part V* (ed. R.C. Moore). Geological Society of America and University of Kansas Press, Lawrence.

Bulman, O.M.B. (1963) On *Glyptograptus dentatus* (Brongniart) and some allied species. *Palaeontology* **6** (4), 665–689.

Bulman, O.M.B. (1964) Lower Palaeozoic plankton. *Quarterly Journal of the Geological Society* **120**, 455–476.

Bulman, O.M.B. (1970a) Graptolithina, in *Treatise on Invertebrate Paleontology, Part V*, 2nd edn (ed. C. Teichert). Geological Society of America and University of Kansas Press, Lawrence.

Bulman, O.M.B. (1970b) A new *Dictyonema* fauna from the Salmien of the Stavelot Massif (with a preface by F. Geukens). *Bulletin de la Societé Belge Geologie, Paleontologie, Hydrologie* **79**, 213–224.

Bulman, O.M.B. and Rickards, R.B. (1966) A revision of Wiman's dendroid and tuboid graptolites. *Bulletin of the Geological Institution of Uppsala* **43** (6), 1–72.

Bulman, O.M.B. and Rickards, R.B. (1968) Some new diplograptids from the Llandovery of Britain and Scandinavia. *Palaeontology* **11** (1), 1–15.

Burdon-Jones, C. (1952) Development and biology of the larva of *Saccoglossus horsti* (Enteropneusta). *Philosophical Transactions of the Royal Society of London B* **236**, 553–589.

Bustin, R.M., Link, C. and Goodarzi, F. (1989) Optical properties and chemistry of graptolite periderm following laboratory simulated maturation. *Organic Geochemistry* **14**, 355–364.

Butterfield, N.J. (1990) Organic preservation of non-mineralizing organisms and the taphonomy of the Burgess Shale. *Paleobiology* **16** (3), 272–286.

Butterfield, N.J., Balthasar, U. and Wilson, L.A. (2007) Fossil diagenesis in the Burgess Shale. *Palaeontology* **50** (3), 537–543.

Calman, W.T. (1908) On a parasitic copepod from *Cephalodiscus. Transactions of the South African Philosphical Society XVII. Marine Investigations in South Africa* **5**, 174–184, pls. 18–19.

Calner, M. (2005) Silurian carbonate platforms and extinction events – ecosystem changes exemplified from Gotland, Sweden. *Facies* **51**, 584–591.

Cameron, C.B., Swalla, B.J. and Garey, J.R. (2000) Evolution of the chordate body plan: new insights from phylogenetic analysis of deuterostome phyla. *Proceedings of the National Academy of Sciences of the United States of America* **97** (9), 4469–4474.

Cannon, J.T., Rychel, A.L., Eccleston, H., Halanych, K.M. and Swalla, B.J. (2009) Molecular phylogeny of Hemichordata, with updated status of deep-sea enteropneusts. *Molecular Phylogenetics and Evolution* **52**, 17–24.

Cannon, J.T., Swalla, B.J. and Halanych, K.M. (2013) Hemichordate molecular phylogeny reveals a novel cold-water clade of Harrimaniid acorn worms. *The Biological Bulletin* **225** (3), 194–204.

Cannon, J.T., Kocot, K.M., Waits, D.S., Weese, D.A., Swalla, B.J., Santos, S.R. and Halanych, K.M. (2014) Phylogenomic resolution of the hemichordate and echinoderm clade. *Current Biology* **24** (23), 2827–2832.

Cantino, P.D. (2004) Classifying species versus naming clades. *Taxon* **53** (3), 795–798.

Cantino, P.D., Bryant, H.N., de Queiroz, K., Donoghue, M.J., Eriksson, T., Hillis, D.M. and Lee, M.S.Y. (1999) Species names in phylogenetic nomenclature. *Systematic Biology* **48** (4), 790–807.

Caron, J.-B. and Rudkin, D. (2009) *A Burgess Shale Primer*. Field Trip Companion Volume – ICCE 2009. Burgess Shale Consortium, Toronto, Canada.

Caron, J.-B., Conway Morris, S. and Shu, D. (2010) Tentaculate fossils from the Cambrian of Canada (British Columbia) and China (Yunnan) interpreted as primitive deuterostomes. *PLoS One* **5** (3), 1–13.

Caron, J.-B., Conway Morris, S. and Cameron, C.B. (2013) Tubicolous enteropneusts from the Cambrian period. *Nature* **495**, 503–506.

Carruthers, W. (1868). A revision of British graptolites, with descriptions of new species, and notes on their affinities. *Geological Magazine* **5**, 64–74, 125–133, pl. V.

Carter, C. (1972) Ordovician (Upper Caradocian) graptolites from Idaho and Nevada. *Journal of Paleontology* **46** (1), 43–49.

Carter, C. (1989) A Middle Ordovician graptolite fauna from near the contact between the Ledbetter Slate and the Metaline Limestone in the Pend Orteille Mine, northeastern Washington State. *US Geological Survey Bulletin* **1860**, A1–A23.

Chang, H. (1933) On the discovery of the graptolite shale from Lientan, Yuenan District in Kwantung Province and its stratigraphic correlation. *Geological Society of China* **12** (2), 249–257.

Chang, H. (1938) Über einige geologische und paläobiologische Probleme der Monograptolithen. *Palaeobiologica* **6**, 190–196.

Chapman, A.J., Rickards, R.B. and Grayson, R.F. (1993) The Carboniferous dendroid graptolites of Britain and Ireland. *Proceedings of the Yorkshire Geological Society* **49** (4), 295–319.

Chapman, A.J., Durman, P.N. and Rickards, R.B. (1996) A provisional classification of the graptolite order Dendroidea. *Paläontologische Zeitschrift* **70** (1/2), 189–202.

Chapman, F. (1917) Report on Cambrian fossils from Knowley East, near Heathcote. *Geological Survey of Victoria, Records* **4** (1), 87–102, pls. 6–7.

Chapman, F. (1919) On some hydroid remains of Lower Palaeozoic age from Monegetta, near Lancefield. *Proceedings of the Royal Society of Victoria, N. S.* **2** (31), 388–393, pls. 19–20.

Chapman, F. and Thomas, D.E. (1936) The Cambrian Hydroidea of the Heathcote and Monegeeta Districts. *Proceedings of the Royal Society of Victoria, N. S.* **48** (2), 193–212, pls. 14–17.

Chen, J.Y., Zhu, M.Y. and Zhou, G.Q. (1995) The early Cambrian medusiform metazoan *Eldonia* from the Chengjiang Lagerstätte. *Acta Palaeontologica Polonica* **40** (3), 213–244.

Chen, J.Y., Zhou, G.Q., Zhu, M.Y. and Yeh, K.Y. (1997) *The Chengjiang Biota – A Unique Window of the Cambrian Explosion*. National Museum of Natural Science, Taichung, Taiwan [in Chinese].

Chen, X. and Mitchell, C.E. (1995) A proposal – the base of the *austrodentatus* zone as a level for global subdivision of the Ordovician System. *Palaeoworld* **5**, 104.

Chen, X. and Zhang, Y.D. (1996) Isograptids of China, in *Centennial Memorial Volume of Prof. Sun Yunzhu (Y. C. Sun): Palaeontology and Stratigraphy* (eds H.Z. Wang and X.L. Wang). China University of Geosciences Press, Beijing, pp. 82–89.

Chen, X., Zhang, Y.D. and Mitchell, C.E. (2001) Early Darriwilian graptolites from central and western China. *Alcheringa* **25**, 191–210.

Chen, X., Melchin, M.J., Fan, J.X. and Mitchell, C.E. (2003) Ashgillian graptolite fauna of the Yangtze eregion and the biogeographical distribution of diversity in the latest Ordovician. *Bulletin de la Societe Geologique de France* **174** (2), 141–148.

Chen, X., Fan, J.X., Melchin, M.J. and Mitchell, C.E. (2005) Hirnantian (latest Ordovician) graptolites from the Upper Yangtze region, China. *Palaeontology* **48**, 235–280.

Chen, X., Bergström, S.M., Zhang, Y.D. and Wang, Z.H. (2013) A regional tectonic event of Katian (Late Ordovician) age across three major blocks of China. *Chinese Science Bulletin, Geology* **58** (34), 4292–4299.

Chlupác, I. and Kukal, Z. (1977) The boundary stratotype at Klonk. The Silurian–Devonian Boundary. *IUGS Series A* **5**, 96–109.

Cisne, J.L. and Chandlee, G.O. (1982) Taconic foreland basin graptolites: age zonation, depth zonation, and use in ecostratigraphic correlation. *Lethaia* **15**, 343–363.

Clark, T.S. (1924) The Paleontology of the Beekmantown Series at Levis, Quebec. *Bulletin of American Paleontology* **10** (41), 1–151.

Conway Morris, S. (1981) Parasites and the fossil record. *Parasitology* **82**, 489–509.

Conway Morris, S. (1982) Parasites in the fossil record. *Trends and Perspectives in Parasitology* **2**, 24–44.

Cooper, R.A. (1970) Tectonic distortion of a syntype of *Isograptus forcipiformis latus* Ruedemann. *Journal of Paleontology* **44** (5), 980–983.

Cooper, R.A. (1973) Taxonomy and evolution of *Isograptus* Moberg in Australasia. *Palaeontology* **16** (1), 45–115.

Cooper, R.A. (1979a) Sequence and correlation of Tremadoc graptolite assemblages. *Alcheringa* **3**, 7–19.

Cooper, R.A. (1979b) Ordovician geology and graptolite faunas of the Aorangi Mine area, north west Nelson, New Zealand. *New Zealand Geological Survey Palaeontological Bulletin* **47**, 1–127.

Cooper, R.A. (1990) Interpretation of tectonically deformed fossils. *New Zealand Journal of Geology and Geophysics* **33** (2), 321–332.

Cooper, R.A. and Fortey, R.A. (1982) The Ordovician graptolites of Spitsbergen. *Bulletin of the British Museum (Natural History), Geology Series* **36** (3), 157–302.

Cooper, R.A. and Fortey, R.A. (1983) Development of the graptoloid rhabdosome. *Alcheringa* **7** (3–4), 201–221.

Cooper, R.A. and Lindholm, K. (1985) The phylogenetic relationship of the graptolites *Tetragraptus phyllograptoides* and *Pseudophyllograptus cor*. *Geologiska Föreningens i Stockholm Förhandlingar* **106** (3), 279–291.

Cooper, R.A. and Lindholm, K. (1990) A precise worldwide correlation of early Ordovician graptolite sequences. *Geological Magazine* **127** (6), 497–525.

Cooper, R.A. and Ni, Y.N. (1986) Taxonomy, phylogeny and variability of *Pseudisograptus* Beavis. *Palaeontology* **29** (2), 313–363.

Cooper, R.A. and Sadler, P.M. (2010) Facies preference predicts extinction risk in Ordovician graptolites. *Paleobiology* **36** (2), 167–187.

Cooper, R.A. and Sadler, P.M. (2012) The Ordovician Period. With a contribution by F.M. Gradstein and O. Hammer, in *The Geologic Time Scale 2012* (eds F.M. Gradstein, J.G. Ogg, M. Schmitz and G. Ogg). Elsevier, Oxford, pp. 489–524.

Cooper, R.A. and Stewart, I. (1979) The Tremadoc graptolite sequence of Lancefield, Victoria. *Palaeontology* **22** (4), 767–797.

Cooper, R.A., Fortey, R.A. and Lindholm, K. (1991) Latitudinal and depth zonation of early Ordovician graptolites. *Lethaia* **24**, 199–218.

Cooper, R.A., Maletz, J., Wang, H.F. and Erdtmann, B.-D. (1998) Taxonomy and evolution of earliest Ordovician graptolites. *Norsk Geologisk Tiddskrift* **78**, 3–32.

Cooper, R.A., Nowlan, G.S. and Williams, S.H. (2001) Global stratotype section and point for the base of the Ordovician. *Episodes* **24** (1), 19–28.

Cooper, R.A., Rigby, S., Loydell, D.K. and Bates, D.E.B. (2012) Palaeoecology of the Graptoloidea. *Earth-Science Reviews* **112**, 23–41.

Cooper, R.A., Sadler, P.M., Munnecke, A. and Crampton, J.S. (2014) Graptoloid evolutionary rates track Ordovician–Silurian global climate change. *Geological Magazine* **151** (2), 349–364.

Crowther, P. (1981) The fine structure of graptolite periderm. *Special Papers in Palaeontology* **26**, 1–119.

Crowther, P. and Rickards, R.B. (1977) Cortical bandages and the graptolite zooid. *Geologica et Palaeontologica* **11**, 9–46.

Czega, W., Hanisch, C., Junge, F., Zerling, L. and Baborowski, M. (2006) Changes in uranium concentration in the Weisse Elster river as a mirror of the remediation in the former WISMUT mining area, in *Uranium in the Environment* (eds B.J. Merkel and A. Hasche-Berger). Springer, Berlin, pp. 875–884.

Damas, D. and Stiasny, O. (1961) Les larves planctoniques d'entéropneustes. *Mémoire de l'' Academie Royale de Belgique C1. Sci.* **15**, 1–68.

Darwin, C. (1859) *On the Origin of Species by Means of Natural Selection, or the Preservation of Favoured Races in the Struggle for Life*. John Murray, London.

Davydov, V.I., Korn, D. and Schmitz, M.D. (2012) The Carboniferous Period, in *The Geologic Time Scale* (eds F. Gradstein, J. Ogg, M. Schmitz and G. Ogg). Elsevier, Oxford, pp. 603–651.

Dawson, D.H. and Melchin, M.J. (2007) A possible transitional stage between the resorption porus and the primary porus in early monograptid graptolites. *Acta Palaeontologica Sinica* **46** (suppl.), 89–94.

Dawydoff, C. (1948) Classe des Ptérobranches, in *Traité de Zoologie, Anatomie, Systématique, Biologie, Vol. 11 (Echinodermes–Stomocordés–Protocordés)* (ed. P.-P. Grassé). Masson et Cie., Paris, pp. 455–532.

Dayrat, B. (2005) Ancestor–descendant relationships and the reconstruction of the Tree of Life. *Paleobiology* **31** (3), 347–353.

Decker, C.E. (1935) The graptolites of the Simpson Group. *Proceedings of the National Academy of Sciences* **21**, 239–243.

Deng, B. (1986) On the morphological characteristics of the *spiralis* group and the stratigraphic significance of the appearance of *Cyrtograptus*. *Geological Society Special Publication* **20**, 191–195.

Deng, G. (1985) *Dictyonema* finds in the Permian System of Hainan Island, Guangdong, China. *Journal of Paleontology* **59** (5), 1323–1324.

De Queiroz, K. (2006) The PhyloCode and the distinction between taxonomy and nomenclature. *Systematic Biology* **55**, 160–162.

De Queiroz, K. (2007) Toward an integrated system of clade names. *Systematic Biology* **56** (6), 956–974.

Dieni, I., Giordano, G., Loydell, D.K. and Sassi, F.P. (2005) Discovery of Llandovery (Silurian) graptolites and probable Devonian corals in the Southalpine Metamorphic Basement of the eastern Alps (Agordo, NE Italy). *Geological Magazine* **142**, 1–5.

Dilly, N.P. (1986) Modern pterobranchs: observations on their behaviour and tube building. *Geological Society Special Publication* **20**, 261–269.

Dilly, N.P. and Ryland, J.S. (1985) An intertidal *Rhabdopleura* (Hemichordata, Pterobranchia) from Fiji. *Journal of Zoology* **205**, 611–623.

Dilly, P.N. (2014) *Cephalodiscus* reproductive biology (Pterobranchia, Hemichordata). *Acta Zoologica (Stockholm)* **95** (1), 111–124.

Dilly, P.N., Welsch, U. and Rehkämper, G. (1986) Fine structure of tentacles, arms, and associated coelomic structures of *Cephalodiscus* (Pterobranchia, Hemichordata). *Acta Zoologica (Stockholm)* **67**, 181–191.

Dobrowolska, K. (2013) Reconstruction of the proximal ends of retiolitid rhabdosomes (Graptolithina) from the Upper Wenlock and the Lower Ludlow. *Paläontologische Zeitschrift* **87**, 1–17.

Durman, P.N. and Sennikov, N.V. (1993) A new rhabdopleurid hemichordate from the Middle Cambrian of Siberia. *Palaeontology* **36** (2), 283–296.

Dyni, J.R. (2005) Geology and resources of some world oil-shale deposits. *USGS Scientific Investigations Report* **2005**, 5294, 1–42.

Egenhoff, S. and Fishman, N.S. (2013) Traces in the dark – sedimentary processes and facies gradients in the Upper Shale member of the Upper Devonian–Lower Mississippian Bakken Formation, Williston Basin, North Dakota, U.S.A. *Journal of Sedimentary Research* **83**, 803–824.

Egenhoff, S. and Maletz, J. (2007) Graptolites as indicators of maximum flooding surfaces in monotonous deep-water shelf successions. *Palaios* **22**, 374–384.

Egenhoff, S. and Maletz, J. (2012) The sediments of the Floian GSSP: depositional history of the Ordovician succession at Mount Hunneberg, Västergötland, Sweden. *GFF* **134**, 237–249.

Eichwald, E.J. (1840) Ueber das silurische Schichtensystem in Esthland. *Zeitschrift für Natur- und Heilkunde der k. medicinisch-chirurgischen Akademie, St. Petersburg* **1** (2), 1–210.

Eichwald, E.J. (1855) Beitrag zur geographischen Verbreitung der fossilen Thiere Russlands. Alte Periode. *Bulletin de la Societé des Naturalistes de Moscou* **28** (4), 433–466.

Eisel, R. (1899) Ueber die Zonenfolge ostthüringischer und vogtländischer Graptolithenschiefer. *39. bis 42. Jahresbericht der Gesellschaft von Freunden der Naturwissenschaften in Gera (Reuss) 1896–1899*, 49–62.

Eisel, R. (1903) Nachtrag zum Fundortsverzeichnissse wie zur Zonenfolge thüringisch-vogtländischer Graptolithen. *43.-45. Jahresbericht der Gesellschaft von Freunden der Naturwissenschaften in Gera (Reuss) 1900–1902*, 25–32.

Eisel, R. (1908) Über die Verdrückungen thüringisch-sächsischer Graptolithenformen. *Zeitschrift für Naturwissenschaften, Organ des naturwissenschaftlichen Vereins für Sachsen und Thüringen, Halle a. S.* 218–221.

Eisel, R. (1912) Über zonenweise Entwicklung der Rastriten und Demirastriten. *53./54. Jahresbericht der Gesellschaft von Freunden der Naturwissenschaften Gera*, 27–43.

Eisenack, A. (1934) Neue Mikrofossilien des baltischen Silurs, III und Neue Mikrofossilien des böhmischen Silurs, I. *Paläontologische Zeitschrift* **16**, 52–76.

Eisenack, A. (1935) Neue Graptolithen aus Geschieben baltischen Silurs. *Paläontologische Zeitschrift* **17**, 73–90.

Eisenack, A. (1937) Was ist *Melanostrophus*? *Zeitschrift für Geschiebeforschung und Flachlandsgeologie* **13** (2), 100–104.

Eisenack, A. (1941) *Epigraptus bidens* n. g. n. sp., eine neue Graptolithenart des baltischen Ordoviziums. *Zeitschrift für Geschiebeforschung und Flachlandsgeologie* **17** (1), 24–28.

Eisenack, A. (1942) Über einige Funde von Graptolithen aus ostpreußischen Silurgeschieben. *Zeitschrift für Geschiebeforschung und Flachlandsgeologie* **18**, 29–42.

Eisenack, A. (1951) Retiolithen aus dem Graptolithengestein. *Palaeontographica* **A100** (5–6), 129–163.

Ekström, G. (1937) Upper Didymograptus Shale in Scania. *Sveriges Geologiska Undersökning, Serie* **C403** (10), 1–53.

Elles, G.L. (1897) The subgenera *Petalograptus* and *Cephalograptus*. *Quarterly Journal of the Geological Society of London* **53**, 186–212.

Elles, G.L. (1922) The graptolite faunas of the British Isles. A study in evolution. *Proceedings of the Geologist's Association* **33** (3), 168–200.

Elles, G.L. and Wood, E.M.R. (1901–1918) A monograph of British Graptolites. 11 parts. *Palaeontographical Society Monograph* I–clxxi, 1–539.

Elles, G.L. and Wood, E.M.R. (1902) A monograph of British Graptolites. Part 2. *Palaeontographical Society Monograph* **56** (265), i–xxviii, 55–102.

Elles, G.L. and Wood, E.M.R. (1904) A monograph of British Graptolites, Part 4. *Palaeontographical Society Monograph* **58** (277), lii–lxxii, 135–180, pls 20–25.

Elles, G.L. and Wood, E.M.R. (1906) A monograph of British graptolites, Part 5. *Palaeontographical Society Monograph* **60** (288), lxxiii–xcvi, 181–216, pls 26–27.

Elles, G.L. and Wood, E.M.R. (1907) A monograph of British Graptolites. Part 6. *Palaeontographical Society Monograph* **61** (297), xcvii–cxx, 217–272.

Elles, G.L. and Wood, E.M.R. (1908) A monograph of British Graptolites. Part 7. *Palaeontographical Society Monograph* **62** (305), cxxxi–cxlviii, 273–358.

Elles, G.L. and Wood, E.M.R. (1911) A monograph of British Graptolites. Part 8. *Palaeontographical Society Monograph* **64** (316), 359–414, pls 36–41.

Elles, G.L. and Wood, E.M.R. (1914) A monograph of British Graptolites. Part 10. *Palaeontographical Society Monograph* **67** (327), 487–526, pls 50–52.

Emig, C.C. (1977) Sur une nouvelle espèce de *Cephalodiscus*, *C.* (*C.*) *caliciformis* n. sp. (Hemichordata, Pterobranchia), récoltée à Madagascar. *Bulletin du Museum d'Histoire Naturelle, Paris* **342**, 1077–1082.

Erdtmann, B.-D. (1976) Middle Silurian dendroid communities in the inter-reefs of the North American platform, in *Graptolites and Stratigraphy* (eds D. Kaljo and T.N. Koren). Academy of Sciences of the Estonian SSR, Institute of Geology, Tallinn, pp. 245–253.

Erdtmann, B.-D. (1982a) A reorganization and proposed phylogenetic classification of planktic Tremadoc (early Ordovician) dendroid graptolites. *Norsk Geologisk Tidsskrift* **62** (2), 121–145.

Erdtmann, B.-D. (1982b) Palaeobiogeography and environments of planktic dictyonemid graptolites during the earliest Ordovician. The Cambrian–Ordovician boundary: sections, fossil distributions, and correlations. *Cardiff, National Museum of Wales, Geological Series* **3**, 9–27.

Erdtmann, B.-D. (1986a) Von *Dictyonema* zu *Rhabdinopora*. *Geschiebekunde aktuell* **2**, 43–48.

Erdtmann, B.-D. (1986b) Comments on some earliest Ordovician (Salmien) graptolites from Solwaster, Massif de Stavelot, Belgian Ardennes. *Aardkundige Mededelingen* **3**, 75–88.

Erdtmann, B.-D. (1988) The earliest Ordovician nematophorid graptolites: taxonomy and correlation. *Geological Magazine* **125** (4), 327–348.

Erdtmann, B.-D. and VandenBerg, A.H.M. (1985) *Araneograptus* gen. nov. and its two species from the late Tremadocian (Lancefieldian, La2) of Victoria. *Alcheringa* **9** (1), 49–63.

Erdtmann, B.-D., Maletz, J. and Gutiérrez-Marco, J.C. (1987) The new Early Ordovician (Hunneberg stage) graptolite genus *Paradelograptus* (Kinnegraptidae), its phylogeny and biostratigraphy. *Paläontologische Zeitschrift* **61**, 109–131.

Finney, S.C. (1978) The affinities of *Isograptus*, *Glossograptus*, *Cryptograptus*, *Corynoides*, and allied graptolites. *Acta Palaeontologica Polonica* **23**, 481–495.

Finney, S.C. (1980) Thamnograptid, dichograptid and abrograptid graptolites from the Middle Ordovician Athens Shale of Alabama. *Journal of Paleontology* **54** (6), 1184–1208.

Finney, S.C. (1985) Nemagraptid graptolites from the Middle Ordovician Athens Shale, Alabama. *Journal of Paleontology* **59** (5), 1100–1137.

Finney, S.C. and Bergström, S.M. (1986) Biostratigraphy of the Ordovician *Nemagraptus gracilis* Zone. *Geological Society Special Publication* **20**, 47–59.

Finney, S.C. and Berry, W.B.N. (1997) New perspectives on graptolite distributions and their use as indicators of platform margin dynamics. *Geology* **25** (10), 919–922.

Finney, S.C. and Berry, W.B.N. (1998) An actualistic model of graptolite biogeography. *Temas Geológico-Mineros* **23**, 183–185.

Finney, S.C. and Berry, W.B.N. (2003) Ordovician to Devonian graptolite distributions along the Cordilleran margin of Laurentia. *INSUGEO, Serie Correlación Geológica* **18**, 27–32.

Flügel, H.W., Mostler, H. and Schönlaub, H.-P. (1993) Erinnerungen an Dozent Dr. rer. Nat. habil. Hermann Jaeger. *Jahrbuch der Geologischen Bundes-Anstalt* **136** (1), 13–17.

Foerste, A.F. (1923) Notes on Medinan, Niagaran, and Chester fossils. *Denison University Bulletin* **20**, 37–120, pls 4–15a

Fortey, R.A. (1971) *Tristichograptus*, a triserial graptolite from the Lower Ordovician of Spitsbergen. *Palaeontology* **14** (1), 188–199.

Fortey, R.A. (1993) Charles Lapworth and the biostratigraphic paradigm. *Journal of the Geological Society, London* **150**, 209–218.

Fortey, R.A. and Bell, A. (1987) Branching geometry and function of multiramous graptoloids. *Paleobiology* **13**, 1–19.

Fortey, R.A. and Cocks, L.R.M. (1986) Marginal faunal belts and their structural implications, with examples from the Lower Palaeozoic. *Journal of the Geological Society of London* **143**, 151–160.

Fortey, R.A. and Cooper, R.A. (1986) A phylogenetic classification of the graptoloids. *Palaeontology*, **29** (4), 631–654.

Fortey, R.A., Beckly, A.J. and Rushton, A.W.A. (1990) International correlation of the base of the Llanvirn Series Ordovician System. *Newsletters on Stratigraphy* **22** (2/3), 119–142.

Fortey, R.A., Zhang, Y.D. and Mellish, C. (2005) The relationships of biserial graptolites. *Palaeontology* **48** (6), 1241–1272.

Franke, D., Gründel, J., Lindert, W., Meissner, B., Schulz, E, Zagora, I. and Zagora, K. (1994) Die Ostseebohrung G 14 – eine Profilübersicht. *Zeitschrift für Geologische Wissenschaften* **22** (1/2), 235–240.

Frech, F. (1897) *Lethaea geognostica oder Beschreibung und Abbildung für die Gebirgs-Formationen bezeichnendsten Versteinerungen. 1. Teil – Lethaea Palaeozoica.* Schweizerbart'sche Verlagshandlung, Stuttgart.

Fu, H.-Y. (1982) Graptolithina, in *The Palaeontological Atlas of Hunan* (ed. Geological Bureau Of Hunan). *Geological Memoirs of the Ministry of Geology and Mineral Resources, People's Republic of China 2(1).* Geological Publishing House, Beijing, pp. 410–479 [in Chinese].

Fu, L.P. (1994) Four evolutionary lineages of cyrtograptids from late Llandovery to early Wenlock, in *Graptolite Research Today* (eds X. Chen, B.-D. Erdtmann and Y. Ni). Nanjing University Press, Nanjing, pp. 256–262, 2 pls.

Ganis, G.R. (2005) Darriwilian graptolites of the Hamburg succession (Dauphin Formation), Pennsylvania, and their geologic significance. *Canadian Journal of Earth Sciences* **42** (5), 791–813.

Ganis, G.R., Williams, S.H. and Repetski, J.E. (2001) New biostratigraphic information from the western part of the Hamburg klippe, Pennsylvania, and its significance for interpreting the depositional and tectonic history of the klippe. *GSA Bulletin* **113**, 109–128.

Gavryushkina, A., Welch, D., Stadler, T. and Drummond, A.J. (2014) Bayesian inference of sampled ancestor trees for epidemiology and fossil calibration. *PLoS Computational Biology* **10** (12), e1003919.

Ge, M.Y., Zheng, Z. and Li, Y. (1990) *Research of the Ordovician and Silurian Graptolites and Graptolite-Bearing Strata from Ningxia and the Neighbouring Districts.* Nanjing University Press, Nanjing [in Chinese].

Gegenbaur, C. (1870) *Grundzüge der vergleichenden Anatomie. Zweite, umgearbeitete Auflage.* Wilhelm Engelmann, Leipzig.

Geh, M.Y. (1964) Some species of *Tetragraptus* from the Ningkuo Shale (Lower Ordovician) of Zheijiang (Chekiang). *Acta Palaeontologia Sinica* **12** (3), 367–410.

Geinitz, H.B. (1842) Über Graptolithen. *Neues Jahrbuch Mineralogie, Geognosie, Geologie und Petrefakten-Kunde, Jahrgang 1842,* 697–701, pl. 10.

Geinitz, H.B. (1852) *Die Versteinerungen der Grauwackenformation in Sachsen und den angrenzenden Länder-Abtheilungen. Heft 1. Die Silurische Formation. Die Graptolithen, ein monographischer Versuch zur Beurtheilung der Grauwackenformation in Sachsen und den angrenzenden Länderabtheilungen sowie der Silurischen Formation überhaupt.* Verlag von Wilhelm Engelmann, Leipzig.

Geinitz, H.B. (1890) Die Graptolithen des K. Mineralogischen Museums in Dresden. *Mitteilungen des k. Mineralogischen-Geologischen und naturhistorischen Museums in Dresden,* 11–33.

Gerhart, J., Lowe, C. and Kirschner, M. (2005) Hemichordates and the origin of chordates. *Current Opinion in Genetics and Development* **15**, 461–467.

Gilchrist, J.D.F. (1908) New forms of the Hemichordata from South Africa. *Transactions of the South African Philosophical Society* **17** (1), 151–176.

Gilchrist, J.D.F. (1923) A form of dimorphism and asexual reproduction in *Ptychodera capensis* (Hemichordata). *Journal of the Linnean Society of London, Zoology* **35** (236), 393–398.

Gingerich, P.D. (1979) The stratophenetic approach to phylogeny reconstruction in vertebrate paleontology, in *Phylogenetic Analysis and Paleontology* (eds J. Cracraft and N. Eldredge). Columbia University Press, New York. pp. 41–77.

Gingerich, P.D. (1990) 5.2.4. Stratophenetics, in *Palaeobiology, a synthesis* (eds D.E.G. Briggs and P.R. Crowther). Blackwell Scientific Publications, Oxford, pp. 437–442.

Goldman, D. (1995) Taxonomy, evolution, and biostratigraphy of the *Orthograptus quadrimucronatus* species group (Ordovician, Graptolithina). *Journal of Paleontology* **69** (3), 516–540.

Goldman, D., Campbell, S.M. and Rahl, J.M. (2002) Three-dimensionally preserved specimens of *Amplexograptus* (Ordovician, Graptolithina) from the North American mid-continent: taxonomic and biostratigraphic significance. *Journal of Paleontology* **76** (5), 921–927.

Goldman, D., Mitchell, C.E., Maletz, J., Riva, J.F., Leslie, S.A. and Motz, G. (2007) Ordovician graptolites and conodonts of the Phi Kappa Formation in the Trail Creek Region of Central Idaho: a revised, integrated biostratigraphy. *Acta Palaeontologica Sinica* Suppl. **46**, 155–162.

Goldman, D., Mitchell, C.E., Melchin, M.J., Fan, J., Wu, S.-Y. and Sheets, H.D. (2011) Biogeography and mass extinction: extirpation and re-invasion of *Normalograptus* species (Graptolithina) in the late Ordovician palaeotropics. *Proceedings of the Yorkshire Geological Society* **58** (4), 227–246.

Goldman, D., Maletz, J., Melchin, M.J. and Fan, J. (2013) Graptolite biogeography. *Geological Society, London, Memoir* **38**, 415–428.

Goodarzi, F. (1990) Graptolite reflectance and thermal maturity of lower Paleozoic rocks, in *Applications of Thermal Maturity Studies to Energy Exploration* (eds V.F. Nuccio, C.E. Barker and S.J. Dyson). Eastwood Print. and Publ., Denver, CO, pp. 19–22.

Goodarzi, F. and Norford, B.S. (1985) Graptolites as indicators of the temperature histories of rocks. *Journal of the Geological Society of London* **142** (6), 1089–1099.

Goodarzi, F. and Norford, B.S. (1989) Variation of graptolite reflectance with depth of burial. *International Journal of Coal Geology* **11**, 127–141.

Göppert, H.R. (1860) Ueber die fossile Flora der Silurischen, der Devonischen und Unteren Kohlenformation oder des sogenannten Uebergangsgebirges. *Nova Acta Academiae Caesareae Leopoldino Carolinae Germanicae Naturae Curiosorum* **27** (8), 1–179.

Gortani, M. (1922) Faune Paleozoiche della Sardegna. I. Le graptoliti di Goni. *Palaeontographia Italica* **28**, 51–67, pls 8–13.

Gradstein, F.M., Ogg, J.G., Schmitz, M.D. and Ogg, G.M. (2012) *The Geologic Time Scale 2012*. Elsevier, Oxford.

Gravier, M.C. (1912) Sur une espèce nouvelle de *Cephalodiscus* (*C. anderssoni* n. sp.) provenant de la deuxième Expédition Antarctique Française. *Bulletin du Museum d'Histoire Naturelle, Paris* **18** (3), 146–150.

Greiling, L. (1958) Graptolithen-Erhaltung in "weißer Kieselsäure". *Senckenbergiana lethaea* **39** (3/4), 289–299.

Gümbel, C.W. (1868) Ueber den Pyrophyllit als Versteinerungsmittel. *Sitzungsberichte der königlich bayerischen Akademie der Wissenschaften zu München, Jahrgang 1868*, **1**, 498–502.

Gümbel, C.W. (1878) Einige Bemerkungen über Graptolithen (Mittheilungen an Professor H.B. Geinitz, 21 Jan. 1878). *Neues Jahrbuch für Mineralogie, Geologie und Palaeontologie, Jahrgang* **1878**, 292–296.

Gupta, N.S. (2014) Chapter 9. Molecular preservation in graptolites, in *Biopolymers: A Molecular Paleontology Approach* (ed. N.S. Gupta). Springer Science+Business Media Dordrecht, The Netherlands: Topics in Geobiology 38, pp. 147–156. doi:10.1007/978-94-007-7936-5_9

Gupta, N.S., Briggs, D.E.G. and Pancost, R.D. (2006) Molecular taphonomy of graptolites. *Journal of the Geological Society* **163**, 897–900.

Gürich, G. (1908) *Leitfossilien. Ein Hilfsbuch zum Bestimmen von Versteinerungen bei geologischen Arbeiten in der Sammlung und im Felde. Erste Lieferung: Kambrium und Silur.* Verlag Gebrüder Bornträger, Berlin.

Gurley, R.R. (1896) North American graptolites, new species and vertical ranges. *Journal of Geology* **4**, 63–102, 291–311.

Gutiérrez-Marco, J.C., Lenz, A.C., Robardet, M. and Piçarra, J.M. (1996) Wenlock–Ludlow graptolite biostratigraphy and extinction: a reassessment from the southwestern Iberian Peninsula (Spain and Portugal). *Canadian Journal of Earth Sciences* **33** (5), 656–663.

Haberfelner, E. (1933) Muscle-scars of Monograptidae. *American Journal of Science* **148** (25), 298–302.

Hadding, A. (1913) Undre Dicellograptus skiffern i Scane. *Lunds Universitets Arsskrift, Nya följe, Afd.* **2**, 9 (15), 1–91.

Hadding, A. (1915) Om *Glossograptus, Cryptograptus* och tvenne dem närstaende graptolitsläkten. *Geologiska Föreningens i Stockholm Förhandlingar* **37**, 303–336.

Haeckel, E. (1866) *Generelle Morphologie der Organismen. Allgemeine Grundzüge der organischen Formen-Wissenschaft, mechanisch begründet durch die von Charles Darwin reformirte Descendenz-Theorie. Erster Band: Allgemeine Anatomie der Organismen.* Verlag Georg Reimer, Berlin.

Haeckel, E. (1868) *Natürliche Schöpfungsgeschichte. Gemeinverständliche wissenschaftliche Vorträge über die Entwicklungslehre im Allgemeinen und diejenige von Darwin, Goethe und Lamarck im Besonderen, über die Anwendung derselben auf den Ursprung des Menschen und andere damit zusammenhängende Grundfragen der Naturwissenschaft.* Verlag Georg Reimer, Berlin.

Hagadorn, J.W. and Belt, E.S. (2008) Stranded in upstate New York: Cambrian medusae from the Potsdam Sandstone. *Palaios* **23**, 424–441.

Hagadorn, J.W., Dott Jr., R.H. and Damrow, D. (2002) Stranded on a Late Cambrian shoreline: medusae from central Wisconsin. *Geology* **30**, 147–150.

Halanych, K.M. (1993) Suspension feeding by the lophophore-like apparatus of the pterobranch hemichordate *Rhabdopleura normani*. *Biological Bulletin* **185**, 417–427.

Halanych, K.M. (1995) The phylogenetic position of the pterobranch hemichordates based on 18S rDNA sequence data. *Molecular Phylogenetics and Evolution* **4**, 72–76.

Halanych, K.M., Cannon, J.T., Mahon, A.R., Swalla, B.J. and Smith, C.R. (2013) Modern Antarctic acorn worms form tubes. *Nature Communications* **4**, 2738. doi:10.1038/ncomms3738.

Hall, J. (1847) *Paleontology of New York, Vol. 1, Containing Descriptions of the Organic Remains of the Lower Division of the New York System (Equivalent of the Lower Silurian Rocks of Europe)*, C. Van Benthuysen Publishers.

Hall, J. (1850) On graptolites, their duration in geological periods, and their value in the identification of strata. *Proceedings of the American Association for the Advancement of Science. Second Meeting, held at Cambridge, August 1849*, 351–352.

Hall, J. (1851) New genera of fossil corals from the report by James Hall, on the Palaeontology of New York. *American Journal of Science and Arts, 2nd Series* **11**, 398–401.

Hall, J. (1865) *Figures and Descriptions of Canadian Organic Remains. Decade II, Graptolites of the Quebec Group*. Dawson Brothers, Montreal.

Hall, T.S. (1892) On a new species of *Dictyonema*. *Proceedings of the Royal Society of Victoria 4*, 7–8, pls 1–2.

Hall, T.S. (1895) The geology of Castlemaine, with a subdivision of part of the Lower Silurian rocks of Victoria and a list of minerals. *Proceedings of the Royal Society of Victoria 7*, 55–58.

Hall, T.S. (1897) Victorian graptolites. Part 1. (a) Ordovician from Matlock. (b) *Dictyonema macgillivrayi*, nom. mut. *Proceedings of the Royal Society of Victoria (New Series)* **10** (1), 13–16.

Hall, T.S. (1899) Victorian graptolites, Part II, The graptolites of the Lancefield beds. *Proceedings of the Royal Society of Victoria (New Series)* **11**, 164–178.

Hall, T.S. (1902) The graptolites of New South Wales in the collection of the Geological Survey. *Records of the Geological Survey of New South Wales 7*, 49–59.

Hall, T.S. (1907) Reports on graptolites. *Records of the Geological Survey of Victoria 2*, 137–143.

Han, N.-R. and Chen, X. (1994) Regeneration in *Cardiograptus*. *Lethaia* **27**(2), 117–118.

Harkness, R. (1851) Description of the graptolites found in the Black Shales of Dumfriesshire. *Quarterly Journal of the Geological Society of London 7*, 58–65, pl. 1.

Harmer, S.F. (1905) The Pterobranchia of the Siboga-Expedition with an account of other species. *Siboga Expedition Monograph* **26**, 1–133.

Harris, W.J. (1933) *Isograptus caduceus* and its allies in Victoria. *Proceedings of the Royal Society of Victoria, New Series* **46**, 79–114.

Harris, W.J. and Keble, R.A. (1928) The *Staurograptus* bed of Victoria. *Proceedings of the Royal Society of Victoria* **40**, 91–95.

Harris, W.J. and Thomas, D.E. (1935) Victorian graptolites (New Series), Part III. *Proceedings of the Royal Society of Victoria* **47** (N.S.), 288–313.

Harris, W.J. and Thomas, D.E. (1938a) Victorian graptolites (New Series), Part V. *Mining and Geological Journal of Victoria* **1** (2), 70–81.

Harris, W.J. and Thomas, D.E. (1938b) A revised classification and correlation of the Ordovician graptolite beds of Victoria. *Mining and Geological Journal of Victoria* **1** (3), 62–72.

Harris, W.J. and Thomas, D.E. (1939) Victorian graptolites (New Series) Part VI, Some multiramous forms. *Mining and Geological Journal of Victoria* **2** (1), 55–60.

Harris, W.J. and Thomas, D.E. (1942) Victorian graptolites (New Series) Part 10, *Clonograptus pervelatus* sp. nov. and *Goniograptus macer*, T.S. Hall and some related forms. *Mining and Geological Journal of Victoria* **2**, 365–366.

Hart, M.W., Miller, R.L., Madin, L.P. (1994) Form and feeding mechanism of a living *Planctosphaera pelagica* (phylum Hemichordata). *Marine Biology* **120**, 521–533.

Harvey, T.H.P., Ortega-Hernández, J., Lin, J.-P., Zhao, Y. and Butterfield, N.J. (2012) Burgess Shale-type microfossils from the middle Cambrian Kaili Formation, Guizhou Province, China. *Acta Palaeontologica Polonica* **57** (2), 423–436.

Haupt, K. (1878) Die Fauna des Graptolithengesteines. *Neues Lausitzer Magazin* **54**, 29–113.

Hegarty, K.A., Birch, W.D., Douglas, J.G., McCann, D., Archbold, N.W., Gleadow, A.J.W., VandenBerg, A.H.M. and Phillips, G.N. (2003) Inspired observations: Examples of how Victorian geology has advanced the Earth Sciences worldwide. *Geological Society of Australia Special Publication* **23**, 687–700.

Heidenhain, F. (1869) Ueber Graptolithen führende Diluvial-Geschiebe der norddeutschen Ebene. *Zeitschrift der Deutschen Geologischen Gesellschaft* **21**, 143–182.

Hennig, W. (1950) *Grundzüge einer Theorie der phylogenetischen Systematik.* Deutscher Zentral-Verlag, Berlin.

Hennig, W. (1965) Phylogenetic systematics. *Annual Review of Entomology* **10**, 97–116.

Hennig, W. (1989) *Phylogenetic Systematics.* University of Illinois Press, Urbane and Chicago.

Herrmann, M.O. (1882) Vorläufige Mittheilung über eine neue Graptolithenart und mehrere noch nicht aus Norwegen gekannte Graptolithen. *Nyt Magasin for Naturvidenskap* **27**, 341–362.

Herrmann, M.O. (1885) Die Graptolithenfamilie Dichograptidae, Lapw., mit besonderer Berücksichtigung von Arten aus dem norwegischen Silur. Inaugural-Dissertation zur Erlangung der philosophischen Doctorwürde an der Universität Leipzig. Kristiania 1885, 1–94 [same in: Herrmann, M.O. (1885) Die Graptolithenfamilie Dichograptidae Lapworth. *Nyt Magazin for Naturvidenskaberne* **29**, 124–214].

Hewitt, R.A. and Birkler, I. (1986) The *Thallograptus* and *Diplospirograptus* from the Silurian Eramosa Member in Hamilton (Ontario, Canada). *Canadian Journal of Earth Sciences* **23**, 849–853.

Hills, E.S. and Thomas, D.E. (1954) Turbidity currents and the graptolitic facies in Victoria. *Journal of the Geological Society of Australia* **1**, 119–133.

Hincks, T. (1880) *History of the British Marine Polyzoa.* Van Voorst, London.

Hind, W. (1907) On the occurrence of dendroid graptolites in British Carboniferous rocks. *Proceedings of the Yorkshire Geological Society* **16**, 155–157.

Hisinger, W. von (1837–1841) *Lethæa suecica seu petrificata Sueciæ iconibus et characteribus illustrata* (med två supplement, 1837–41). Norstedat et Filii.

Hoffknecht, A. (1991) Mikropetrographische, organisch-geochemische, mikrothermometrische und mincralogische Untersuchungen zur Bestimmung der organischen Reife von Graptolithen-Periderm. *Göttinger Arbeiten zur Geologie und Paläontologie* **48**, 1–98.

Hoffmann, N., Jödicke, H., Fluche, B., Jording, A. and Müller, W. (1998) Modellvorstellungen zur Verbreitung potentieller präwestfälischer Erdgas-Muttergesteine in Norddeutschland. *Zeitschrift für Angewandte Geologie* **44**, 140–158.

Holland, C.H. (1985) Series and Stages of the Silurian System. *Episodes* **8** (2), 101–103.

Holm, G. (1881) Bidrag till kännedomen om Skandinaviens graptoliter I. *Pterograptus*, ett nytt graptolitslägte. *Öfversigt af Konglika Vetenskaps-Akademiens Förhandlingar* 1881 (4), 71–84.

Holm, G. (1890) Gotlands Graptoliter. *Svenska Vetenskaps.-Academiens Handlingar, Bihang* **16**, 1–34.

Holm, G. (1895) Om *Didymograptus, Tetragraptus* och *Phyllograptus*. *Geologiska Föreningens i Stockholm Förhandlingar* **17**, 319–359. [English translation: Holm, G. (1895) On *Didymograptus, Tetragraptus* and *Phyllograptus*. *Geological Magazine* **11**, 433–441, 481–492.]

Hopkinson, J. (1869) On British graptolites. *Journal of the Quekett Microscopical Club* **1**, 151–166, pl. 158.

Hopkinson, J. (1870) On the structure and affinities of the genus *Dicranograptus*. *Geological Magazine* **7**, 353–359, pl. 16.

Hopkinson, J. (1872) On some new species of graptolites from the South of Scotland. *Geological Magazine* **9**, 501–509.

Hopkinson, J. and Lapworth, C. (1875) Descriptions of the graptolites of the Arenig and Llandeilo rocks of St. David's. *Quarterly Journal of the Geological Society* **31**, 631–672, pls 33–37.

Horstig, G. von (1952) Neue Graptolithen-Funde in gotlandischen Lyditen des Frankenwaldes und ihre Erhaltung in weißer Kieselsäure. *Senckenbergiana* **33** (4/6), 345–351.

Hou, X.-G., Aldridge, R., Bergström, J., Siveter, D.J., Siveter, D. and Feng, X.-H. (2004) *The Cambrian Fossils of Chengjiang, China: The Flowering of Early Animal Life.* John Wiley & Sons Ltd, London.

Hou, X.-G., Aldridge, R.J., Siveter, D.J., Siveter, D.J., Williams, M., Zalasiewicz, J. and Ma, X.-Y. (2011) An Early Cambrian hemichordate zooid. *Current Biology* **21**, 1–5.

Howe, M.P.A. (1983) Measurements of thecal spacing in graptolites. *Geological Magazine* **120** (6), 635–638.

Hsü, S.C. (1934) The graptolites of the Lower Yangtze Valley. *Academia Sinica. Monograph of the National Research Institute of Geology* **A4**, 1–106.

Hsü, S.C. (1959) A newly discovered *Climacograptus* with a particular basal appendage. *Acta Palaeontologica Sinica* **7** (5), 346–352.

Hsü, S.C. and Ma, C.T. (1948) The Ichang formation and the Ichangian fauna. *Bulletin of the National Research Institute of Geology, Academia Sinica* **8**, 1–51.

Hu, S., Steiner, M., Zhu, M., Erdtmann, B.-D., Luo, H., Chen, L. and Weber, B. (2007) Diverse pelagic predators from the Chengjiang Lagerstätte and the establishment of modern-style pelagic ecosystems in the early Cambrian. *Palaeogeography, Palaeoclimatology, Palaeoecology* **254**, 307–316.

Hundt, R. (1924) *Die Graptolithen des deutschen Silurs.* Verlag Max Weg, Leipzig.

Hundt, R. (1934) Kieselsäureerhaltung bei obersilurischen Graptolithen aus dem gemengten Diluvium Mitteldeutschlands. *Zeitschrift für Geschiebeforschung* **10** (2), 101–104.

Hundt, R. (1935) Die Graptolithenfauna des obersten Obersilurs Thüringens (mit einem Beitrag über Graptolithen in Roteisensteinknollen). *Zeitschrift für Naturwissenschaften* **91** (1), 1–34.

Hundt, R. (1936) Neue Forschungen über mitteldeutsche Graptolithen (*Diplograptus, Retiolites*, Großkolonien). *Paläontologische Zeitschrift* **18** (1–2), 39–48.

Hundt, R. (1939) *Das Mitteldeutsche Graptolithenmeer.* Martin Boerner Verlag, Halle.

Hundt, R. (1946) In Kieselsäure erhaltene Graptolithen aus dem Ostthüringer Gotlandium. *Ostthüringer Geologie, Heft* **2**, 62–63. [Published by Geologischer Verein für Gera und Umgebung] Verlag Josef Wiroth, Gera.

Hundt, R. (1965) *Aus der Welt der Graptolithen.* Commercia Verlag, Berlin.

Huo, S.C., Fu, P.P. and Shu, D.G. (1986) A mathematical study of the *Cyrtograptus sakmaricus* lineage with discussions of the evolutionary trends in this lineage. *Geological Society Special Publication* **20**, 197–205.

Hutt, J.E. (1974a) The development of *Clonograptus tenellus* and *Adelograptus hunnebergensis*. *Lethaia* **7** (1), 79–92.

Hutt, J.E. (1974b) A new group of Llandovery biform monograptids. *Special Papers in Palaeontology* **13**, 189–203.

Hutt, J. and Rickards, R.B. (1970) The evolution of the earliest Llandovery monograptids. *Geological Magazine* **107** (1), 67–77.

Hutt, J.E., Rickards, R.B. and Skevington, D. (1970) Isolated Silurian graptolites from the Bollerup and Klubbudden Stages of Dalarna, Sweden. *Geologica et Palaeontologica* **4**, 1–23.

Hutt, J.E., Rickards, R.B. and Berry, W.B.N. (1972) An outline of the evolution of Silurian–early Devonian graptoloids. *International Geological Congress, Abstracts – Congres Geologique Internationale, Resumes* **24** (24), 226–227.

Hyman, L.H. (1959) Chapter 17. The enterocoelous coelomates – Phylum Hemichordata, in *The Invertebrates: Smaller Coelomate Groups, vol. 5.* McGraw-Hill Book Company, Inc., New York.

ICZN (1954) Opinion 198. Suppression under the plenary powers, of the generic names "*Lomatoceras*" Bronn, 1834, and "*Monoprion*" Barrande, 1850 (Class Graptolithina) and validation of the generic name "*Monograptus*" Geinitz, 1852. *Opinions and declarations rendered by the International Commission on Zoological Nomenclature* **3** (17), 217–228.

Jaanusson, V. (1960) Graptoloids from the Ontikan and Viruan (Ordov.) Limestones of Estonia and Sweden. *Bulletin of the Geological Institutions of the University of Uppsala* **38** (3–4), 289–366.

Jackson, D.E. (1964) Observations on the sequence and correlation of Lower and Middle Ordovician graptolite faunas of North America. *Bulletin of the Geological Society of America* **75** (6), 523–534.

Jackson, D.E. (1966) On the occurrence of *Parabrograptus* and *Sinograptus* from the Middle Ordovician of Western Canada. *Geological Magazine* **103** (3), 263–268.

Jackson, D.E. (1967) *Psigraptus*, a new graptolite genus from the Tremadocian of Yukon, Canada. *Geological Magazine* **104** (4), 317–321.

Jackson, D.E. (1974) Tremadoc graptolites from Yukon Territory, Canada. *Special Papers in Palaeontology* **13**, 35–58.

Jackson, D.E. and Lenz, A. (1999) Occurrences of *Psigraptus* and *Chigraptus* gen. nov. in the Tremadoc of the Yukon territory, Canada. *Geological Magazine* **136** (2), 153–157.

Jackson, D.E. and Lenz, A.C. (2000) Some graptolites from the late Tremadoc and early Arenig of Yukon, Canada. *Canadian Journal of Earth Sciences* **37**, 1177–1193.

Jackson, D.E. and Lenz, A.C. (2003) Taxonomic and biostratigraphical significance of the Tremadoc graptolite fauna from northern Yukon Territory, Canada. *Geological Magazine* **140** (2), 131–156.

Jackson, J.B.C. (1979) Morphological strategies of sessile animals, in *Biology and Systematics of Colonial Organisms*, Systematics Association Special Volume **11** (eds G. Larwood and B.R. Rosen). Academic Press, London, pp. 499–555.

Jackson, J.W. and Alkins, W.E. (1919) Obituary – John Hopkinson. *Geological Magazine (Decade VI)* **6** (9), 431–432.

Jaeger, H. (1959) Graptolithen und Stratigraphie des jüngsten Thüringer Silurs. *Abhandlungen der deutschen Akademie der Wissenschaften zu Berlin, Klasse für Chemie, Geologie und Biologie* **1959** (2), 1–197.

Jaeger, H. (1978) Late graptoloid faunas and the problem of graptoloid extinction. *Acta Palaeontologica Polonica* **23** (4), 497–521.

Jaeger, H. (1988) Devonian Graptoloidea, *in Devonian of the World. Vol. III. Proceedings of the Second International Symposium on the Devonian System* (eds. M.J. McMillan, A.F. Embry and D.J. Glass). Canadian Society of Petroleum Geologists, Calgary, Alberta, pp. 431–438.

Jaeger, H. (1991) Neue Standard-Graptolithenzonenfolge nach der "Großen Krise" an der Wenlock/Ludlow-Grenze (Silur). *Neues Jahrbuch für Geologie und Paläontologie, Abhandlungen* **182** (3), 303–354.

Jaekel, O. (1889) Über das Alter des sog. Graptolithengesteins mit besonderer Berücksichtigung der in demselben enthaltenen Graptolithen. *Zeitschrift der Deutschen Geologischen Gesellschaft* **41**, 653–716.

Jarochowska, E., Tonarová, P., Munnecke, A., Ferrová, L., Sklenář, J. and Vodrážková, S. (2013) An acid-free method of microfossil extraction from clay-rich lithologies using the surfactant Rewoquat. *Palaeontologia Electronica* **16** (3), 7T. 1–16.

Jenkins, C.J. (1980) *Maeandrograptus schmalenseei* and its bearing on the origin of the diplographtids. *Lethaia* **13** (4), 289–302.

Jenkins, C.J. (1982) *Isograptus gibberulus* (Nicholson) and the Isograptids of the Arenig Series (Ordovician) of England and Wales. *Proceedings of the Yorkshire Geological Society* **44** (11, Part 2), 219–248.

Jenkins, C.J. (1987) The Ordovician graptoloid *Didymograptus murchisoni* in South Wales and its use in three-dimensional absolute strain analysis. *Transactions of the Royal Society of Edinburgh: Earth Sciences* **78** (2), 105–114.

Jeppsson, L. (1998) Silurian oceanic events. Summary of general characteristics. *New York State Museum Bulletin* **491**, 239–257.

Jiao, X. [Qiao, X.] (1977) *Kalpinograptus*, a new graptolite genus from the Saergan Formation in Kalpin of Xinjiang. *Acta Palaeontologica Sinica* **16** (2), 287–292, pl. 1.

Jones, H., Zalasiewicz, J.A. and Rickards, R.B. (2002) Clingfilm preservation of spiraliform graptolites: Evidence of organically sealed Silurian seafloors. *Geology* **30**(4), 343–346.

Jones, W.D.V. and Rickards, R.B. (1967) *Diplograptus penna* Hopkinson, 1869, and its bearing on viesicular structures. *Paläontologische Zeitschrift* **41** (3/4), 173–185.

Kämpf, H., Bankwitz, P., Schauer, M. and Lange, G. (1995). Uranlagerstätten der Gera-Jachymov-Störungszone. *Zeitschrift für Geologische Wissenschaften* **23**, 516–808.

Kann, J., Raukas, A. and Siirde, A. (2013) About the gasification of kukersite oil shale. *Oil Shale* **30** (2S), 283–293.

Kayser, E. (1878) Die Fauna der ältesten Devon-Ablagerungen des Harzes Mit einem Atlas von

36 lithographischen Tafeln. *Abhandlungen zur geologischen Specialkarte von Preussen und den Thüringischen Staaten* **2** (4), 1–295.

Keble, R.A. and Benson, W.N. (1939) Graptolites of Australia: bibliography and history of research. *Memoir of the National Museum of Melbourne* **11**, 11–99.

Keble, R.A. and Harris, W.J. (1934) Graptolites of Victoria; New species and additional records. *Memoirs of the National Museum, Melbourne* **8**, 166–183.

Kendrick, P., Kvacek, Z. and Bengtson, S. (1999) Semblant land plants from the Middle Ordovician of the Prague Basin reinterpreted as animals. *Palaeontology* **42**, 991–1002.

Kim, J.Y., Cho, H.S. and Erdtmann, B.-D. (2006) *Psigraptus jacksoni* from the early Ordovician Mungok Formation, Yeongwol area, Korea: taxonomy and development. *Alcheringa* **30**, 11–22.

Kirk, N.H. (1969) Some thoughts on the ecology, mode of life and evolution of the Graptolithina. *Proceedings of the Geological Society of London* **1659**, 273–292.

Kirk, N.H. (1973) More thoughts on bithecae, budding and branching in the Graptolithina. *University College of Wales Aberystwyth Department of Geology Publications* **2** (2), 1–12.

Kirk, N.H. (1974) Some thoughts on convergence and divergence in the Graptolithina. *University College of Wales Aberystwyth Department of Geology Publications* **5**, 29 pp.

Kirk, N.H. (1978) Mode of life of graptolites. *Acta Palaeontologica Polonica* **23** (4), 533–555.

Kirk, N.H. (1979) Thoughts on coloniality in the Graptolithina. *Systematics Association Special Volume* **11**, 411–432.

Kirk, N.H. (1990) Juvenile sessility, vertical automobility, and passive lateral transport as factors in graptoloid evolution. *Modern Geology* **14** (3), 153–187.

Kistinger, S. (1999) Reclamation strategy at the Ronneburg uranium mining site before flooding the mine. *International Mine Water Association, Proceedings*, 299–303.

Klapper, G. and Furnish, W.M. (1962) Devonian-Mississippian Englewood Formation in Black Hills, South Dakota. *Bulletin of the American Association of Petroleum Geologists* **46** (11), 2071–2078.

Kobell, F. von (1870) Der Gümbelit, ein neues Mineral von Nordhalben bei Steben in Oberfranken. *Sitzungsberichte der königlich bayerischen Akademie der Wissenschaften zu München, Jahrgang 1870*, **1**, 294–296.

Kolata, D.R., Frost, J.K. and Huff, W.D. (1986) K-bentonites of the Ordovician Decorah Subgroup, upper Mississippi Valley: correlation by chemical fingerprinting. *Illinois State Geological Survey, Circular* **537**, 1–30.

Kolata, D.R., Huff, W.D. and Bergström, S.M. (1996) Ordovician K-bentonites of eastern North America. *Geological Society of America Special Paper* **313**, 1–84.

Komai, T. (1949) Internal structure of the pterobranch *Atubaria heterolopha* Sato, with an appendix on the homology of the notochord. *Proceedings of the Japan Academy* **25**, 19–24.

Koren, T.N. (1974) The phylogeny of some Lower Devonian monograptids. *Special Papers in Palaeontology* **13**, 249–260, pls 25–26.

Koren, T.N. (1979) Late monograptid faunas and the problem of graptolite extinction. *Acta Palaeontologica Polonica* **24**, 79–106.

Koren, T.N. (1987) Graptolite dynamics in Silurian and Devonian time. *Bulletin of the Geological Society of Denmark* **35**, 149–159.

Koren, T.N. (1991a) Evolutionary crisis of the Ashgill graptolites. *Geological Survey of Canada*, Paper **90–9**, 157–164.

Koren, T.N. (1991b) The *C. lundgreni* extinction event in Central Asia and its bearing on graptolite biochronology within the Homerian. *Proceedings of the Estonian Academy of Science, Geology* **40**, 74–78.

Koren, T.N. (1992) Novye. Pozdnevenlokskio Monograpti Alaiskogo reta. *Paleontological Zhurnal* **2**, 21–33.

Koren, T.N. (1993) Osnovnye rubezhi v evolyutsii ludlovskikh graptolitov. *Stratigrafia. Geologicheskaya Korrelyatsiya* **1**, 44–52.

Koren, T.N. (1994a) Diversity dynamics of graptolites at the Wenlock–Ludlow boundary. *Stratigraphy. Geological Correlation* **2** (1), 39–45 [in Russian].

Koren, T.N. (1994b) The Homerian monograptid fauna of central Asia: zonation, morphology and phylogeny, in *Graptolite Research Today* (eds X. Chen, B.-D. Erdtmann and Y. Ni). Nanjing University Press, Nanjing, pp. 140–148.

Koren, T. and Bjerreskov, M. (1997) Early Llandovery monograptids from Bornholm and the southern Urals: taxonomy and evolution. *Bulletin of the Geological Society of Denmark* **44**, 1–43.

Koren, T.N. and Rickards, R.B. (1996) Taxonomy and evolution of Llandovery biserial graptolites from the southern Urals, western Kazakhstan. *Special Papers in Palaeontology* **54**, 1–103.

Koren, T.N. and Rickards, R.B. (2004) An unusually diverse Llandovery (Silurian) diplograptid fauna from the southern Urals of Russia and its evolutionary significance. *Palaeontology* **47**, 859–918.

Koren, T.N. and Suyarkova, A.A. (1997) Late Ludlow and Pridoli monograptids from the Turkestan-Alai Mountains, South Tien-Shan. *Palaeontographica* **A247**, 59–90.

Koren, T.N. and Suyarkova, A.A. (1998) Specialized thecal structures in some Ludlow monograptids, Upper Silurian, Central Asia. *Temas Geológico-Mineros ITGE* **23**, 198–201.

Koren, T.N. and Suyarkova, A.A. (2004) The Ludlow (Late Silurian) neocucullograptid fauna from the southern Tien Shan, Kyrghizstan. *Alcheringa* **28** (2), 333–387.

Koren, T.N. and Urbanek, A. (1994) Adaptive radiation of monograptids after the late Wenlock crisis. *Acta Palaeontologica Polonica* **39** (2), 137–167.

Koren, T.N., Lenz, A.C., Loydell, D.K., Melchin, M.J., Štorch, P. and Teller, L.A. (1996) Silurian standard graptolite zonation. *Lethaia* **29**, 59–60.

Kozłowska, A. (2015) Evolutionary history of the *Gothograptus* lineage of the Retiolitidae (Graptolithina). *Estonian Journal of Earth Sciences* **64** (1), 56–61.

Kozłowska, A. and Bates, D.E.B. (2014) The simplest retiolitid (Graptolithina) species *Plectodinemagraptus gracilis* from the Ludlow of Poland. *GFF* **136** (1), 147–152.

Kozłowska, A., Lenz, A.C. and Melchin, M.J. (2009) Evolution of the retiolitid *Neogothograptus* (Graptolithina) and its new species from the Upper Wenlock of Poland, Baltica. *Acta Palaeontologica Polonica* **54** (3), 423–434.

Kozłowska, A., Dobrowolska, K. and Bates, D.E.B. (2013) A new type of colony in Silurian (upper Wenlock) retiolitid graptolite *Spinograptus* from Poland. *Acta Palaeontolgica Polonica* **58** (1), 85–92.

Kozłowska-Dawidziuk, A. (1990) The genus *Gothograptus* (Graptolithina) from the Wenlock of Poland. *Acta Palaeontologica Polonica* **35** (3–4), 191–209.

Kozłowska-Dawidziuk, A. (1995) Silurian retiolitids of the East European Platform. *Acta Palaeontologica Polonica* **40** (3), 261–326.

Kozłowska-Dawidziuk, A. (1997) Retiolitid graptolite *Spinograptus* from Poland and its membrane structures. *Acta Palaeontologica Polonica* **42** (3), 391–412.

Kozłowska-Dawidziuk, A. (2001) Phylogenetic relationships within the Retiolitidae (Graptolithina) and a new genus, *Cometograptus*. *Lethaia* **34**, 84–96.

Kozłowska-Dawidziuk, A. (2004) Evolution of retiolitid graptolites – a synopsis. *Acta Palaeotologica Polonica* **49**, 505–518.

Kozłowska-Dawidziuk, A. and Lenz, A.C. (2001) Evolutionary developments in the Silurian Retiolitidae (Graptolites). *Journal of the Czech Geological Society* **46** (3–4), 227–238.

Kozłowska-Dawidziuk, A., Lenz, A.C., Bates, D.E.B. (2003) A new classification of ancorate diplograptids. *INSUGEO, Serie Correlación Geológica* **18**, 49–53.

Kozłowski, R. (1938) Informations preliminaires sur les Graptolithes du Tremadoc de la Pologne et sur leur portee theorique. *Annales Musei Zoologici Polonici* **13** (16), 183–196.

Kozłowski, R. (1947) Les affinites des Graptolithes. *Biological Reviews* **22** (2), 93–108.

Kozłowski, R. (1949) Les graptolithes et quelques nouveaux groups d'animaux du Tremadoc de la Pologne. *Palaeontologia Polonica* **3**, 1–235.

Kozłowski, R. (1951) [often cited as 1952] O niezwyklym graptolicie ordowickim. *Acta Geologica Polonica* **2** (3), 291–299. [French translation: Sur un remarquable graptolite ordovicien. *Acta Geologica Polonica* **2**, 86–93.]

Kozłowski, R. (1953) Badania nad nowym gatunkiem z rodzaju *Corynoides* (Graptolithina). *Acta Geologica Polonica* **3**, 193–209.

Kozłowski, R. (1954) Sur la structure de certain Dichograptids. *Acta Geologica Polonica* **4**, 118–135.

Kozłowski, R. (1960) *Calyxdendrum graptoloides* n. gen., n. sp. – a graptolite intermediate between the Dendroidea and the Graptoloidea. *Acta Palaeontologica Polonica* **5** (2), 107–124.

Kozłowski, R. (1962) Crustoidea – nouveau groupe de graptolites. *Acta Palaeontologica Polonica* **7** (1–2), 3–52, pls 1–4.

Kozłowski, R. (1963) Le développement d'un graptolite tuboïde. *Acta Palaeontologica Polonica* **8** (2), 103–134.

Kozłowski, R. (1965) Oeufs fossiles des cephalopodes? *Acta Palaeontologica Polonica* **10** (1), 3–9.

Kozłowski, R. (1967) Sur certains fossiles ordoviciens à test organique. Acta *Palaoentologica Polonica* **12** (2), 99–132.

Kozłowski, R. (1970) Tubotheca – a peculiar morphological element in some graptolites. *Acta Palaeontologica Polonica* **15** (4), 393–410.

Kozłowski, R. (1971) Early development stages and the mode of life of graptolites. *Acta Palaeontologica Polonica* **16** (4), 313–342.

Kraatz, R. (1958) Stratigraphische und paläontologische Untersuchungen (besonders im Gotlandium) im Gebiet zwischen Wieda und Zorge (südl. West-Harz). *Zeitschrift der Deutschen Geologischen Gesellschaft* **110**, 22–70.

Kraft, P. (1926) Ontogenetische Entwicklung und Biologie von *Diplograptus* und *Monograptus*. *Paläontologische Zeitschrift* **7**, 207–249.

Kraft, P. (1932) Neue optische Wege in der Mikrophotographie und Mikroskopie im Dienste der Geologie und Paläontologie. *Zeitschrift der Deutschen Geologischen Gesellschaft* **84**, 651–652.

Kraft, P. and Kraft, J. (2003) Middle Ordovician graptolite fauna from Praha – Červený vrch (Prague Basin, Czech Republic). *Bulletin of Geosciences* **78** (2), 129–139.

Kraft, P. and Kraft, J. (2007) Planktic dendroids – subversives of graptolite taxonomy. *Wissenschaftliche Mitteilungen, Institut für Geologie, Freiberg* **36**, 72–73.

Kriz, J., Jaeger, H., Paris, F. and Schönlaub, H.P. (1986) Pridoli – the fourth subdivision of the Silurian. *Jahrbuch der geologischen Bundesanstalt, Wien* **129** (2), 291–360.

Kühne, W.G. (1955) Unterludlow-Graptolithen aus Berliner Geschieben. *Neues Jahrbuch für Geologie und Paläontologie. Abhandlungen* **100** (3), 350–401.

Kurck, C. (1882) Några nya graptolitarter från Skåne. *Geologiska Föreningens i Stockholm Förhandlingar* **6**, 294–304, pl. 14.

Lacépède, B.G.E. (1801) *Histoire Naturelle des Poissons, 3*. Plassan, Paris.

Lankester, E.R. (1877) Notes on the embryology and classification of the animal kingdom; comprising a revision of speculations relative to the origin and significance of the germlayers. *Quarterly Journal of Microscopical Science, N.S.* **17**, 339–454, pl. 25.

Lankester, E.R. (1884) A contribution to the knowledge of *Rhabdopleura*. *Quarterly Journal of Microscopical Science* **24**, 622–647, pls 37–41.

Lapworth, C. (1873a) Notes on the British graptolites and their allies. 1. On an improved classification of the Rhabdophora, part 1. *Geological Magazine* **10**, 500–504.

Lapworth, C. (1873b) Notes on the British graptolites and their allies. 1. On an improved classification of the Rhabdophora, part 1. *Geological Magazine* **10**, 555–560.

Lapworth, C. (1876) The Silurian System in the South of Scotland, in *Catalogue of the Western Scottish Fossils* (eds J. Armstrong, J. Young and D. Robertson). Blackie & Son, Glasgow, pp. 1–9.

Lapworth, C. (1877) The graptolites of County Down. *Proceedings of the Belfast Naturalists' Field Club (Appendix)* 1876–77, 125–144.

Lapworth, C. (1878) The Moffat Series. *Quarterly Journal of the Geological Society of London* **34**, 240–343, pls 11–13.

Lapworth, C. (1879) On the tripartite classification of the Lower Palaeozoic rocks. *Geological Magazine; New Series, Decade II*, **6**, 1–15.

Lapworth, C. (1880a) On new British graptolites. *Annals and Magazine of Natural History* **5** (5), 149–177.

Lapworth, C. (1880b) On the geological distribution of the Rhabdophora. Part III. Results (continued from vol. VI, p. 29). *Annals and Magazine of Natural History* **5** (6), 185–207.

Lapworth, C. (1897) in Walter, J. 1897. Die Lebensweise der Graptolithen. *Zeitschrift der Deutschen Geologischen Gesellschaft* **49**, 238–258.

Lee, C.K. (1961) Graptolites from the Dawan Formation (Lower Ordovician) of W. Hupeh and S. Kueichou. *Acta Palaeontologica Sinica* **9**, 48–79.

Legg, D.P. (1976) Ordovician trilobites and graptolites from the Canning Basin, Western Australia. *Geologica et Palaeontologica* **10**, 1–58.

Legrand, P. (1964) Deux nouvelles espèces du genre *Adelograptus* (Graptolites) dans L'Ordovicien inférieur du Sahara algérien. *Bulletin de la Societé Géologique de France, Compte Rendu Sommaire* **7** (6), 295–304.

Legrand, P. (1974) Development of rhabdosomes with four primary branches in the group *Dictyonema flabelliforme* (Eichwald). *Special Papers in Palaeontology* **13**, 19–34.

Lenz, A.C. (1974) A membrane-bearing *Cyrtograptus* and an interpretation of the hydrodynamics of cyrtograptids. *Special Papers in Palaeontology* **13**, 205–214.

Lenz, A.C. (1977) Some Pacific Faunal Province graptolites from the Ordovician of northern Yukon, Canada. *Canadian Journal of Earth Sciences* **14**, 1946–1952.

Lenz, A.C. (1988a) Revision of Upper Silurian and Lower Devonian graptolite biostratigraphy and morphological variation in *Monograptus yukonensis* and related Devonian graptolites, northern Yukon, Canada, in *Devonian of the World, vol. III* (eds J. McMillan, A.F. Embry and D.J. Glass). *N. Canadian Society of Petroleum Geology Memoir* **14**, 439–447.

Lenz, A.C. (1988b) Upper Silurian and Lower Devonian graptolites and graptolite biostratigraphy, northern Yukon, Canada. *Canadian Journal of Earth Sciences* **25** (3), 355–369.

Lenz, A.C. (1990) Ludlow and Pridoli (Upper Silurian) graptolite biostratigraphy of the central Arctic Islands: a preliminary report. *Canadian Journal of Earth Sciences* **27** (8), 1074–1083.

Lenz, A.C. (1993) Late Wenlock and Ludlow (Silurian) Plectograptinae (retiolitid graptolites), Cape Phillips Formation, Arctic Canada. *Bulletin of American Paleontology* **104** (342), 1–52.

Lenz, A.C. (1994a) Uppermost Wenlock and Lower Ludlow plectograptine graptolites, Arctic Islands, Canada: new isolated material. *Journal of Paleontology* **68** (4), 851–860.

Lenz, A.C. (1994b) A sclerotized retiolitid, and its bearing on the origin and evolution of Silurian retiolitid graptolites. *Journal of Paleontology* **68** (6), 1344–1349.

Lenz, A.C. (1994c) The graptolites *Pristiograptus praedeubeli* (Jaeger) and *Pristiograptus ludensis* (Murchison) (uppermost Wenlock, Silurina) from Arctic Canada: taxonomy and evolution. *Canadian Journal of Earth Sciences* **31**, 1419–1426.

Lenz, A.C. (1995) Upper Homerian (Wenlock, Silurian) graptolites and graptolite biostratigraphy, Arctic Archipelago, Canada. *Canadian Journal of Earth Sciences* **32** (9), 1378–1392.

Lenz, A.C. (2013) Early Devonian graptolites and graptolite biostratigraphy, Arctic Islands, Canada. *Canadian Journal of Earth Sciences* **50**, 1097–1115.

Lenz, A.C. and Jackson, D.E. (1986) Arenig and Llanvirn graptolite biostratigraphy, Canadian Cordillera. *Geological Society Special Publication* **20**, 27–45.

Lenz, A.C. and Kozłowska, A. (2006) Graptolites from the *lundgreni* biozone (Lower Homerian: Silurian), Arctic Islands, Canada: new species and supplementary material. *Journal of Paleontology* **80** (4), 616–637.

Lenz, A.C. and Kozłowska-Dawidziuk, A. (1998) Sicular annuli and thickened interthecal septa in Silurian graptolites: new information. *Temas Geológico-Mineros ITGE* **23**, 212–214.

Lenz, A.C. and Kozłowska-Dawidziuk, A. (2001) Upper Wenlock (Silurian) graptolites of Arctic Canada: pre-extinction, *lundgreni* Biozone fauna. *Palaeontographica Canadiana* **20**, 1–21.

Lenz, A.C. and Kozłowska-Dawidziuk, A. (2004) *Ludlow and Pridoli (Upper Silurian) graptolites from the Arctic Islands, Canada*. NRC Research Press, Ottawa, Ontario, Canada.

Lenz, A.C. and Melchin, M.J. (1987a) Peridermal and interthecal tissue in Silurian retiolitid graptolites: with examples from Sweden and Arctic Canada. *Lethaia* **20** (4), 353–359.

Lenz, A.C. and Melchin, M.J. (1987b) Silurian retiolitids from the Cape Phillips Formation, Arctic Islands, Canada. *Bulletin of the Geological Society of Denmark* **35**, 161–170.

Lenz, A.C. and Melchin, M.J. (1989) *Monograptus spiralis* and its phylogenetic relationship to early cyrtograptids. *Journal of Paleontology* **63** (3), 341–348.

Lenz, A.C. and Melchin, M.J. (1991) Wenlock (Silurian) graptolites, Cape Phillips Formation, Canadian Arctic Islands. *Transactions of the*

Royal Society of Edinburgh: Earth Sciences **82** (3), 211–237.

Lenz, A.C. and Melchin M.J. (1997) Phylogenetic analysis of the Silurian Retiolitidae. *Lethaia* **29** (4), 301–309.

Lenz, A.C. and Melchin, M.J. (2008) Convergent evolution in two Silurian graptolites. *Acta Palaeontologica Polonica* **53** (3), 449–460.

Lenz, A.C. and Thorsteinsson, R. (1997) Fusellar banding in Silurian retiolitid graptolites. *Journal of Paleontology* **71** (5), 917–920.

Lenz, A.C., Noble, P.J., Masiak, M., Poulson, S.R. and Kozłowska, A. (2006). The *lundgreni* Extinction Event: integration of paleontological and geochemical data from Arctic Canada. *GFF* **128**, 153–158.

Lenz, A.C., Senior, S.H., Kozłowska, A. and Melchin, M.J. (2012) Graptolites from the mid-Wenlock (Silurian) middle and upper Sheinwoodian, Arctic Canada. *Palaeontographica Canadiana* **32**, 1–93.

Lester, S.M. (1985) *Cephalodiscus* sp. (Hemichordata: Pterobranchia): observations of functional morphology, behavior and occurrence in shallow water around Bermuda. *Marine Biology* **85**, 263–268.

Lester, S.M. (1988a) Ultrastructure of adult gonads and development and structure of the larva of *Rhabdopleura normani* (Hemichordata: Pterobranchia). *Acta Zoologica (Stockholm)* **69**, 95–109.

Lester, S.M. (1988b) Settlement and metamorphosis of *Rhabdopleura normani* (Hemichordata: Pterobranchia). *Acta Zoologica (Stockholm)* **69**, 111–120.

Li, J.J. (1984) Some early Ordovician graptolites from Chongyi, Jiangxi. *Acta Palaeontologica Sinica* **23** (5), 578–585.

Li, J.J. (1987) On the origin of dimorphograptids. *Bulletin of the Geological Society of Denmark* **35**, 171–177.

Lidgard, S. (2008) Predation on marine bryozoan colonies: taxa, traits and trophic groups. *Marine Ecology Progress Series* **359**, 117–131.

Lin, Y.-K. (1981) New materials of Graptodendroids with special reference to the classification of the Graptodendroidea. *Bulletin of the Nanjing Institute of Geology and Palaeontology, Academia Sinica* **3** (5), 241–262 [259–261: English abstract].

Lin, Y.-K. (1986) A new planktonic graptolite fauna, in *Aspects of Cambrian–Ordovician Boundary in Dayangcha, China* (ed. J.Y. Chen). China Prospect Publishing House, Beijing, pp. 224–254, pls 55–66.

Lin, Y.-K. (1988) On proximal tufts of threads in *Dictyonema*. *Acta Palaeontologica Sinica* **27** (2), 218–237.

Lindholm, K. (1991) Ordovician graptolites from the early Hunneberg of southern Scandinavia. *Palaeontology* **34** (2), 283–327.

Lindholm, K. and Maletz, J. (1989) Intraspecific variation and relationships of some Lower Ordovician species of the dichograptid, Clonograptus. *Palaeontology* **32** (4), 711–743.

Linnæus, C. (1735) *Systema naturæ, sive regna tria naturæ systematice proposita per classes, ordines, genera, and species*. 1–12. Lugduni Batavorum. (Haak). Available at http://www.biodiversitylibrary.org/item/15373

Linnæus, C. (1751) *Skånska resa, på hœga œfwerhetens befallning fœrrættad år 1749kåMed rœn och anmærkningar uti oeconomien, naturalier, antiquiteter, seder, lefnadssætt. Med tilhœrige figurer*. Uplagd på L. Salvii kostnad, Stockholm. Available at http://hdl.handle.net/2027/nyp.33433006610202

Linnæus, C. (1758) *Systema naturæ per regna tria naturæ, secundum classes, ordines, genera, species, cum characteribus, differentiis, synonymis, locis*. Tomus I. Editio decima, reformata, pp. [1–4], 1–824. Holmiæ. (Salvius). Available at http://reader.digitale-sammlungen.de/resolve/display/bsb10076014.html; http://www.biodiversitylibrary.org/item/10277

Linnæus, C. (1768) *Systema naturæ per regna tria naturæ, secundum classes, ordines, genera, species, cum characteribus & differentiis*. Tomus III, pp. 1–236 [1–20], Tab. I–III. Holmiæ. (Salvius). Available at http://gdz.sub.uni-goettingen.de/dms/load/img/?PPN=PPN362053855&IDDOC=215259

Linnarsson, J.G.O. (1871) Om nagra försteningar fran Sveriges och Norges Primordialzon. *Öfversigt af Kongl. Vetenskaps-Akademiens Förhandlingar* **6**, 789–797.

Linnarsson, J.G.O. (1876) On the vertical range of the graptolitic types in Sweden. *Geological Magazine* **3**, 241–245.

LoDuca, S.T. (1990) *Medusaegraptus mirabilis* Ruedemann as a noncalcified dasyclad alga. *Journal of Paleontology* **64**, 469–474.

LoDuca, S.T., Melchin, M.J. and Verbruggen, H. (2011) Complex noncalcified macroalgae from the Silurian of Cornwallis Island, Arctic Canada. *Journal of Paleontology* **85** (1), 111–121.

Longrich, N.R. (2014) The horned dinosaurs *Pentaceratops* and *Kosmoceratops* from the upper Campanian of Alberta and implications for dinosaur biogeography. *Cretaceous Research* **51**, 292–308.

Loxton, J., Melchin, M.J., Mitchell, C.E. and Senior, S.J.H. (2011) Ontogeny and astogeny of the graptolite genus *Appendispinograptus* (Li and Li, 1985). *Proceedings of the Yorkshire Geological Society* **58** (4), 253–260.

Loydell, D.K. (1990) *Sinodiversograptus*: Its occurrence in Australia and northern Canada. *Journal of Paleontology* **64** (5), 847–849.

Loydell, D.K. (1991a) Isolated graptolites from the Llandovery of Kallholen, Sweden. *Palaeontology* **34** (3), 671–693.

Loydell, D.K. (1991b) The biostratigraphy and formational relationships of the upper Aeronian and lower Telychian (Llandovery, Silurian) formations of western mid-Wales. *Geological Journal* **26** (3), 209–244.

Loydell, D.K. (1992) Upper Aeronian and Lower Telychian (Llandovery) graptolites from western Mid-Wales. Part 1. *Monograph of the Palaeontographical Society* **146** (589), 1–55, pl. 1.

Loydell, D.K. (1993) Upper Aeronian and Lower Telychian (Llandovery) graptolites from western Mid-Wales, Part 2. *Monograph of the Palaeontographical Society* **147** (592), 56–180, pls. 2–5.

Loydell, D.K. (2012) Graptolite biozone correlation charts. *Geological Magazine* **149** (1), 124–132.

Loydell, D.K. and Cave, R. (1996) The Llandovery–Wenlock boundary and related stratigraphy in eastern mid-Wales with special reference to the Banwy River section. *Newsletter on Stratigraphy* **34** (1), 39–64.

Loydell, D.K. and Loveridge, R.F. (2001) The world's longest graptolite? *Geological Journal* **36**, 55–57.

Loydell, D.K. and Maletz, J. (2004) The Silurian graptolite genera *Streptograptus* and *Pseudostreptograptus*. *Journal of Systematic Palaeontology* **2** (2), 65–93.

Loydell, D.K. and Maletz, J. (2009) Isolated graptolites from the *Lituigraptus convolutus* Biozone (Silurian, Llandovery) of Dalarna, Sweden. *Palaeontology* **52** (2), 273–296.

Loydell, D.K. and Nestor, V. (2006) Isolated graptolites from the Telychian (Upper Llandovery, Silurian) of Latvia and Estonia. *Palaeontology* **49** (3), 585–619.

Loydell, D.K., Štorch, P. and Melchin, M.J. (1993) Taxonomy, evolution and biostratigraphical importance of the Llandovery graptolite *Spirograptus*. *Palaeontology* **36** (4), 909–926.

Loydell, D.K., McKenniff, J. and Lenz, A.C. (1997) Graptolites of the genus *Mediograptus* Bouček and Přibyl from the uppermost Llandovery of Cornwallis Island, Arctic Canada. *Canadian Journal of Earth Sciences* **34**, 765–769.

Loydell, D.K., Zalasiewicz, J. and Cave, R. (1998) Predation on graptoloids: new evidence from the Silurian of Wales. *Palaeontology* **41** (3), 423–427.

Loydell, D.K., Orr, P.J. and Kearns, S. (2004) Preservation of soft tissues in Silurian graptolites from Latvia. *Palaeontology* **47** (3), 503–513.

Loydell, D.K., Butcher, A. and Al-Juboury, A. (2013a) Biostratigraphy of a Silurian 'hot' shale from western Iraq. *Stratigraphy* **10** (4), 249–255.

Loydell, D.K., Butcher, A. and Frýda, J. (2013b) The middle Rhuddanian (Lower Silurian) 'hot' shale of North Africa and Arabia: an atypical hydrocarbon source rock. *Palaeogeography, Palaeoclimatology, Palaeoecology* **386**, 233–256.

Lukasik, J.J. and Melchin, M.J. (1994) *Atavograptus primitivus* (Li) from the earliest Silurian of Arctic Canada: Implications for monograptid evolution. *Journal of Paleontology* **68** (5), 1159–1163.

Lukasik, J.J. and Melchin, M.J. (1997) Morphology and classification of some early Silurian monograptids (Graptoloidea) from the Cape Phillips formation, Canadian Arctic Islands. *Canadian Journal of Earth Sciences* **34** (8), 1128–1149.

Lüning, S., Craig, J., Loydell, D.K., Štorch, P. and Fitches, B. (2000) Lower Silurian 'hot shales' in North Africa and Arabia: regional distribution and depositional model. *Earth Science Reviews* **49**, 121–200.

Lüning, S., Shahin, Y.M., Loydell, D.K., Al-Rabi, H.T., Masri, A., Tarawneh, B. and Kolonic, S. (2005) Anatomy of a world-class source rock: distribution and depositional model of Silurian organic-rich shales in Jordan and implications for hydrocarbon potential. *Bulletin of the American Association of Petroleum Geologists* **89**, 1397–1427.

Maletz, J. (1992a) The proximal development in anisograptids (Graptoloidea, Anisograptidae). *Paläontologische Zeitschrift* **66** (3/4), 297–309.

Maletz, J. (1992b) The Arenig/Llanvirn boundary in the Quebec Appalachians. *Newsletters on Stratigraphy* **26** (1), 49–64.

Maletz, J. (1993) A possible abrograptid graptolite (Abrograptidae, Graptoloidea) from western Newfoundland. *Paläontologische Zeitschrift* **67** (3/4), 323–329.

Maletz, J. (1994a) Pendent Didymograptids (Graptoloidea, Dichograptina), in *Graptolite Research Today* (eds X. Chen, B.-D. Erdtmann and Y. Ni). Nanjing University Press, Nanjing, pp. 27–43.

Maletz, J. (1994b) The rhabdosome architecture of *Pterograptus* (Graptoloidea, Dichograptidae). *Neues Jahrbuch für Geologie und Paläontologie, Abhandlungen* **191**, 345–356.

Maletz, J. (1995) *Dicranograptus clingani* in einem Geschiebe von Nienhagen (Mecklenburg). *Geschiebekunde aktuell* **11**, 33–35.

Maletz, J. (1997a) Arenig biostratigraphy of the Pointe-de-Lévy slice, Québec Appalachians, Canada. *Canadian Journal of Earth Sciences* **34**, 733–752.

Maletz, J. (1997b) Graptolites from the *Nicholsonograptus fasciculatus* and *Pterograptus elegans* Zones (Abereiddian, Ordovician) of the Oslo region, Norway. *Greifswalder Geowissenschaftliche Beiträge* **4**, 5–100.

Maletz, J. (1997c) The rhabdosome structure of a *Saetograptus* species (Graptoloidea, Monograptacea) from a North German glacial boulder. *Paläontologische Zeitschrift* **71**, 247–255.

Maletz, J. (1998a) Die Graptolithen des Ordoviziums von Rügen (Norddeutschland, Vorpommern). *Paläontologische Zeitschrift* **72**, 351–372.

Maletz, J. (1998b) *Undulograptus dicellograptoides* n. sp., an abnormal diplograptid from the Late Arenig of western Newfoundland, Canada. *Paläontologische Zeitschrift* **72**, 111–116.

Maletz, J. (2001a) Graptolite research in Germany. *Geologica Saxonica, Abhandlungen des Staatlichen Museums für Mineralogie und Geologie Dresden* **46/47**, 169–180.

Maletz, J. (2001b) Graptolite biostratigraphy of the Rügen wells. *Neues Jahrbuch für Geologie und Paläontologie, Abhandlungen* **222**, 55–72.

Maletz, J. (2003) Genetically controlled cortical tissue deposition in *Normalograptus scalaris* (Hisinger, 1837). *Paläontologische Zeitschrift* **77**, 471–476.

Maletz, J. (2006a) Dendroid graptolites from the Devonian of Germany. *Paläontologische Zeitschrift* **80**, 221–229.

Maletz, J. (2006b) The graptolite genus *Hunnegraptus* in the early Ordovician of Texas, USA. *Journal of Paleontology* **80** (3), 423–429.

Maletz, J. (2008) Retiolitid graptolites from the collection of Hermann Jaeger in the Museum für Naturkunde, Berlin (Germany). I. *Neogothograptus and Holoretiolites*. *Paläontologische Zeitschrift* **82** (3), 285–307.

Maletz, J. (2009) *Holmograptus spinosus* and the Middle Ordovician (Darriwilian) graptolite biostratigraphy at Les Méchins (Quebec, Canada). *Canadian Journal of Earth Sciences* **46**, 739–755.

Maletz, J. (2010a) *Xiphograptus* and the evolution of the virgella-bearing graptoloids. *Palaeontology* **53** (2), 415–439.

Maletz, J. (2010b) Retiolitid graptolites from the collection of Hermann Jaeger II; *Cometograptus, Spinograptus* and *Plectograptus*. *Paläontologische Zeitschrift* **84**, 501–522.

Maletz, J. (2011a) The identity of the Ordovician (Darriwilian) graptolite *Fucoides dentatus* Brongniart, 1828. *Palaeontology* **54** (4), 851–865.

Maletz, J. (2011b) The identity of *Climacograptus pungens* Ruedemann, 1904. *Canadian Journal of Earth Sciences* **48**, 1355–1367.

Maletz, J. (2011c) The proximal development of the Middle Ordovician graptolite *Skanegraptus janus* from the Krapperup drill core of Scania, Sweden. *GFF* **133**, 49–56.

Maletz, J. (2011d) Scandinavian isograptids (Graptolithina; Isograptidae): Biostratigraphy and taxonomy. *Proceedings of the Yorkshire Geological Society*, **58** (4), 267–280.

Maletz, J. (2014a) The classification of the Graptolithina Bronn, 1849. *Bulletin of Geosciences* **89** (3), 477–540.

Maletz, J. (2014b) Fossil Hemichordata (Pterobranchia, Enteropneusta). *Palaeogeography, Palaeoclimatology, Palaeoecology* **398**, 16–27.

Maletz, J. (2015) Graptolite reconstructions and interpretations. *Paläontologische Zeitschrift* **89**, 271–286.

Maletz, J. and Ahlberg, P. (2011a) The Lerhamn drill core and its bearing for the graptolite biostratigraphy of the Ordovician Tøyen Shale in Scania, southern Sweden. *Lethaia* **44**, 350–368.

Maletz, J. and Ahlberg, P. (2011b) Darriwilian (Ordovician) graptolite faunas and biogeography of the Tøyen Shale in the Krapperup drill core (Scania, southern Sweden). *Cuadernos del Museo Geominero* **14**, 327–332.

Maletz, J. and Egenhoff, S.O. (2001) Late Tremadoc to early Arenig graptolite faunas of southern Bolivia and their implications for a worldwide biozonation. *Lethaia* **34**, 47–62.

Maletz, J. and Egenhoff, S.O. (2005) Dendroid graptolites in the Elnes Formation (Middle Ordovician), Oslo Region, Norway. *Norwegian Journal of Geology* **85**, 217–221.

Maletz, J. and Kozłowska, A. (2013) Dendroid graptolites from the Lower Ordovician (Tremadocian) of the Yichang area, Hubei, China. *Paläontologische Zeitschrift* **87**, 445–454.

Maletz, J. and Mitchell, C.E. (1996) Evolution and phylogenetic classification of the Glossograptidae and Arienigraptidae (Graptoloidea): new data and remaining questions. *Journal of Paleontology* **70** (4), 641–655.

Maletz, J. and Steiner, M. (2014) The earliest graptolites (Pterobranchia) and their recognition, in *Abstract Volume, 4th International Palaeontological Congress, Mendoza* (ed. E. Cerdeño), p. 570.

Maletz, J. and Steiner, M. (2015) Graptolite (Pterobranchia, Graptolithina) preservation and identification in the Cambrian Series 3. *Palaeontology* **58** (6), 1073–1107.

Maletz, J. and Zhang, Y.-D. (2003) The proximal structure and development in the Ordovician graptolite *Parisograptus* Chen and Zhang, 1996. *Palaeontology* **46** (2), 295–306.

Maletz, J., Königshof, P., Meco, S. and Schindler, E. (1998) Late Wenlock to Early Ludlow grapto-lites from Albania. *Senckenbergiana lethaea* **78** (1/2), 141–151.

Maletz, J., Egenhoff, S. and Erdtmann, B.-D. (1999) Late Tremadoc to early Arenig graptolite succession of southern Bolivia. *Acta Universitatis Carolinae – Geologica* **43** (1/2), 29–32.

Maletz, J., Suarez-Soruco, R. and Egenhoff, S. (2002) Silurian (Wenlock–Ludlow) graptolites from Bolivia. *Palaeontology* **45** (2), 327–341.

Maletz, J., Goldman, D. and Cone, M. (2005a) The Early to Middle Ordovician graptolite faunal succession of the Trail Creek section, Central Idaho, U.S.A. *Geologica Acta* **3** (4), 395–409.

Maletz, J., Steiner, M. and Fatka, O. (2005b) Middle Cambrian pterobranchs and the question: What is a graptolite? *Lethaia* **38**, 73–85.

Maletz, J., Carlucci, J. and Mitchell, C.E. (2009) Graptoloid cladistics, taxonomy and phylogeny. *Bulletin of Geosciences* **84**, 7–19.

Maletz, J., Reimann, C., Spiske, M., Bahlburg, H. and Brussa, E.D. (2010) Darriwilian (Middle Ordovician) graptolites of southern Peru. *Geological Journal* **45** (4), 397–411.

Maletz, J., Egenhoff, S., Böhme, M., Asch, R., Borowski, K., Höntzsch, S., Kirsch, M. and Werner, M. (2011) A tale of both sides of Iapetus – Upper Darriwilian (Ordovician) grapto-lite faunal dynamics on the edges of two continents. *Canadian Journal of Earth Sciences* **48**, 841–859.

Maletz, J., Bates, D.E.B., Brussa, E.D., Cooper, R.A., Lenz, A.C., Riva, J.F., Toro, B.A. and Zhang, Y.D. (2014) Treatise on Invertebrate Paleontology, Part V, revised, Chapter 12: Glossary of the Hemichordata. *Treatise Online* **62**, 1–23.

Malinconico, M.A.L. (1992) Graptolite reflectance in the prehnite-pumpelliyite zone of Northern Maine, U.S.A. *Organic Geochemistry* **18** (3), 263–271.

Malinconico, M.A.L. (1993) Reflectance cross-plot analysis of graptolites from the anchi-metamorphic region of northern Maine, U.S.A. *Organic Geochemistry* **20** (2), 197–207.

Manck, E. (1923). Untersilurische Graptolithenarten der Zone 10 des Obersilurs, ferner *Diversograptus* gen. nov. sowie einige neue Arten anderer Gattungen. *Natur, Leipzig* **14**, 282–289.

Markham, J.C. (1971) On a species of *Cephalodiscus* collected during Operation Deep

Freeze, 1957–59. *Antarctic Research Series, Washington* **17**, 83–110.

Marr, J.E. (1925) Conditions of deposition of the Stockdale Shales of the Lake District. *Quarterly Journal of the Geological Society of London* **81**, 113–131.

Martin, E.L.O., Pittet, B., Gutiérrez-Marco, J.C., Vannier, J., El Hariri, K., Lerosey-Aubril, R., Masrour, M., Nowak, H., Servais, T., Vandenbroucke, T.R.A., Van Roy, P., Vaucher, R. and Lefebre, B. (2015) The Lower Ordovician Fezouata Konservat-Lagerstätte from Morocco: Age, environment and evolutionary perspective. *Gondwana Research* **34**, 274–283.

Masterman, A.T. (1897) On the 'notochord' of Cephalodiscus. *Zoologischer Anzeiger* **20**, 443–450.

McCrady, J. (1859) Remarks on the zoological affinities of graptolites. *Proceedings of the Elliott Society of Natural History* **1**, 229–236.

M'Coy, F. (1850) On some new genera and species of Silurian Radiata in the Collection of the University of Cambridge. *Annals and Magazine of Natural History, Series* **2** (6), 270–290.

M'Coy, F. (1874) Prodromus of the Palaeontology of Victoria; or, Figures and descriptions of Victorian organic remains, Decade 1. *Geological Survey of Victoria*, 1–16.

M'Coy, F. (1875) Prodromus of the Palaeontology of Victoria; or, Figures and descriptions of Victorian organic remains, Decade 1. *Geological Survey of Victoria*, 28–37.

M'Coy, F (1876) On a new Victorian graptolite. *Annales and Magazine of Natural History Series* **4** (18), 128–130.

Melchin, M.J. (1987) *Late Ordovician and Early Silurian grapolites, Cape Phillips Formation, Canadian Arctic Archipelago*. University of Western Ontario, London, Ontario, 762 pp (PhD thesis, unpublished).

Melchin, M.J. (1998) Morphology and phylogeny of some early Silurian 'diplograptid' genera from Cornwallis Island, Arctic Canada. *Palaeontology* **41** (2), 263–315.

Melchin, M.J. (1999) Origin of the Retiolitidae: insights from a new graptolite genus from the early Silurian of Arctic Canada. *Lethaia* **32**, 261–269.

Melchin, M.J. (2003) Restudying a global stratotype for the base of the Silurian: a progress report.

INSUGEO, Serie Correlación Geológica **18**, 147–149.

Melchin, M.J. and Anderson, A.J. (1998) Infrared video microscopy for the study of graptolites and other organic-walled fossils. *Journal of Paleontology* **72** (2), 397–400.

Melchin, M.J. and DeMont, M.E. (1995) Possible propulsion modes in Graptoloidea: a new model for graptoloid locomotion. *Paleobiology* **21** (1), 110–120.

Melchin, M.J. and Doucet, K.M. (1996) Modelling flow patterns in conical dendroid graptolites. *Lethaia* **29** (1), 39–46.

Melchin, M.J. and Koren, T.N. (2001) Morphology and phylogeny of some early Silurian (mid-Rhuddanian) monograptid graptolites from the South Urals of Russia. *Journal of Paleontology* **75** (1), 165–185.

Melchin, M.J. and Lenz, A.C. (1986) Uncompressed specimens of *Monograptus turriculatus* (Barrande, 1850) from Cornwallis Island, Arctic Canada. *Canadian Journal of Earth Sciences* **23** (4), 579–582.

Melchin, M.J. and Mitchell, C.E. (1991) Late Ordovician extinction in the Graptoloidea. *Geological Survey of Canada Paper* **90-9**, 143–156.

Melchin, M.J., Naczk-Cameron, A. and Koren, T.N. (2003) New insights into the phylogeny of Rhuddanian (Lower Llandovery) graptolites. *INSUGEO, Serie Correlación Geológica* **18**, 67–68.

Melchin, M.J., Mitchell, C.E., Naczak-Cameron, A., Fan, J.X. and Loxton, J. (2011) Phylogeny and adaptive radiation of the Neograptina (Graptoloidea) during the Hirnantian mass extinction and Silurian recovery. *Proceedings of the Yorkshire Geological Society* **58** (4), 281–309.

Melchin, M.J., Sadler, P.M. and Cramer, B.D. (2012) The Silurian Period, in *The Geological Time Scale 2012* (eds F.M. Gradstein, J.G. Ogg, M. Schmitz and G. Ogg). Elsevier, Oxford, pp. 525–559.

Merkel, B.J. and Hasche-Berger, A. (eds) (2006) *Uranium in the Environment*. Springer, Berlin.

Mierzejewski, P. (1986) Ultrastructure, taxonomy and affinities of some Ordovician and Silurian organic microfossils. *Palaeontologia Polonica* **47**, 129–220, pls 19–37.

Mierzejewski, P. (1991) *Estoniocaulis* Obut et Rotsk, 1958 and *Rhadinograptus* Obut, 1960 are not graptolites. *Acta Palaeontologica Polonica* **36** (1), 77–81, pls 13–14.

Mierzejewski, P. (2000) On the nature and development of graptoblasts. *Acta Palaeontologica Polonica* **45**, 227–238.

Mierzejewski, P. and Kulicki, C. (2003a) Cortical fibrils and secondary deposits in periderm of the hemichordate *Rhabdopleura* (Graptolithoidea). *Acta Palaeontologica Polonica* **48** (1), 99–111.

Mierzejewski, P. and Kulicki, C. (2003b) *Helicotubulus*, a replacement name for *Helicosyrinx* Kozłowski, 1967 (?Phoronoidea) preoccupied by *Helicosyrinx* Baur, 1864 (Mollusca, Gastropoda). *Acta Palaeontologica Polonica* **48**, 446.

Mierzejewski, P. and Urbanek, A. (2004) The morphology and fine structure of the Ordovician *Cephalodiscus*-like genus *Melanostrophus*. *Acta Palaeontologica Polonica* **49** (4), 519–528.

Miller, F.X. (1977) The graphic correlation method in biostratigraphy, in *Concepts and Methods of Biostratigraphy* (eds E.G. Kauffman and J.E. Hazel). Dowden, Hutchinson and Ross Inc., Stroudsburg, pp. 165–186.

M'Intosh, W.C. (1882) Preliminary notice of *Cephalodicus*, a new type allied to Prof. Allman's *Rhabdopleura*, dredged in H.M.S. 'Challenger'. *Annals and Magazine of Natural History (5th Series)* **10**, 337–348.

M'Intosh, W.C. (1887) Report on *Cephalodiscus dodecalophus*, M'Intosh, a new type of the Polyzoa, procured on the voyage of H.M.S. Challenger during the years 1873–76 (with an appendix, pp. 38–47, by S. Harmer on the anatomy of *Cephalodiscus*). *Challenger Reports, Zoology* **20**, 1–47.

Mitchell, C.E. (1987) Evolution and phylogenetic classification of the Diplograptacea. *Palaeontology* **30** (2), 353–405.

Mitchell, C.E. (1988) The morphology and ultrastructure of *Brevigraptus quadrithecatus* n. gen., n. sp. (Diplograptacea), and its convergence upon *Dicaulograptus hystrix* (Bulman). *Journal of Paleontology* **62** (3), 448–463.

Mitchell, C.E. (1990) Directional macroevolution of the diplograptacean graptolites: a product of astogenetic heterochrony and directed specia-tion, in *Major Evolutionary Radiations* (eds P.D. Taylor and G.P. Larwood). Clarendon Press, Oxford, pp. 235–264.

Mitchell, C.E. (1994) Astogeny and rhabdosome architecture of graptolites of the *Undulograptus austrodentatus* species group, in *Graptolite Research Today* (eds X. Chen, B.-D. Erdtmann and Y. Ni). Nanjing University Press, Nanjing, pp. 49–60.

Mitchell, C.E. and Carle, K.J. (1986) The nematularium of *Pseudoclimacograptus scharenbergi* (Lapworth) and its secretion. *Palaeontology* **29** (2), 373–390, pls 28–29.

Mitchell, C.E. and Maletz, J. (1995) Proposal for adoption of the base of the *Undulograptus austrodentatus Biozone* as a global Ordovician stage and series boundary level. *Lethaia* **28**, 317–331.

Mitchell, C.E., Wilson, M.A. and St. John, J.M. (1993) In situ crustoid graptolite colonies from an Upper Ordovician hardground, southwestern Ohio. *Journal of Paleontology* **67** (6), 1011–1016.

Mitchell, C.E., Maletz, J. and Zhang, Y.D. (1995) Evolutionary origins of the Diplograptina. *The Pacific Section Society for Sedimentary Geology (SEPM) Book* **77**, 401–404.

Mitchell, C.E., Chen, X., Bergström, S.M., Zhang, Y.D., Wang, Z.H., Webby, B.D. and Finney, S.C. (1997). Definition of a global boundary stratotype for the Darriwilian Stage of the Ordovician System. *Episodes* **20** (3), 158–166.

Mitchell, C.E., Brussa, E.D. and Astini, R.A. (1998) A diverse Da2 fauna preserved within an altered volcanic ash fall, Eastern Precordillera, Argentina: implications for graptolite paleoecology. *Temas Geológico–Mineros ITGE* **23**, 222–223.

Mitchell, C.E., Adhya, S., Bergström, S.M., Joy, M.P. and Delano, J.W. (2004) Discovery of the Ordovician Millbrig K-bentonite Bed in the Trenton Group of New York State: implications for regional correlation and sequence stratigraphy in eastern North America. *Palaeogeography, Palaeoclimatology, Palaeoecology* **210**, 331–346.

Mitchell, C.E., Goldman, D., Klosterman, S.L., Maletz, J., Sheets, H.D. and Melchin, M.J. (2007a) Phylogeny of the Diplograptoidea. *Acta Palaeontologica Sinica* **46** (Suppl.), 332–339.

Mitchell, C.E., Chen, C. and Finney, S.C. (2007b) The structure and possible function of 'basal membranes' in the spinose climacograptid graptolite *Appendispinograptus* Li and Li 1985. *Journal of Paleontology* **81** (5), 1122–1127.

Mitchell, C.E., Brussa, E.D. and Maletz, J. (2008) A mixed isograptid-didymograptid graptolite assemblage from the Middle Ordovician of west Gondwana (NW Bolivia). *Journal of Paleontology* **82**, 1114–1126.

Mitchell, C.E., Maletz, J. and Goldman, D. (2009) What is *Diplograptus*? *Bulletin of Geosciences* **84**, 27–34.

Mitchell, C.E., Melchin, M.J., Cameron, C.B. and Maletz, J. (2013) Phylogeny of the tube-bearing Hemichordata reveals that *Rhabdopleura* is an extant graptolite. *Lethaia* **46**, 34–56.

Moberg, J.C. (1890) Om en afdelning inom Ölands Dictyonemaskiffern såsom motsvarighet till Ceratopygeskiffern i Norge samt anteckningar om Ölands Ortocerkalk. *Sveriges Geologiske Undersökning, Afhandling och Uppsatser C* **109**, 1–22.

Moberg, J.C. (1892) Om några nya graptoliter från Skånes Undre graptolitskiffer. *Geologiska Föreningens i Stockholm Förhandlingar* **14** (4), 339–350.

Moberg, J.C. (1900) Nya bidrag till utredning af frågan om gränsen mellan Undersilur och Kambrium. *Geologiska Föreningens i Stockholm Förhandlingar* **22** (7), 523–540.

Monsen, A. (1925) Über eine neue ordovizische Graptolithenfauna. *Norsk Geologisk Tidsskrift* **8**, 147–187.

Monsen, A. (1937) Die Graptolithenfauna im Unteren Didymograptus Schiefer (Phyllograptusschiefer) Norwegens. *Norsk Geologisk Tidsskrift* **16** (2–4), 57–267.

Moors, H.T. (1968) An attempted statistical appraisal of the graptolite fauna of Willey's Quarry, Victoria, Australia. *Proceedings of the Royal Society of Victoria* **81** (2), 137–141.

Moors, H.T. (1969) The position of graptolites in turbidites. *Sedimentary Geology* **3** (4), 241–261.

Moors, H.T. (1970) Current orientation of graptolites: its significance and interpretation. *Sedimentary Geology* **4** (2), 117–134.

Morris, W.G. (1988) *A systematic survey of Lancefield graptolites from Victoria, Australia.* Unpublished PhD thesis, Cambridgeshire College of Arts and Technology/Sedgewick Museum, University of Cambridge.

Mottequin, B., Poty, E. and Prestianni, C. (2015) Catalogue of the types and illustrated specimens recovered from the 'black marble' of Denée, a marine conservation-Lagerstätte from the Mississippian of southern Belgium. *Geologica Belgica* **18** (1), 1–14.

Mu, A.T. (1946) Graptolite faunas from the Wufeng shale. *Bulletin of the Geological Society of China* **25** (1–4), 201–209.

Mu, A.T. (1948) Silurian succession and graptolite fauna of Lientan. *Bulletin of the Geological Society of China* **28**, 207–231.

Mu, A.T. (1950) On the evolution and classification of graptolites. *Geological Review* **15**, 171–183.

Mu, A.T. (1957) Some new or little known graptolites from the Ningkuo Shale (Lower Ordovician) of Changshan, western Chekiang. *Acta Palaeontologica Sinica* **5** (3), 369–437.

Mu, A.T. (1958) *Abrograptus*, a new graptolite genus from the Hulo Shale (Middle Ordovician) of Kiangshan, western Chekiang. *Acta Palaeontologica Sinica* **6** (3), 259–267 [English text: 263–265].

Mu, A.T. (1963) On the complication of graptolite rhabdosome. *Acta Palaeontologica Sinica* **11** (3), 346–374 [English text: 374–377].

Mu, E.Z. (1980) Researches on the Graptolithina of China. *Acta Palaeontologica Sinica* **19** (2), 143–151.

Mu, E.Z. (1987) Graptolite taxonomy and classification. *Bulletin of the Geological Society of Denmark* **35** (3–4), 203–207.

Mu, A.T. and Chen, X. (1962) *Sinodiversograptus multibrachiatus* gen. et sp. nov. and its developmental stages. *Acta Palaeontologica Sinica* **10**, 143–154.

Mu, A.T. and Lee, C. K. (1958) Scandent graptolites from the Ningkuo Shale of Kiangshan–Changshan area, western Chekiang. *Acta Palaeontologica Sinica* **6** (4), 391–427.

Mu, A.T. and Qiao, X.D. (1962) New materials of Abrograptidae. *Acta Palaeontologica Sinica* **10** (1), 1–8.

Mu, A.T., Lee, C.K. and Geh, M.Y. (1960) Ordovician graptolites from Xinjiang (Sikiang). *Acta Palaeontologica Sinica* **8** (1), 27–39.

Mu, A.T., Li, J.J., Ge, M.Y., Chen, X., Ni, Y.N., Lin, Y.K. and Mu, X. (1974) Graptolites, in *A*

Handbook of the Stratigraphy and Palaeontology of Southwest China (ed. NIGPAS). Nanjing Institute of Geology and Palaeontology, Nanjing, pp. 154–164, pls 67–70.

Mu, A.T., Ge, M.Y., Chen, X., Ni, Y.N. and Lin, Y.K. (1979) Lower Ordovician graptolites of Southwest China. *Palaeontologica Sinica (New Series B)* **156** (13), 1–192.

Mu, E.-Z., Ge, M.-J. and Chen, X. (1981) Lower Carboniferous graptolites from Shilu, Hainan Island. *Acta Palaeontologica Sinica* **20** (3), 185–187.

Mu, E., Li, J., Ge, M., Lin, Y. and Ni, Y. (2002) *Fossil Graptolites of China.* Nanjing University Press, Nanjing.

Muir, L. (1999) A cladistic analysis of some Llandovery (Silurian) Monograptidae (Graptolithina). University of Edinburgh [unpublished PhD dissertation].

Müller, A.H. (1975) Über das tierische Grossplankton (Graptoloidea) der silurischen Meere mit einigen allgemeinen Angaben über Graptolithina (Hemichordata). *Biologische Rundschau* **13**, 325–344.

Müller, A.H. and Schauer, M. (1969) Über Schwebeeinrichtungen bei Diplograptidae (Graptolithina) aus dem Silur. *Freiberger Forschungshefte* **C245**, 5–26.

Münch, A. (1928) Die Entwicklung der Monograptuskolonie. 22. *Bericht der Naturwissenschaftlichen Gesellschaft zu Chemnitz*, 24–28.

Münch, A. (1931) *Retiolites mancki.* Ein neuer Retiolites aus dem norddeutschen Geschiebe. *23. Bericht der Naturwissenschaftlichen Gesellschaft zu Chemnitz*, 35–42.

Münch, A. (1952) Die Graptolithen aus dem anstehenden Gotlandium Deutschlands und der Tschechoslowakei. *Geologica, Schriftenreihe der Geologischen Institute der Universtäten Berlin, Greifswald, Halle, Rostock* **7**, 1–157.

Munnecke, A. and Servais, T. (2008) Palaeozoic calcareous plankton: evidence from the Silurian of Gotland. *Lethaia* **41**, 185–194.

Murchison, R.I. (1839) *The Silurian System, founded on geological researches in the counties of Salop, Hereford, Radnor, Montgomery, Caermarthen, Brecon, Pembroke, Monmouth, Gloucester, Worcester, and Stratford; with descriptions of the coal-fields and overlying formations. Part 1.* John Murray, London.

Newell, G.E. (1952) The homology of the stomochord of the Enteropneusta. *Proceedings of the Zoological Society of London* **121** (4), 741–746.

Newman, A. (1998) Pyrite oxidation and museum collections: A review of theory and conservation treatments. *The Geological Curator* **6** (10), 363–371.

Ni, Y.N. (1978) Lower Silurian graptolites from Yichang, western Hubei. *Acta Palaeontologica Sinica* **17**(4), 387–416. [in Chinese with English abstract]

Ni, Y.N. and Cooper, R.A. (1994) The graptolite *Glossograptus* Emmons and its proximal structure. *Alcheringa* **18** (1–2), 161–167.

Ni, Y.N. and Xiao, C.X. (1994) On the genus *Apiograptus* Cooper and McLaurin, 1974, in *Graptolite Research Today* (eds X. Xu, B.-D. Erdtmann and Y. Ni). Nanjing University Press, Nanjing, pp. 14–19.

Nicholson, H.A. (1867a) On a new genus of graptolites, with notes on reproductive bodies. *Geological Magazine* **4**, 256–263, pl. 211.

Nicholson, H.A. (1867b) On some fossils from the Lower Silurian rocks of the south of Scotland. *Geological Magazine* **1** (4), 107–113.

Nicholson, H.A. (1868a) Notes on the distribution in time of the various British species and genera of graptolites. *Annals and Magazine of Natural History* **4** (2), 347–357.

Nicholson, H.A. (1868b). On the graptolites of the Coniston Flags; with notes on the British species of the genus Graptolites. *Quarterly Journal of the Geological Society of London* **24**, 521–545.

Nicholson, H.A. (1869) On some new species of graptolites. *Annales and Magazine of Natural History, London, Series* **4** (4), 231–242.

Nicholson, H.A. (1872a) *A Monograph of the British Graptolitidae. Part 1. General Introduction.* Williams Blackwood and Sons, Edinburgh and London.

Nicholson, H.A. (1872b) Migration of the Graptolites. *Quarterly Journal of the Geological Society of London* **28**, 217–232.

Nicholson, H.A. (1872c) *A Manual of Palaeontology for the Use of Students with a General Introduction on the Principles of Palaeontology.* Blackwood & Sons, Edinburgh & London.

Nicholson, H.A. and Marr, J.E. (1895) Notes on the phylogeny of graptolites. *Geological Magazine* **11**, 529–539.

Nixon, K.C., Carpenter, J.M. and Stevenson, D.W. (2003) The PhyloCode is fatally flawed, and the 'Linnaean' System can easily be fixed. *The Botanical Review* **69** (1), 111–120.

Noble, P.J., Lenz, A.C., Holmden, C., Masiak, M., Zimmerman, M.K., Poulson, S.R. and Kozłowska, A. (2012) Isotope geochemistry and plankton response to the Ireviken (earliest Wenlock) and *Cyrtograptus lundgreni* extinction events, Cape Phillips Formation, Arctic Canada, in *Earth and Life* (ed. J.A. Talent). Springer Science and Business Media, pp. 631–652.

Norman, A.M. (1869) Shetland Final Dredging Report. Part 2. On the Crustacea, Tunicata, Polyzoa, Echinodermata, Actinozoa, Hydrozoa and Porifera. *Reports of the British Association of the Advancement of Science* **38**, 247–336.

Obrehl, J. (1959) Ein Landpflanzenfund im mittelböhmischen Ordovizium. *Geologie* **8**, 535–541.

Obst, K., Böhnke, A., Maletz, J. and Katzung, G. (2002) Pb–Pb zircon dating for tuff horizons in the *Cyrtograptus* Shale (Wenlock, Silurian) of Bornholm, Denmark. *Bulletin of the Geological Society of Denmark* **48**, 1–8.

Obut, A.M. (1960) Korrelyatsiya nekotorykh chastei razreza ordovikskikh i siluriiskikh otlozhenii Estonskoi SSR po graptolitam. *Eesti NSV Teaduste Akad., Geol. Inst., Uurimused* **5** Serial 143–157. [Correlation of some parts of Estonian Ordovician and Silurian deposits according to graptolites. *Eesti NSV Geoloogia Instituudi Uurimused*, **5**, 143–158; in Russian.]

Obut, A.M. (1964) Podtip Stomochordata. Stomokhordovye, in *Osnovy paleontogii: Echinodermata, Hemichordata, Pogonophora, Chaetognatha* (ed. Y.A. Orlov). Nedra Press, Moscow, pp. 279–337.

Obut, A.M. and Rotsk, G.V. (1958) Ordovikskie i siluriyskie dendroidei Estonii. *Eesti NSV Geoloogia Instituudi Uurimused* **3**, 125–144 [in Russian].

Obut, A.M. and Sobolevskaya, R.F. (1962) Problemi neftegazonosti Sovjetskoj Arktiki: Paleontologija i biostratigrafija: Graptoliti rannego Ordovika na Taimyre. *Trudy Nautshno Issledowatelskogo Instituta Geologii, arktiki* **127** (3), 65–96 [in Russian].

Obut, A.M. and Sobolevskaya, R.F., in Obut, A.T., Sobolevskaya, R.F., Bondarev A.M., et al. (1965) Silurian graptolites of Taymyr. *Akademia Nauk SSSR*, 1–120 [in Russian].

Obut, A.M. and Sobolevskaya, R.F. (1967) Nekotorye stereostolonaty pozdnego kembriya i ordovika Noril'skogo rayona [Some stereostolonates of the late Cambrian and Ordovician of the Norilsk Region], in *Novye dannye po biostratigrafii nizhnego paleozoya Sibirskoy platformy* (eds A.B. Ivanovskiy and B.S. Sokolov). Nauka, Moskva, pp. 45–64 [in Russian].

Ohtsuka, S., Kitazawa, K. and Boixhall, G.A. (2010) A new genus of endoparasitic copepods (Cyclopoida: Enterognathidae), forming a gall in the calyx of deep-sea crinoids. *Zoological Science* **27** (8), 689–696.

Öpik, A. (1933) Über einen kambrischen Graptolithen aus Norwegen. *Norsk Geologisk Tidsskrift* **13**, 8–10.

Osborn, K.J., Kuhnz, L.A., Priede, I.G., Urata, M., Gebruk, A.V. and Holland, N.D. (2012) Diversification of acorn worms (Hemichordata, Enteropneusta) revealed in the deep sea. *Proceedings of the Royal Society* **B279**, 1646–1654.

Page, A., Gabbott, S.E., Wilby, P.R. and Zalasiewicz, J.A. (2008) Ubiquitous Burgess Shale-style "clay-templates" in low-grade metamorphic mudrock. *Geology* **36** (11), 855–858.

Palmer, D. and Rickards, B. (1991) *Graptolites – Writing in the Rocks*. The Boydell Press, Suffolk.

Pannell, C.L., Clarkson, E.N.K. and Zalasiewicz, J.A. (2006) Fine-scale biostratigraphy within the *Stimulograptus sedgwickii* Zone (Silurian: Llandovery) at Dob's Linn, Southern Uplands. *Scottish Journal of Geology* **42**, 59–64.

Paul, M., Gengnagel, M., Vogel, D. and Kuhn, W. (2002) Four years of flooding WISMUT's Ronneburg uranium mine – a status report, in *Uranium in the Aquatic Environment 4* (eds B.J. Merkel, B. Planer-Friedrich and Ch. Wolkersdorfer). Springer, Heidelberg, pp. 783–792.

Perner, J. (1895) *Etudes sur les graptolites de Boheme. IIieme Partie: Monographie des graptolites de L'etage D*. Prague, en commission chez Raimund Gerhard, Leipzig.

Perner, J. (1897) *Études sur les graptolithes de Bohême, 3a*, 1–25, pls 9–13. Palaeontographica Bohemiae, Prague.

Perner, J. (1899) *Études sur les graptolithes de Bohême, 3b*, 1–24, pls 14–17. Palaeontographica Bohemiae, Prague.

Petersen, H.I., Schovsbo, N.H. and Nielsen, A.T. (2013) Reflectance measurements of zooclasts and solid bitumen in Lower Paleozoic shales, southern Scandinavia: correlation into vitrinate reflectance. *International Journal of Coal Geology* **114**, 1–18.

Peterson, K.J., Su, Y.-H., Arnone, M.I., Swalla, B.J. and King, B.L. (2013) MicroRNAs support the monophyly of enteropneust hemichordates. *Journal of Experimental Zoology, Part B, Molecular and Developmental Evolution* **9999**, 1–7.

Philippe, H., Brinkmann, H., Copley, R.R., Moroz, L.L., Nakano, H., Poustka, A.J., Wallberg, A., Peterson, K.J. and Telford, M.J. (2011) Acoelomorph flatworms are deuterostomes related to *Xenoturbella*. *Nature* **470**, 255–260.

Philippot, A. (1950) Les graptolites du Massif Armoricain. (Thèses présentées a la Faculté des Sciences de L`Université de Rennes). *Mémoires de la Societé Géologique et Minéralogique de Bretagne* **8**, 1–295.

Phillips, N.G. and Hughes, M.J. (1996) The geology and gold deposits of the Victorian gold province. *Ore Geology Reviews* **11**, 255–302.

Podhalańska, T. (2013) Graptolites – stratigraphic tool in the exploration of zones prospective for the occurrence of unconventional hydrocarbon deposits. *Przeglad Geologiczny* **61**, 621–629.

Pohlmann, J. (1887) Fossils from the Waterlime Group near Buffalo, N.Y. *Bulletin of the Buffalo Society of Natural Sciences* **5**, 23–32.

Porębska, E. (1984) Latest Silurian and Early Devonian grapolites from Zdanow section, Bardo Mts. (Sudetes). *Annales Societatis Geologorum Poloniae* **52**, 89–209.

Porębska, E. and Sawłowicz, Z. (1997) Palaeoceanographic linkage of geochemical and graptolite events across the Silurian–Devonian boundary in the Bardzkie Mountains (southwest Poland). *Palaeogeography, Palaeoclimatology, Palaeoecology* **132**, 343–354.

Porębska, E., Kozłowska-Dawidziuk, A. and Masiak, M. (2004) The *lundgreni* event in the Silurian of the East European Platform, Poland. *Palaeogeography, Palaeoclimatology, Palaeoecology* **213**, 271–294.

Portlock, J.E. (1843) *Report on the Geology of the county of Londonderry, and of parts of Tyrone and Fermanagh. Examined and Described Under the Authority of the Master General and Board of Ordnance.* Alexander Thom, Dublin.

Přibyl, A. (1940) Die Graptolithenfauna des mittleren Ludlows von Böhmen. *Vestnik statniho geologickeho Ustavu C.S.R.* **16**, 63–73.

Přibyl, A. (1941) O ceskych a cizich zastupcich rodu *Rastrites* Barrande, 1850 (Von böhmischen und fremden Vertretern der Gattung Rastrites Barrande 1850). *Rozpravy II. Tridy Ceske Akademie* **51** (6), 1–22.

Přibyl, A. (1943) Revise zastupcu rodu *Pristiograptus*, ze skupiny *P. dubius* a *P. vulgaris* z ceskeho a ciziho siluru. *Rozpravy II. Tridy Ceske Akademie* **53** (4), 1–48. (German version: Přibyl, A. (1943) Revision aller Vertreter der Gattung *Pristiograptus* aus der Gruppe *P. dubius* und *P. vulgaris* aus dem böhmischen und ausländischen Silur. *Academie tcheque des Sciences, Bulletin International* 44, 33–81).

Přibyl, A. and Münch, A. (1942) Revise středoevropských zástupců rodu *Demirastrites* Eisel. *Rozpravy II. Třídy České Akademie* **51**, 1–29. (German version: Pribyl, A. and Münch, A. (1942) Revision der mitteleuropäischen Vertreter der Gattung *Demirastrites* Eisel. *Rozpravy II. Třídy České Akademie* 52 (30), 1–26).

Pritchard, G.B. (1892) On a new species of Graptolitidae (*Temnograptus magnificus*). *Proceedings of the Royal Society of Victoria (New Series)* **4**, 56–58.

Pritchard, G.B. (1895) Notes on some Lancefield graptolites. *Proceedings of the Royal Society of Victoria, New Series* **7**, 27–30.

Prout, H.A. (1851) Description of a new graptolite found in the lower Silurian rocks near the falls of St. Croix River. *American Journal of Science* **11**, 187–191.

Quenstedt, F.A. (1840) Über die vorzüglichsten Kennzeichen der Nautileen. *Neues Jahrbuch für Mineralogie, Geognosie, Geologie und Petrefakten-Kunde, Jahrgang* 1840, 253–291.

Quoy, J.R.C. and Gaimard, P. (1825) Zoologie. 2.ᶜ Partie, in *Voyage autour du Monde, Entrepris par Ordre du Roi, sous le Ministere et Conformement aux instructions de S. Exc. M. Le Vicomte du Bouchage, … exécuté sur les corvettes de L. M. "L'Uranie" et "La Physicienne," pendant les années 1817, 1818, 1819 et 1820* (ed.

L. de Freycinet). Paris. pp. 192–401 [1–328 in 1824; 329–616 in 1825], Atlas pls. 43–65.

Radzevičius, S. (2003) *Pristiograptus* (Graptoloidea) from the *perneri-lundgreni* biozones (Silurian) of Lithuania. *Carnets de Geologie*, Article 2003/07.

Radzevičius, S. (2006) Late Wenlock biostratigraphy and the *Pristiograptus virbalensis* group (Graptolithina) in Lithuania and the Holy Cross Mountains. *Geological Quarterly* **50** (3), 333–344.

Radzevičius, S. (2007) The genus *Pristiograptus* in Wenlock of east Baltic and the Holy Cross Mountains. *Dissertationes Geologicae Universitatis Tartuensis* **20**, 1–131.

Radzevičius, S., Schopf, J.W. and Kudryavtsev, A.B. (2013) Bacterial epibionts encrusting Silurian graptolites, in *Proceedings of the 3rd IGCP 591 Annual Meeting – Lund, Sweden* (eds A. Lindskog and K. Mehlqvist). Lund University, pp. 267–269.

Raiswell, R. and Berner, R.A. (1985) Pyrite formation in euxinic and semi-euxinic sediments. *American Journal of Science* **285**, 710–724.

Raiswell, R. and Canfield, D.E. (1998) Sources of iron for pyrite formation in marine sediments. *American Journal of Science* **298** (3), 219–245.

Richter, R. (1853) Thüringische Graptolithen. *Zeitschrift der Deutschen Geologischen Gesellschaft* **5**, 439–464, pl. 12.

Richter, R. (1871) Aus dem Thüringischen Schiefergebirge. *Zeitschrift der Deutschen Geologischen Gesellschaft* **23**, 231–256.

Richter, R. (1875) Aus dem Thüringischen Schiefergebirge. *Zeitschrift der Deutschen Geologischen Gesellschaft* **27**, 261–273, pl. 8.

Rickards, R.B. (1973) Bipolar monograptids and the Silurian genus *Diversograptus* Manck. *Paläontologische Zeitschrift* **47** (3–4), 175–187.

Rickards, R.B. (1975) Palaeoecology of the Graptolithina, an extinct class of the phylum Hemichordata. *Biological Reviews* **50**, 397–436.

Rickards, R.B. (1990) 1.7.2 Plankton, in *Palaeobiology: A Synthesis* (eds D.E.G. Briggs and P.R. Crowther). Blackwell Science, pp. 49–52.

Rickards, R.B. (1996) The graptolite nema: problem to all our solutions. *Geological Magazine* **133** (3), 343–346.

Rickards, R.B. (1999) A century of graptolite research in Cambridge. *The Geological Curator* 7(2), 71–76.

Rickards, R.B. and Bulman, O.M.B. (1965) The development of *Lasiograptus harknessi* (Nicholson, 1867). *Palaeontology* **8** (2), 272–280.

Rickards, R.B. and Chapman, A. (1991) Bendigonian graptolites (Hemichordata) of Victoria. *Memoirs of the Museum of Victoria* **52** (1), 1–135.

Rickards, R.B. and Durman, P.N. (2006) Evolution of earliest graptolites and other hemichordates. *Cardiff, National Museum of Wales, Geological Series* **25**, 5–92.

Rickards, R.B. and Hutt J.E. (1970) The earliest monograptid. *Proceedings of the Geological Society of London* **1663**, 115–119.

Rickards, R.B. and Koren, T.N. (1974) Virgellar meshwork and sicular spinosity in Llandovery graptoloids. *Geological Magazine* **111** (3), 193–204.

Rickards, R.B. and Lane, P.D. (1997) Two new coremagraptid graptolites from the type Llandovery (Silurian) district, and a review of the genus *Coremagraptus*. *Neues Jahrbuch für Geologie und Paläontologie, Abhandlungen* **204** (2), 171–183.

Rickards, R.B. and Stait, B.A. (1984) *Psigraptus*, its classification, evolution and zooid. *Alcheringa* **8**, 101–111.

Rickards, R.B. and Wright, A.J. (1999) Systematics, biostratigraphy and evolution of the late Ludlow and Pridoli (Late Silurian) graptolites of the Yass District, New South Wales, Australia. *Records of the Australian Museum* **51**, 187–214.

Rickards, R.B., Hutt, J.E. and Berry, W.B.N. (1977) Evolution of Silurian and Devonian graptoloids. *Bulletin of the British Museum of Natural History, Geological Series* **28** (1), 1–120.

Rickards, R.B., Baillie, P.W. and Jago, J.B. (1990) An Upper Cambrian (Idamean) dendroid assemblage from near Smithton, northwestern Tasmania. *Alcheringa* **14**, 207–232.

Rickards, R.B., Partridge, P.L. and Banks, M.R. (1991) *Psigraptus jacksoni* – systematics, reconstruction, distribution and preservation. *Alcheringa* **15**, 243–254.

Rickards, R.B., Hamedi, M.A. and Wright, A.J. (1994) A new Arenig (Ordovician) graptolite fauna from the Kerman District, east-central Iran. *Geological Magazine* **131** (1), 35–42.

Rickards, R.B., Packham, G.H., Wright, A.J. and Williamson, P.L. (1995) Wenlock and Ludlow graptolite faunas and biostratigraphy of the Quarry Creek district, New South Wales. *Association of Australasian Paleontologists, Memoir* **17**, 1–68.

Ridewood, W.G. (1907) Pterobranchia. *Cephalodiscus. National Antarctic Expedition 1901–1904. Natural History. Vol. II. Zoology* (*Vertebrata: Mollusca: Crustacea*), 1–67.

Ridewood, W.G. and Fantham, H.B. (1907) On *Neurosporidium cephalodisci*, n. g., n. sp., a Sporozoon from the nervous system of *Cephalodiscus nigrescens. Quarterly Journal of Microscopical Sciences* **51**, 81–100.

Riediger, C., Goodarzi, F. and MacQueen, R.W. (1989) Graptolites as indicators of regional maturity in Lower Palaeozoic sediments, Selwyn Basin, Yukon and Northwest Territories, Canada. *Canadian Journal of Earth Sciences* **26**, 2003–2015.

Rigby, S. (1991) Feeding strategies in graptoloids. *Palaeontology* **34** (4), 797–815.

Rigby, S. (1992) Graptoloid feeding efficiency, rotation and astogeny. *Lethaia* **25**, 51–68.

Rigby, S. (1993a) Population analysis and orientation studies of graptoloids from the Middle Ordovician Utica Shale, Quebec. *Palaeontology* **36**, 267–282.

Rigby, S. (1993b) Graptolite functional morphology; a discussion and critique. *Modern Geology* **17** (4), 271–287.

Rigby, S. (1994) Erect tube growth in *Rhabdopleura compacta* (Hemichordata, Pterobranchia) from Start Point, Devon. *Journal of Zoology, London* **233**, 449–455.

Rigby, S. and Dilly, N.P. (1993) Growth rates of pterobranchs and the lifespan of graptolites. *Paleobiology* **19** (4), 459–475.

Rigby, S. and Milsom, C.V. (1996) Benthic origins of zooplankton: An environmentally determined macroevolutionary effect. *Geology* **24** (1), 52–54.

Rigby, S. and Milsom, C.V. (2000) Origins, evolution, and diversification of zooplankton. *Annual Review of Ecology and Systematics* **31**, 293–313.

Rigby, S. and Rickards, R.B. (1989) New evidence for the life habitat of graptoloids from physical modelling. *Paleobiology* **15** (4), 402–413.

Riva, J.F. (1974) A revision of some Ordovician graptolites of eastern North America. *Palaeontology* **17** (1), 1–40.

Riva, J.F. (1976) *Climacograptus bicornis bicornis* (Hall), its ancestor and likely descendants, in *The Ordovician System: Proceedings of a Palaeontological Association Symposium, Birmingham, September, 1974* (ed. M.G. Bassett). University of Wales and National Museum of Wales, Cardiff, pp. 589–619.

Riva, J.F. (1994) *Yutagraptus mantuanus* Riva in Rickards 1994, a pendent xiphograptid from the Lower Ordovician of Utah, in *Graptolite Research Today* (eds X. Chen, B.-D. Erdtmann and Y. Ni). Nanjing University Press, Nanjing, pp. 1–13.

Riva, J.F. and Kettner, K.B. (1989) Ordovician graptolites from the northern Sierra de Cobachi, Sonora, Mexico. *Transactions of the Royal Society of Edinburgh: Earth Sciences* **80** (2), 71–90.

Roemer, F.A. (1855) Graptolithen am Harze. *Neues Jahrbuch für Mineralogie, Geognosie, Geologie und Petrefaktenkunde* 1855, 540–542, pl. 7.

Roemer, F. (1861) *Die fossile Fauna der Silurischen Diluvial-Geschiebe von Sadewitz bei Oels in Niederschlesien*. Breslau, Robert Nischkowski.

Röttinger, E. and Lowe, C.J. (2012) Evolutionary crossroads in developmental biology: hemichordates. *Development* **139**, 2463–2475.

Roychoudhury, A.N., Kostka, J.E. and van Cappellen, P. (2003) Pyritization: a palaeoenvironmental and redox proxy reevaluated. *Estuarine, Coastal and Shelf Science* **57**, 1183–1193.

Ruedemann, R. (1895) Development and the mode of growth of *Diplograptus* M'Coy. *New York State Geological Survey Annual Report for* **1894**, 219–249.

Ruedemann, R. (1897) Evidence of current action in the Ordovician of New York. *American Geologist* **21**, 367–391.

Ruedemann, R. (1898) Synopsis of recent progress in the study of graptolites. *The American Naturalist* **32** (373), 1–16.

Ruedemann, R. (1904) Graptolites of New York, part I. Graptolites of the lower beds. *New York State Museum Memoir* **7**, 455–807.

Ruedemann, R. (1908) Graptolites of New York, Part 2. *New York State Museum Memoir* **11**, 1–583.

Ruedemann, R. (1911) Stratigraphic significance of the wide distribution of graptolites. *Bulletin* of the Geological Society of America **22**, 231–237.

Ruedemann, R. (1912) The Lower Siluric Shales of the Mohawk Valley. *New York State Bulletin* **162** (525), 1–151.

Ruedemann, R. (1925) Some Silurian (Ontarian) faunas of New York. *New York State Museum Bulletin* **265**, 1–134.

Ruedemann, R. (1937) A new North American graptolite faunule. *American Journal of Science* **33** (193), 57–62.

Ruedemann, R. (1947) Graptolites of North America. *Geological Society of America Memoir* **19**, 1–652.

Ruedemann, R. and Lochmann, C. (1942) Graptolites from the Englewood Formation (Mississippian) of the Black Hills, South Dakota. *Journal of Paleontology* **16** (5), 657–659.

Ruppert, E.E. (2005) Key characters uniting hemichordates and chordates: homologies or homoplasies? *Canadian Journal of Zoology* **83**, 3–23.

Rushton, A.W.A. (2011) Deflexed didymograptids from the Lower Ordovician Skiddaw Group of northern England. *Proceedings of the Yorkshire Geological Society* **58**, 319–327.

Sachanski, V., Özgül, N. and Arpat, E. (2006) The graptolite species *Hunnegraptus copiosus* Lindholm, 1991 from the Lower Ordovician of Central Taurus, Turkey. *Geosciences* **2006**, 49–52.

Sadler, P.M. (2004) Quantitative biostratigraphy: achieving finer resolution in global correlation. *Annual Review of Earth and Planetary Sciences* **32**, 187–213.

Sadler, P.M., Cooper, R.A. and Melchin, M.J. (2009) High-resolution, early Paleozoic (Ordovician–Silurian) time scales. *Geological Society of America Bulletin* **121**, 887–906.

Sadler, P.M., Cooper, R.A. and Melchin, M.J. (2011) Sequencing the graptoloid clade: building a global diversity curve from local range charts, regional composites and global time-lines. *Proceedings of the Yorkshire Geological Society* **58** (4), 329–343.

Salter, J.W. (1852) Description of some graptolites from the south of Scotland. *Quarterly Journal of the Geological Society of London* **8**, 388–392.

Salter, J.W. (1863) Notes on the Skiddaw Slate Fossils. *Quarterly Journal of the Geological Society of London* **19**, 135–140.

Sars, G.O. (1872) *On some remarkable forms of animal life from the great deeps off the Norwegian coast: I. Partly from posthumous manuscripts of the late Professor Dr. Michael Sars.* University Programme for the 1st half-year 1869. Brogger and Bristiff, Christiania.

Sars, G.O. (1874) On *Rhabdopleura mirabilis* (M. Sars). *Quarterly Journal of Microscopical Science, New Series* **14**, 23–44.

Sars, M. (1868) Fortsatte Bemærkninger over det dyriske livs udbredning i havets dybder. *Særskilt aftryt af Videnskabers-selskap forhandlinger for 1868*, 245–275 [separate pagination in reprint: 1–32].

Sato, T. (1936) Vorläufige Mitteilung über *Atubaria heterolopha* gen. nov. sp. nov., einen in freiem Zustand aufgefundenen Pterobranchier aus dem Stillen Ozean. *Zoologischer Anzeiger* **115**, 97–106.

Saunders, K.M., Bates, D.E.B., Kluessendorf, J., Loydell, D.K. and Mikulic, D.G. (2009) *Desmograptus micronematodes*, a Silurian dendroid graptolite, and its ultrastructure. *Palaeontology*, **52** (3), 541–559.

Scharenberg, W. (1851) *Ueber Graptolithen mit besonderer Berücksichtigung der bei Christiania vorkommenden Arten.* Dissertation. Robert Nischkowsky, Breslau.

Schauer, M. (1967) Biostratigraphie und Taxionomie von *Rastrites* (Pterobranchiata, Graptolithina) aus dem anstehenden Silur Ostthüringens und des Vogtlandes. *Freiberger Forschungshefte C* **213**, 171–199.

Schauer, M. (1971) Biostratigraphie und Taxionomie der Graptolithen des tieferen Silurs unter besonderer Berücksichtigung der tektonischen Deformation. *Freiberger Forschungshefte* **C273**, 1–185.

Scheltema, R.S. (1970) Two new records of *Planctosphaera* larvae (Hemichordata: Planctosphaeroidea). *Marine Biology* **7**, 47–48.

Schepotieff, A. (1905) Ueber die Stellung der Graptolithen im zoologischen System. *Neues Jahrbuch für Mineralogie, Geologie und Paläontologie 2, Abhandlungen* **1**, 78–98.

Schepotieff, A. (1906) Die Pterobranchier. Anatomische und histologische Untersuchungen über *Rhabdopleura normani* Allman und *Cephalodiscus dodecalophus* M'Int. 1. Teil. *Rhabdopleura normani* Allman. Die Anatomie von *Rhabdopleura*. *Zoologische Jahrbücher. Abteilung für Anatomie und Ontogenie der Tiere* **23**, 463–534, pls 25–33.

Schepotieff, A. (1907a) Die Pterobranchier. Anatomische und histologische Untersuchungen über *Rhabdopleura normani* Allman und *Cephalodiscus dodecalophus* M'Int. 1. Teil. *Rhabdopleura normani* Allman. 2. Abschnitt. Knospungsprozess und Gehäuse von *Rhabdopleura*. *Zoologische Jahrbücher. Abteilung für Anatomie und Ontogenie der Tiere* **24**, 193–238, pls 17–23.

Schepotieff, A. (1907b) Die Pterobranchier. Anatomische und histologische Untersuchungen über *Rhabdopleura normani* Allman und *Cephalodiscus dodecalophus* M'Int. 2. Teil. *Cephalodiscus dodecalophus* M'Int. 1. Abschnitt. Die Anatomie von *Cephalodiscus*. *Zoologische Jahrbücher. Abteilung für Anatomie und Ontogenie der Tiere* **24**, 553–609, pls 38–48.

Schleiger, N. (1968) Orientation distribution patterns of graptolite rhabdosomes from Ordovician sediments in central Victoria, Australia. *Journal of Sedimentary Petrology* **38** (2), 462–472.

Schovsbo, N.H., Nielsen, A.T. and Gautier, D.L. (2014) The Lower Palaeozoic shale gas play in Denmark. *Geological Survey of Denmark and Greenland Bulletin* **31**, 19–22.

Sdzuy, K. (1974) Mittelkambrische Graptolithen aus NW-Spanien. *Paläontologische Zeitschrift* **48** (1–2), 110–139.

Sell, B.K., Leslie, S.A. and Maletz, J. (2011) New U–Pb zircon data for the GSSP for the base of the Katian in Atoka, Oklahoma, USA and the Darriwilian in Newfoundland, Canada. *Cuadernos del Museo Geominero* **14**, 537–546.

Sennikov, M., Tolmacheva, T. and Obut, O. (2011) Obituary: Tatyana N. Koren, 1935–2010. *Newsletter of the Subcommission on Devonian Stratigraphy* **26**, 2–4.

Sewera, L.J. (2011). Determining the composition of the dwelling tubes of Antarctic pterobranchs. Senior Honors thesis, Illinois Wesleyan University. Paper 48. Available at http://digitalcommons.iwu.edu/bio_honproj/48

Shaw, A.B. (1964) *Time in Stratigraphy*. McGraw-Hill, New York.

Sherwin, L. (1975) Silurian graptolites from the Forbes group, New South Wales. *Records of the Geological Survey of New South Wales* **16** (3), 227–237.

Signor, P.W. and Vermeij, G.J. (1994) The plankton and the benthos: origins and early history of an evolving relationship. *Paleobiology* **20**, 297–319.

Skevington, D. (1963a) Graptolites from the Ontikan Limestones (Ordovician) of Öland, Sweden. I: Dendroidea, Tuboidea, Camaroidea, and Stolonoidea. *Bulletin of the Geological Institution of Uppsala* **42** (6), 1–62.

Skevington, D. (1963b) A correlation of Ordovician graptolite-bearing sequences. *Geologiska Föreningens i Stockholm Förhandlingar* **85** (3), 2298–2319.

Skevington, D. (1965) Graptolites from the Ontikan limestones (Ordovician) of Öland, Sweden. 2. Graptoloidea and Graptovermida. *Bulletin of the Geological Institutions of the University of Upsala* **43** (3), 1–74.

Skevington, D. (1967) Probable instance of genetic polymorphism in the graptolites. *Nature* **213** (5078), 810–812.

Skevington, D. (1968) British and North American Lower Ordovician correlation: Discussion. *Geological Society of America Bulletin* **79**, 1259–1264.

Skevington, D. (1973) Ordovician graptolites, in *Atlas of Palaeobiogeography* (ed. A. Hallam). Elsevier Scientific Publishing Company, Amsterdam, pp. 27–35.

Skevington, D. (1974) Controls influencing the composition and distribution of Ordovician graptolite faunal provinces. *Special Papers in Palaeontology* **13**, 59–73.

Skwarko, S.K. (1974) Some graptolites from the Canning Basin, Western Australia 2. Graptolites from the Goldwyer No. 1 well. *Bulletin of the Bureau of Mineral Resources, Geology and Geophysics of Australia* **150**, 43–56.

Slavík, L. and Carls, P. (2012) Post-lau event (late Ludfordian, Silurian) recovery of conodont faunas of Bohemia. *Bulletin of Geosciences* **87** (4), 815–832.

Smith, J.D.D. (1957) Graptolites with associated sedimentary grooving. *Geological Magazine* **94** (5), 425–428.

Smith Jr., K.L., Holland, N.D. and Ruhl, H.A. (2005) Enteropneust production of spiral fecal trails on the deep-sea floor observed with time-lapse photography. *Deep-Sea Research I* **52**, 1228–1240.

Spencer, J.W.W. (1884) Niagara fossils. Part 1. Graptolitidae of the Upper Silurian System. *Transactions of the Academy of Science of Saint Louis* **4**, 555–593, pls 1–6.

Spengel, J. (1932) *Planktosphaera pelagica. Report on the Scientific Results of the 'Michael Sars' North Atlantic Deep Sea Expedition* **5**, 1–28.

Spjeldnaes, N. (1963) Some Upper Tremadocian graptolites from Norway. *Palaeontology* **6** (1), 121–131.

Stebbing, A.R.D. (1970a) Aspects of the reproduction and life cycle of *Rhabdopleura compacta* (Hemichordata). *Marine Biology* **5**, 205–212.

Stebbing, A.R.D. (1970b) The status and ecology of *Rhabdopleura compacta* (Hemichordata) from Plymouth. *Journal of the Marine Biological Association of the United Kingdom* **50**, 209–221.

Steiner, M. (1994) Die neoproterozoischen Megaalgen Südchinas. *Berliner Geowissenschaftliche Abhandlungen* **E15**, 1–146.

Stenzel, S.R., Knight, I. and James, N.P. (1990) Carbonate platform to foreland basin: revised stratigraphy of the Table Head Group (Middle Ordovician) western Newfoundland. *Canadian Journal of Earth Sciences*, **27** (1), 14–26.

Stiasny, G. (1910) Zur Kenntnis der Lebensweise von *Balanoglossus clavigerus* Delle Chiaje. *Zoologischer Anzeiger* **34**, 561–565, 633.

Štorch, P. (1994) Graptolite biostratigraphy of the Lower Silurian (Llandovery and Wenlock) of Bohemia. *Geological Journal* **29** (2), 137–165.

Štorch, P. (1995) Biotic crises and post-crisis recoveries recorded by Silurian planktonic graptolite faunas of the Barrandian area (Czech Republic). *Geolines* **3**, 59–70.

Štorch, P. (1998a) Biostratigraphy, palaeobiogeographical links and environmental interpretation of the Llandovery and Wenlock graptolite faunas of peri-Gondwanan Europe. *Temas Geologico-Mineros ITGE* **23**, 126–129.

Štorch, P. (1998b) Graptolites of the *Pribylograptus leptotheca* and *Lituigraptus convolutus* biozones of Tman (Silurian, Czech Republic). *Journal of the Czech Geological Society* **43** (3), 209–272.

Štorch, P. (1998c) New data on Telychian (Upper Llandovery, Silurian) graptolites from Spain. *Journal of the Czech Geological Society* **43** (3), 113–141.

Štorch, P. and Loydell, D.K. (1992). Graptolites of the *Rastrites linnei* Group from the European Llandovery (Lower Silurian). *Neues Jahrbuch für Geologie und Paläontologie, Abhandlungen* **184** (1), 63–86.

Štorch, P., Fatka, O. and Kraft, P. (1993) Lower Palaeozoic of the Barrandian area (Czech republic) – a review. *Coloquios de Paleontologia* **45**, 163–191.

Štorch, P., Mitchell, C.E., Finney, S.C. and Melchin, M.J. (2011) Uppermost Ordovician (upper Katian–Hirnantian) graptolites of north-central Nevada, U.S.A. *Bulletin of Geosciences* **86** (2), 301–386.

Štorch, P., Manda, S. and Loydell, D.K. (2014) The early Ludfordian *leintwardinensis* graptolite event and the Gorstian–Ludfordian boundary in Bohemia (Silurian, Czech republic). *Palaeontology* **57** (5), 1003–1043.

Størmer, L. (1933) A floating organ in *Dictyonema*. *Norsk Geologisk Tidsskrift* **13**, 102–112, pl. 1.

Strachan, I. (1954) The structure and development of *Peiragraptus fallax*, gen. et sp. nov. A new graptolite from the Ordovician of Canada. *Geological Magazine* **41** (6), 509–513.

Strachan, I. (1985) The significance of the proximal end of *Cryptograptus tricornis* (Carruthers) (Graptolithina). *Geological Magazine* **122** (2), 151–155.

Strachan, I. (1990) A new genus of abrograptid graptolite from the Ordovician of Southern Scotland. *Palaeontology* **33** (4), 933–936.

Strandmark, J.E. (1902) Undre Graptolitskiffer vid Fagelsang. *Geologiska Föreningens i Stockholm Förhandlingar* **23**, 548–557.

Stürmer, W. (1951) Zur Technik der Auffindung von Graptolithen-Kieselschiefern in den Main-Geröllen. *Senckenbergiana* **32** (1–4), 157–159.

Stürmer, W. (1952) Zur Technik an Graptolithen und Radiolarien in Main-Kieselschiefern, 2. *Senckenbergiana* **32** (5–6), 351–355.

Sudbury, M. (1958) Triangulate Monograptids from the *Monograptus gregarius* zone (Lower Llandovery) of the Rheidol Gorge (Cardiganshire). *Philosophical Transactions of the Royal Society of London, Series B. Geological Sciences* **241** (685), 485–555.

Suess, E. (1851) Über böhmische Graptolithen. *Naturwissenschaftliche Abhandlungen von W. Haidinger* **4** (4), 87–134.

Sun, Y.Z. (1931) Graptolite-bearing strata of China. *Bulletin of the Geological Society of China* **10**, 291–299.

Sun, Y.Z. (1933) Ordovician and Silurian grapto-lites from China. *Palaeontologia Sinica, Series* **B14**, 1–70.

Sun, Y.Z. (1935) Lower Ordovician graptolite fauna of north China. *Palaeontologica Sinica, Series* **B14**, 1–16.

Sutton, M.D., Briggs, D.E.G., Siveter, D.J. and Siveter, D.J. (2001) Methodologies for the visualization and reconstruction of three-dimensional fossils from the Silurian Herefordshire lagerstätte. *Palaeontologica Electronica* **4** (1).

Swalla, B.J. and Smith, A.B. (2008) Deciphering deuterostome phylogeny: molecular, morphological and palaeontological perspectives. *Philosophical Transactions of the Royal Society B* **363**, 1557–1568.

Taylor, P.D., Berning, B. and Wilson, M. (2013) Reinterpretation of the Cambrian "bryozoan" *Pywackia* as an Octocoral. *Journal of Paleontology* **88** (6), 984–990.

Teichmüller, M. (1978) Nachweis von Graptolithen-Periderm in geschieferten Gesteinen mit Hilfe kohlenpetrologischer Methoden. *Neues Jahrbuch fur Geologie und Palaeontology, Monatshefte* **1978** (7), 430–447.

Teichmüller, M. (1982) Application of coal petrological methods in geology including oil and natural gas prospecting, in *Textbook of Coal Petrology* (eds E. Stach, M.-Th. Mackowsky, R. Teichmüller, G.H. Taylor, D. Chandra and R. Teichmüller). Gebrüder Borntraeger, Berlin, Stuttgart, pp. 316–331.

Teller, L. (1964) Graptolite fauna and stratigraphy of the Ludlovian deposits from Chelm borehole, Eastern Poland. *Studia Geologica Polonica* **13**, 1–88.

Teller, L. (1966) Two new species of Monograptidae from the Upper Ludlowian of Poland. *Bulletin de L'Académie Polonaise des Sciences, Série des Sciences Biologiques* **14**, 553–558.

Teller, L. (1976) Morphology of some Upper Wenlockian Cyrtograptinae from Zawada 1 profile (NE Poland). *Acta Geologica Polonica* **26** (4), 469–484.

Teller, L. (1986) Morphology of selected Monograptidae from the Wenlock of NE Poland. *Palaeontographica* **A192** (1–3), 51–73.

Teller, L. (1994) Astogeny in a *Cyrtograptus* colony, in *Graptolite Research Today* (eds X. Chen, B.-D. Erdtmann and Y. Ni). Nanjing University Press, Nanjing, pp. 128–132.

Teller, L. (1997a) Graptolites and stratigraphy of the Pridoli Series in the East European Platform. *Palaeontologia Polonica* **56**, 59–70.

Teller, L. (1997b) Revision of certain Pridoli monograptids from the Chelm Keysection (EEP). *Palaeontologia Polonica* **56**, 71–85.

Teller, L. (1999) Abnormalities in the development of monograptid colonies. *Geological Quarterly* **43** (3), 347–352.

Thorsteinsson, R. (1955) The mode of cladial generation in *Cyrtograptus*. *Geological Magazine* **92** (1), 37–49, pls 3–4.

Tinn, O., Meidla, T., Ainsaar, L. and Pani, T. (2009) Thallophytic algal flora from a new Silurian Lagerstätte. *Estonian Journal of Earth Sciences* **58** (1), 38–42.

Toghill, P. (1968) The graptolite assemblages and zones of the Birkhill shales (Lower Silurian) at Dobb's Linn. *Palaeontology* **11** (5), 654–668.

Törnquist, S.L. (1879) Några iakttagelser öfver Dalarnes Graptolitskiffrar. *Geologiska Föreningen i Stockholm Förhandlingar* **4**, 446–457.

Törnquist, S.L. (1881) Om några graptolitarter från Dalarne. *Geologiska Föreningens i Stockholm Förhandlingar* **5**, 434–445, pl. 17.

Törnquist, S.L. (1887) Anteckningar om de äldre paleozoiska leden i Ostthüringen och Voigtland. *Geologiska Föreningens i Stockholm Förhandlingar* **9**, 471–491.

Törnquist, S.L. (1890) Undersökningar öfver Siljansområdets graptoliter. I. *Lunds Universitets Årsskrift* **26**, 1–33, pls 1–2.

Törnquist, S.L. (1892) Undersökningar öfver Siljanområdets graptoliter 2. *Lunds Universitets Årsskrift* **28**, 1–47.

Törnquist, S.L. (1893) Observations on the structure of some Diprionidæ. *Lunds Universitets Årsskrift* **29**, 1–12.

Törnquist, S.L. (1894) Nagra anmärkningar om graptoliternas terminologi. *Geologiska Föreningens i Stockholm Förhandlingar* **16**, 375–379.

Törnquist, S.L. (1897) On the Diplograptidae and the Heteroprionidae of the Scanian Rastrites beds. *Acta Regiae Societatis Physiographicae Lundensis* **8**, 1–22.

Törnquist, S.L. (1899) Researches into the Monograptidae of the Scanian Rastrites Beds. *Lunds Universitets Arsskrift* **35** (1), 1–25.

Törnquist, S.L. (1901) Researches into the Graptolites of the Lower Zones of the Scanian and Vestrogothian Phyllo-Tetragraptus beds 1. *Lunds Universitets Arsskrift* 37, Afdeln. **2** (5), 1–26. [also *Kongl. Fysiografiska Sallskapets Handlingar* 12, 5].

Törnquist, S.L. (1904) Researches into the Graptolites of the Lower Zones of the Scanian and Vestrogothian Phyllo-Tetragraptus beds 2. *Lunds Universitets Arsskrift* 40, *Afdeln.* **2** (2), 1–29.

Törnquist, S.L. (1907) Observations on the genus *Rastrites* and some allied species of *Monograptus*. *Lunds Universitete Årsskrift (N.F.) Afd.* 2, **3** (5), 1–22.

Toro, B.A. and Maletz, J. (2008) The proximal development in *Cymatograptus* (Graptoloidea) from Argentina and its relevance for the early evolution of the Dichograptacea. *Journal of Paleontology* **82** (5), 974–983.

Towe, R.M. and Urbanek, A. (1972) Collagen-like structures in Ordovician graptolite periderm. *Nature* **237**, 443–445.

Tsai, C.-H. and Fordyce, R.E. (2015) Ancestor-descendant relationships in evolution: origin of the extant pygmy right whale, *Caperea marginata*. *Biology Letters* **11** (1). doi:10.1098/rsbl.2014.0875

Tsegelniuk, P.D. (1976) Late Silurian and early Devonian monograptids from the south-western margin of the East European Platform (in Russian), in *Palaeontology and Stratigraphy of the Upper Precambrian and Lower Paleozoic of the South-West part of the east European Platform* (ed. P.L. Shulga). Naukova Dumka, Kiev, pp. 91–133.

Tullberg, S.A. (1880) Några Didymograptus-arter I undre graptolitskiffer vid Kiviks-Esperöd. *Geologiska Föreningens i Stockholm Förhandlingar* **5**, 39–43, pl. 2.

Tullberg, S.A. (1882) On the Graptolites described by Hisinger and the older Swedish authors. *Bihang till Kongliga Svenska Vetenskaps Akademiens Handlingar* **6** (13), 3–24.

Tullberg, S.A. (1883) Skånes Graptoliter 2. Graptolitfaunorna i Cardiolaskiffern och Cyrtograptusskiffrarne. *Sveriges Geolokiska Undersökning* **C55**, 1–43.

Ubaghs, G. (1941) Les graptolithes dendroïdes du marbre noir de Denee (Viséen inférieur). *Bulletin du Musée royal d'Histoire naturelle de Belgique* **17** (2), 1–30.

Ulrich, E.O. and Ruedemann, R. (1931) Are the graptolites bryozoans? *Bulletin of the Geological Society of America* **42**, 589–604.

Ulst, R.Z. (1974) The sequence of pristiograptids and coterminous deposits of Wenlock and Ludlow of the Middle Pribaltic, in *Graptolites of the USSR* (ed. A.E. Obut). Publishing House, Nauka', Siberian branch, Novosibirsk, 105–122.

Underwood, C.J. (1992) Graptolite preservation and deformation. *Palaios* **7** (2), 178–186.

Underwood, C.J. (1993) The position of graptolites within Lower Palaeozoic planktic ecosystems. *Lethaia* **26** (3), 189–202.

Underwood, C.J. (1995) Interstipe webbing in the Silurian graptolite *Cyrtograptus murchisoni*. *Palaeontology* **38** (3), 619–625.

Underwood, C.J. (1998) Population structure of graptolite assemblages. *Lethaia* **31**, 33–41.

Urbanek, A. (1958) Monograptidae from erratic boulders of Poland. *Acta Palaeontologica Polonica* **9**, 1–105.

Urbanek, A. (1959) Studies on graptolites. 2. On the development and structure of graptolite genus *Gymnograptus* Bulman. *Acta Palaeontologica Polonica* **4** (3), 279–338.

Urbanek, A. (1960) An attempt at biological interpretation of evolutionary changes in graptolite colonies. *Acta Palaeontologica Polonica* **5** (2), 127–234.

Urbanek, A. (1963) On generation and regeneration of cladia in some Upper Silurian monograptids. *Acta Palaeontologica Polonica* **8** (2), 135–254.

Urbanek, A. (1966) On the morphology and evolution of the Cullograptinae (Monograptidae, Graptolithina). *Acta Palaeontologica Polonica* **11** (3–4), 291–544.

Urbanek, A. (1970) Neocucullograptinae n. subfam. (Graptolithina) their evolutionary and stratigraphic bearing. *Acta Palaeontologica Polonica* **15** (2–3), 163–388.

Urbanek, A. (1973) Organization and evolution of graptolite colonies, in *Animal Colonies – Development and Function Through Time* (eds R.S. Boardman, A.H. Cheetham and W.A. Oliver, Jr.). Dowden, Hutchinson & Ross, Inc., Stroudsburg, PA, pp. 441–514.

Urbanek, A. (1976) The problem of graptolite affinities in the light of ultrastructural studies on peridermal derivatives in Pterobranchs. *Acta Palaeontologica Polonica* **21** (1), 3–36.

Urbanek, A. (1978) Some remarks on colony organization in graptolites. *Acta Palaeontologica Polonica* **23** (4), 631–633.

Urbanek, A. (1986) The enigma of graptolite ancestry: lesson from a phylogenetic debate, in *Enigmatic Fossil Taxa* (eds A. Hoffman and M. Nitecki). Oxford University Press, Oxford, pp. 184–226.

Urbanek, A. (1995). Phyletic evolution of the latest Ludlow spinose monograptids. *Acta Palaeontologica Polonica* **40** (1), 1–17.

Urbanek, A. (1997a) Late Ludfordian and early Pridoli monograptids from the Polish lowland. *Palaeontologia Polonica* **56**, 87–231.

Urbanek, A. (1997b) The emergence and evolution of linograptids. *Palaeontologia Polonica* **56**, 233–269.

Urbanek, A. (2004) Morphogenetic gradients in graptolites and bryozoans. *Acta Palaeontologica Polonica* **49** (4), 485–504.

Urbanek, A. and Mierzejewski, P. (1982) Ultrastructure of the tuboid graptolite tubotheca. *Paläontologische Zeitschrift* **56** (1–2), 87–93.

Urbanek, A. and Mierzejewski, P. (2009) The ultrastructure and building of graptolite dissepiments. *Acta Palaeontologica Polonica* **54** (2), 243–252.

Urbanek, A. and Teller, L. (1997) Graptolites and stratigraphy of the Wenlock and Ludlow Series in the east European Platform. *Palaeontologia Polonica* **56**, 23–57.

Urbanek, A., Koren, T.N. and Mierzejewski, P. (1982) The fine structure of the virgellar apparatus in *Cystograptus vesiculosus*. *Lethaia* **15**, 207–228.

Urbanek, A., Radzevičius, S., Kozłowska, A. and Teller, L. (2012) Phyletic evolution and iterative speciation in the persistent *Pristiograptus dubius* lineage. *Acta Palaeontologica Polonica* **57** (3), 589–611.

VandenBerg, A.H.M. (1990) The ancestry of *Climacograptus spiniferus* Ruedemann. *Alcheringa* **14**, 39–51.

VandenBerg, A.H.M. and Cooper, R.A. (1992) The Ordovician graptolite sequence of Australasia. *Alcheringa* **16** (1–2), 33–85.

Van der Horst, C.J. (1934) The burrow of an enteropneust. *Nature* **134**, 852.

Van der Horst, C.J. (1936) Planctosphaera and Tornaria. *The Quarterly Journal of Microscopical Science* s2-78, 605–613.

Vannier, J., Garcia-Bellido, D.C., Hu, S.-X. and Chen, A.-L. (2009) Arthropod visual predators in the early pelagic ecosystem: evidence from the Burgess Shale and Chengjiang biotas. *Proceedings of the Royal Society B* **276**, 2567–2574.

Van Phuc, N. (1998) *Vietnamograptus*: a new diplograptid genus from the *Monograptus hercynicus* zone of the Muongxen area, northwest part of Central Vietnam. *Temas Geológico-Mineros* **23**, 286.

Van Roy, P., Daley, A.C. and Briggs, D.E.G. (2015) Anomalocaridid trunk limb homology revealed by giant filter-feeder with paired flaps. *Nature* **522**, 77–80.

Varol, Ö.N., Demirel, I.H., Rickards, R.B. and Günay, Y. (2006) Source rock characteristics and biostratigraphy of the Lower Silurian (Telychian) organic-rich shales at Akyaka, central Taurus region, Turkey. *Marine and Petroleum Geology* **23**, 901–911.

Veski, R. and Palu, V. (2003) Investigation of *Dictyonema* oil shale and its natural and artificial transformation products by a vankrevelenogram. *Oil Shale* **20** (3), 265–281.

Vidal, G. and Knoll, A.H. (1983) Proterozoic plankton. *Memoir of the Geological Society of America* **161**, 265–277.

Vinther, J., Stein, M., Longrich, N.R. and Harper, D.A.T. (2014) A suspension-feeding anomalocarid from the Early Cambrian. *Nature* **507**, 496–499.

Wahlenberg, G. (1821) Petrificata telluris Suecana examinata. *Nova Acta Regiae societatis Scientarum Upsaliensis* **5** (8), 1–116.

Walch, J.E.I. (1771) *Die Naturgeschichte der Versteinerungen zur Erläuterung der Knorrischen Sammlung von Merkwürdigkeiten der Natur. Dritter Theil. Vorrede*, 1–235 [in Walch, J.E.I. 1768–1773. Die Naturgeschichte der Versteinerungen, 4 Theile in 5 Bänden, 272 plates in map] [ETH-Bibliothek Zürich, Rar 399, doi:10.3931/e-rara-10246].

Walcott, C.D. (1883) Fossils of the Utica Slate. *Transactions of the Albany Institute* **10**, 18–38.

Walcott, C.D. (1911a) Middle Cambrian holothurians and medusae. *Smithsonian Miscellaneous Collections* **57** (3), 41–68.

Walcott, C.D. (1911b) Middle Cambrian annelids. *Smithsonian Miscellaneous Collections* **57** (2), 109–144.

Walcott, C.D. (1919) Cambrian geology and paleontology IV. Middle Cambrian Algae. *Smithsonian Miscellaneous Collections* **67** (5), 217–260.

Walker, M. (1953) The development of *Monograptus dubius* and *Monograptus chimaera*. *Geological Magazine* **90** (5), 362–373.

Wang, J., Meng, Y., Wang, X., Huang, H.-P., Fu, L.-P. and Zhang, X. (2011) Further studies of *Cyrtograptus robustus*. *Geological Bulletin of China* **30** (8), 1233–1237 [in Chinese].

Wang, X.F. and Chen, X.H. (1996) Astogeny, evolution and classification of psigraptids – a critical review, in *Centennial Memorial Volume of Prof. Sun Yunzhu: Palaeontology and Stratigraphy* (eds H. Wang and X. Wang). China University of Geosciences Press, Beijing, pp. 98–103.

Wang, X.-F. and Jin, Y.-Q. (eds) (1977) *Handbook to Palaeontology of Central-South China. Pt. 1. Early Palaeozoic Era.* Geological Press, Beijing [in Chinese].

Webby, B.D. (2004) Introduction, in *The Great Ordovician Biodiversification Event* (eds B.D. Webby, F. Paris, M.L. Droser and I.G. Percival). Columbia University Press, New York, pp. 1–37.

Webby, B.D., Paris, F., Droser, M.L. and Percival, I.G. (2004a) *The Great Ordovician Biodiversification Event.* Columbia University Press, New York.

Webby, B.D., Cooper, R.A., Bergström, S.M. and Paris, F. (2004b) Stratigraphic framework and time slices, in *The Great Ordovician Biodiversification Event* (eds B.D. Webby, F. Paris, M.L. Droser and I.G. Percival). Columbia University Press, New York, pp. 41–47.

Whitfield, R.P. (1902) Notice of a new genus of marine algæ, fossil in the Niagara Shale. *American Museum of Natural History, Bulletin* **16**, 399–400.

Whittington, H.B. (1954) A new Ordovician graptolite from Oklahoma. *Journal of Paleontology* **28** (5), 613–621.

Whittington, H.B. (1955) Additional new Ordovician graptolites and a chitinozoan from Oklahoma. *Journal of Paleontology* **29** (5), 837–851.

Whittington, H.B. (1985) *The Burgess Shale*. Yale University Press, New Haven.

Whittington, H.B. and Rickards, R.B. (1969) Development of *Glossograptus* and *Skiagraptus*, Ordovician graptoloids from Newfoundland. *Journal of Paleontology* **43**, 800–817.

Wilde, P. and Berry, W.B.N. (1984) Destabilization of the oceanic density structure and its significance to marine "extinction" events. *Palaeogeography, Palaeoclimatology, Palaeoecology* **48**, 143–162.

Wilde, P., Quinby-Hunt, M.S., Berry, W.B.N. and Orth, C.J. (1989) Palaeo-oceanography and biogeography in the Tremadoc (Ordovician) Iapetus Ocean and the origin of the chemostratigraphy of *Dictyonema flabelliforme* black shales. *Geological Magazine* **126** (1), 19–27.

Williams, S.H. (1981) Form and mode of life of *Dicellograptus* (Graptolithina). *Geological Magazine* **118** (4), 401–408.

Williams, S.H. (1983) The Ordovician–Silurian boundary graptolite fauna of Dob's Linn, southern Scotland. *Palaeontology* **26**, 605–639.

Williams, S.H. (1988) Dob's Linn – the Ordovician–Silurian boundary stratotype. *Bulletin of British Museum of Natural History, Geology* **43**, 17–30.

Williams, S.H. (1990) Computer-assisted graptolite studies, in *Microcomputers in Palaeontology* (eds D.L. Bruton and D.A.T. Harper). Contributions from the Palaeontology Museum, University of Oslo, Oslo, pp. 46–55.

Williams, S.H. and Bruton, D.K. (1983) The Caradoc-Ashgill boundary in the central Oslo region and associated graptolite faunas. *Norsk Geologisk Tidsskrift* **63**, 147–191.

Williams, S.H. and Clarke, L.C. (1999) Structure and secretion of the graptolite prosicula, and its application for biostratigraphical and evolutionary studies. *Palaeontology* **42**, 1003–1015.

Williams, S.H. and Stevens, R.K. (1988) Early Ordovician (Arenig) graptolites from the Cow Head Group, western Newfoundland. *Palaeontographica Canadiana* **5**, 1–167.

Williams, S.H. and Stevens, R.K. (1991) Late Tremadoc graptolites from western Newfoundland. *Palaeontology* **34**, 1–47.

Williams, S.H., Bashford, A.R. and Dilly, N.P. (1997) Growth rates and skeletal secretion of siculae in Early Ordovician (Arenig) graptolites from Western Newfoundland: Implications for development and paleoecology of graptolites. *Palaios* **12** (6), 591–597.

Wiman, C. (1893a) Über Diplograptidae Lapworth. *Bulletin of the Geological Institute of the University of Uppsala* **1**, 97–104.

Wiman, C. (1893b) Über Monograptus Geinitz. *Bulletin of the Geological Institute of the University of Uppsala* **1**, 113–117.

Wiman, C. (1895) Über die Graptolithen. *Bulletin of the Geological Institute of the University of Uppsala* **2** (4), 239–316, pls 9–15.

Wiman, C. (1896a) The structure of the grapto-lites. *Natural Science* **9**, 186–192, 240–249.

Wiman, C. (1896b) Über *Dictyonema cavernosum* n. sp. *Bulletin of the Geological Institution of Uppsala* **3** (1), 1–13.

Wiman, C. (1897) Über den Bau einiger gotländis-cher Graptolithen. *Bulletin of the Geological Institute of the University of Uppsala* **3**, 352–368.

Wiman, C. (1901) Über die Borkholmer Schicht im mittelbaltischen Silurgebiet. *Bulletin of the Geological Institute of the University of Uppsala* **5**, 149–222.

Winchell, C.J., Sullivan, J., Cameron, C.B., Swalla, B.J. and Mallatt, J. (2002) Evaluating hypotheses of deuterostome phylogeny and chordate evolu-tion with new LSU and SSU ribosomal DNA data. *Molecular and Biological Evolution* **19** (5), 762–776.

Wood, E.M.R. (1900) The Lower Ludlow Formation and its graptolite fauna. *Quarterly Journal of the Geological Society of London* **56**, 415–492.

Woodward, S.P. (1854–1856) *Manual of the Mollusca; or, Rudimentary Treatise of Recent and Fossil Shells*. John Weale, London.

Woodwick, K.H. and Sensenbaugh, T. (1985) *Saxipendium coronatum*, new genus, new spe-cies (Hemichordata: Enteropneusta): the unusual spaghetti worms of the Galápagos Rift hydro-thermal vents. *Proceedings of the Biological Society Washington* **98**, 351–365.

Worsaae, K., Sterrer, W., Kaul-Strehlow, S., Hay-Schmidt, A. and Giribet, G. (2012) An anatomical description of a miniaturized acorn worm (Hemichordata, Enteropneusta) with asexual repro-duction by paratomy. *PLOS One* **7** (11), 1–19.

Wyatt, A.R.E. (1997) Nancy Kirk: turning the world of graptolites upside down, in *The Role of Women in the History of Geology* (eds C.V.

Burek and B Higgs). *Geological Society, London, Special Publications* **281**, 325–333.

Xiao, C.X., Xia, T.L. and Wang, Z.Y. (1985) New materials of Cardiograptidae from S. Jiangxi and their evolutionary relationship. *Acta Palaeontologica Sinica* **24** (4), 429–439.

Yin, T.H. and Mu, A.T. (1945) Lower Silurian grap-tolites from Tungtze. *Bulletin of the Geological Society of China* **25**, 211–220.

Yolkin, E.A., Kim, A.I., Weddige, K., Talent, J.A. and House, M.R. (1997) Definition of the Pragian/Emsian stage boundary. *Episodes* **20** (4), 235–240.

Young, G.A. and Hagadorn, J.W. (2010) The fossil record of cnidarian medusae. *Palaeoworld* **19**, 212–221.

Yu, J.-H. and Fang, Y.-T. (1981) *Arienigraptus*, a new graptolite genus from the Ningkuo Formation (Lower Ordovician) of South China. *Acta Palaeontologica Sinica* **20** (1), 27–32.

Zalasiewicz, J.A., Rushton, A.W.A. and Owen, A.W. (1995) Late Caradoc graptolitic faunal gra-dients across the Iapetus Ocean. *Geological Magazine* **132** (5), 611–617.

Zalasiewicz, J.A., Russel, C., Snelling, A.M. and Williams, M. (2011) The systematic relationship of the monograptid species *acinaes* Törnquist, 1899 and *rheidolensis* Jones, 1909. *Proceedings of the Yorkshire Geological Society* **58** (4), 351–356.

Zalasiewicz, J.A., Page, A., Rickards, R.B., Williams, M., Wilby, P.R., Howe, M.P.A. and Snelling, A.M. (2013) Polymorphic organization in a planktonic graptoloid (Hemichordata: Pterobranchia) colony of Late Ordovician age. *Geological Magazine* **150** (1), 143–152.

Zessin, W. and Puttkamer, K.F. von (1994) *Melanostrophus fokini* Öpik (Graptolithina, Stolonoidea) – Fund einer vollständigen Kolonie in einem ordovizischen Geschiebe von Rendsburg, Schleswig-Holstein. *Archiv für Geschiebekunde* **1** (10), 563–572.

Zhang, Y.D. (1994) Wenlock isolated graptolites from Cornwallis Island, Arctic Canada, in *Graptolite Research Today* (eds X. Chen, B.-D. Erdtmann and Y. Ni). Nanjing University Press, Nanjing, pp. 113–127.

Zhang, Y.D. and Chen, X. (2007) Palaeobiogeographic distribution of *Pseudisograptus* and early biserials in South China and its implication for the origination of major graptolite faunas in the

Ordovician. *Acta Palaeontologica Sinica,* suppl. **46**, 530–536.

Zhang, Y.D. and Fortey, R.A. (2001) The proximal development and thecal structure of the Ordovician graptolites *Tylograptus* and *Sinograptus. Palaeontology* **44** (3), 553–573.

Zhang, Y.D., Chen, X., Goldman, D., Zhang, Y., Cheng, J.F. and Song, Y.Y. (2010) Diversity and paleobiogeographic distribution patterns of Early and Middle Ordovician graptolites in distinct depositional environments of South China. *Science China, Earth Sciences* **53**, 1811–1827.

Zhao, X. and Zhang, S. (1985) Reclined graptolites of the Xinchangian. *Journal of Changchun College of Geology* **2** (40), 13–26.

Index

Note: page numbers in italics refer to figures

Graptolite Paleobiology, First Edition. Jörg Maletz.
© 2017 John Wiley & Sons Ltd. Published 2017 by John Wiley & Sons Ltd.

diversity, 118, 147, *155*, 168, 193, 228
Diversograptus, 228
Dob's Linn, (GSSP), *100*
dome, 42–43, *43*, *143*
dorsal fold *see* prothecal fold
drawing, 250, *251*, 258
drawing mirror (camera lucida), 250
Dycoryne, 28

ecology, 6, 17, 51
ecosystem, 69
ectocortex, 33, *34*
Egregiograptus, 238, *238*
Eiseligraptus, 214, *215*
Eldonia, 140
Elnes Formation, 58
Emsian, 98, *99*
encrusting, 125
endemic, *62*, 104, 164
endocortex, 33, *34*
endoparasites, 71
Englewood Formation, 136
Enteropneusta, 17, *17*, *18*, 114
    deep-sea, 21
Eocephalodiscus, 28
Eoglyptograptus, 188, 190
epibiont, 72, *73*
Epigraptus, 143, *143*
epipelagic, 53, *54*, *57*, 61, *62*
erect, 124–126
Essex Fauna, 18
Estoniocaulis, 138
Etagraptus, *160*
evolution, 3, 119, 148
    convergent, 120
    lineages, 119
Exigrapus, 183, 188, *189*, *191*
Expansograptus, *11*, 104, *146*, *156*, *159*, 261
    coalification, *87*
    contact metamorphosis, 88
    proximal development, 168, *169*
    thecal terminology, *39*
exploration, 107
    gold, 107–108
    oil & gas, 108
    uranium, 109
extinction events, 118, *119*, 135, 236, 240–242
    final, 242–243
extrathecal development, 45, 267

FAD *see* first appearance datum (FAD)
faunal provincialism, 62–64
faunal succession, 97
feeding 69, *70*
feeding efficiency, 55
Fenestella, 141
Fezouata Biota, 70, 150, Plate 4
fibrils, 33
first appearance datum (FAD), 99, *100*
float, 145 *see also* nematularium
Floian *63*, *103*, 261
    biogeography, 10
    biostratigraphy, 7, *64*
    GSSP, *64*, *103*
folding (thecae), 42, *42*
food chain, 68
fool's gold *see* pyrite
free-floating, 53
Fucoides, 255, *255*
fungus growth, *249*
funicle, 54
fusellum, 32, *85*, *249*, 267
    attenuation, 177–178, 211 *see also* reduction
    maturation (coalification), 86, *87*
    preservation, 79
    reflectance, 108
    replacement, 88
    Retiolitidae, 208–210
fusellus, fuselli, 32, *33*

Galaeplumosus, 22
Galapagos Rift, 20
Galeograptus, *131*
Gangliograptus, 239
gas window, 91
Geinitz, Hanns Bruno, *257*
Geniculograptus, *62*, 73, *73*, *191*, *194*
geniculum, *164*, *191*, 204, *212*
    hood, 216, *218*
geological mapping, 107
glacial erratics (glacial boulder), 95, 132, 235, 255
    monograptids, 235, 236
    retiolitids, 215, *217*, *218*
Global stratotype section and point (GSSP), *63*,
    102, *103*
Glossograptidae, *177*
    origin, 176
Glossograptina, 114, 171
    evolution, *175*